Research in Mathematics Education

This series is designed to produce thematic volumes, allowing researchers to access numerous studies on a theme in a single, peer-reviewed source. Our intent for this series is to publish the latest research in the field in a timely fashion. This design is particularly geared toward highlighting the work of promising graduate students and junior faculty working in conjunction with senior scholars. The audience for this monograph series consists of those in the intersection between researchers and mathematics education leaders—people who need the highest quality research, methodological rigor, and potentially transformative implications ready at hand to help them make decisions regarding the improvement of teaching, learning, policy, and practice. With this vision, our mission of this book series is:

1. To support the sharing of critical research findings among members of the mathematics education community;
2. To support graduate students and junior faculty and induct them into the research community by pairing them with senior faculty in the production of the highest quality peer-reviewed research papers; and
3. To support the usefulness and widespread adoption of research-based innovation.

More information about this series at http://www.springer.com/series/13030

Nicholas H. Wasserman
Editor

Connecting Abstract Algebra to Secondary Mathematics, for Secondary Mathematics Teachers

Springer

Editor
Nicholas H. Wasserman
Department of Mathematics, Science and Technology
Teachers College, Columbia University
New York, NY, USA

ISSN 2570-4729 ISSN 2570-4737 (electronic)
Research in Mathematics Education
ISBN 978-3-319-99213-6 ISBN 978-3-319-99214-3 (eBook)
https://doi.org/10.1007/978-3-319-99214-3

Library of Congress Control Number: 2018962925

This Springer imprint is published by the registered company Springer Nature Switzerland AG
The registered company address is: Gewerbestrasse 11, 6330 Cham, Switzerland

Foreword

"I just want to teach high school mathematics, why should I take abstract algebra?" "Why would taking abstract algebra help me become an effective high school mathematics teacher?" For those of us with experience in secondary mathematics teacher education, these are not unfamiliar questions. This volume can help answer these questions. Indeed, secondary teachers' mathematical knowledge, and, more specifically, their lack of mathematical knowledge, is a common but naïvely cited reason for the myriad problems we encounter as we try to improve mathematics teaching and learning in high schools. Although merely taking more advanced courses has long been shown to be ineffective at improving student learning outcomes, this volume focuses on particularly useful connections which can be made from the study of abstract algebra—a mathematics course several years beyond what high school teachers are expected to teach—and teachers' knowledge of and disciplinary practices related to the mathematics required in their high school classes.

Some ideas in this book are theoretical and speculative. The historical development of mathematical ideas, for example, is illuminative for those of us interested in the epistemology of mathematical understanding. These ideas highlight what *could* be possible if teachers' collegiate and professional education were focused on the "higher standpoint." Teachers *may* develop new tasks, stories, and discussions that introduce students to some of these major leaps of understanding and their associated practices, but only if they understand them and their application to the content they are expected to teach.

The main body of the book, therefore, is empirical, drawing upon the experience and careful research and development conducted by the authors. One of the key innovations of this book is the attention to not just how abstract algebra informs high school content (though it does this, in our estimation, better than any other volume on the subject) but how the situations and dilemmas that arise from teaching high school mathematics can provide a meaningful structure for learning advanced mathematics. Moreover, this mutual impact—making connections from advanced mathematics to high school math and from high school math to advanced mathematics—is not merely focused on content or history but on the ways in which people have grappled with algebraic ideas that connect *learning* and *instruction* through the practices and habits of mind that mark our field as unique.

Mathematics and its teaching *is* unique in the world of academic subjects, and it is often only in grappling with advanced mathematics—stretching our own understanding of content *and* mathematical practices—that we can see beyond the common conception of mathematics as a subject that exists primarily to serve engineering and the applied sciences. Perhaps by advancing the ideas promoted in this book we can help high school students see this as well.

Because of this duality and because of the rigorous theoretical and empirical standpoint taken by the authors, the examples and instructional implications in this volume are of immense practical utility for those of us tasked with the preparation and professional learning experiences of secondary mathematics teachers. As such, this book epitomizes one of the primary rationales that we, as co-editors, provided when we proposed this series: to increase the utility of mathematics education research for those who would impact practice at the cutting edge. We wish to express our great appreciation to the authors of the chapters in this volume and, of course, the editor, Nick Wasserman, for creating such a melding of research and practice.

Department of Mathematics
University of Delaware
Newark, DE, USA

Jinfa Cai

Ira A. Fulton Schools of Engineering
School for Engineering of Matter,
Transport & Energy
Arizona State University
Tempe, AZ, USA

James A. Middleton

Acknowledgments

The work that goes into the creation of a volume is both in conceptualization and realization, and I would like to acknowledge people who played a role in each. First, I would like to thank the many colleagues (too many to list, although several of whom are authors of chapters) who have engaged with me in thinking about advanced mathematics and secondary teacher education and who have challenged me and pushed me to think more deeply and more critically in this area. Moreover, this volume would not exist without the initial encouragement from Rina Zazkis to bring together some of the current work around, and the existing contributions about, abstract algebra and secondary teaching. To all of these colleagues, thank you for your role in helping conceptualize this volume and in shaping many of the ideas in it. Second, I am grateful for the work of all the authors, commentary authors, reviewers, series editors, copyeditors, etc., for helping bring this volume into existence. In this regard, in addition to the authors you see within the volume, I would like to thank Jonathan Bostic, Shawn Broderick, Jinfa Cai, Paul Dawkins, Ben Dickman, Tim Fukawa-Connelly, Steven Greenstein, Will McGuffey, Melissa Mills, Teo Paoletti, Melissa Renga, Jason Samuels, Julianna Stockton, Megan Wawro, Ann Wheeler, Sean Yee, and Bill Zahner.

Contents

List of Contributors

Authors

James A. Mendoza Álvarez Department of Mathematics, The University of Texas at Arlington, Arlington, TX, USA

Erin E. Baldinger Department of Curriculum and Instruction, College of Education and Human Development, University of Minnesota, Saint Paul, MN, USA

Marita Barabash Department of Mathematics, Achva Academic College, Arugot, Israel

Department of Science Teaching, Weizmann Institute of Science, Rehovot, Israel

Hyman Bass Department of Mathematics and School of Education, University of Michigan, Ann Arbor, MI, USA

Anna Marie Bergman Fariborz Maseeh Department of Mathematics and Statistics, Portland State University, Portland, OR, USA

John Caughman Fariborz Maseeh Department of Mathematics and Statistics, Portland State University, Portland, OR, USA

John Paul Cook Department of Mathematical Sciences, Oklahoma State University, Stillwater, OK, USA

James Cummings Department of Mathematical Sciences, Carnegie Mellon University, Pittsburgh, PA, USA

Al Cuoco Education Development Center, Waltham, MA, USA

Allan Donsig Department of Mathematics, University of Nebraska-Lincoln, Lincoln, NE, USA

Ruhama Even Department of Science Teaching, Weizmann Institute of Science, Rehovot, Israel

Joshua Fagan Department of Mathematics, Texas State University, San Marcos, TX, USA

Patrick Galarza Department of Mathematics, Science and Technology, Teachers College, Columbia University, New York, NY, USA

Erin Glover College of Education, Oregon State University, Corvallis, OR, USA

M. Kathleen Heid Department of Curriculum and Instruction, The Pennsylvania State University, University Park, PA, USA

Yvonne Lai Department of Mathematics, University of Nebraska-Lincoln, Lincoln, NE, USA

Sean Larsen Fariborz Maseeh Department of Mathematics and Statistics, Portland State University, Portland, OR, USA

Younhee Lee Department of Curriculum and Instruction, The Pennsylvania State University, University Park, PA, USA

Elise Lockwood Department of Mathematics, Oregon State University, Corvallis, OR, USA

Ami M. Mamolo Faculty of Education, University of Ontario Institute of Technology, Oshawa, ON, Canada

Ofer Marmur Faculty of Education, Simon Fraser University, Burnaby, BC, Canada

William McCallum Department of Mathematics, The University of Arizona, Tucson, AZ, USA

Kathleen Melhuish Department of Mathematics, Texas State University, San Marcos, TX, USA

Eileen Murray Department of Mathematical Sciences, College of Science and Mathematics, Montclair State University, Montclair, NJ, USA

Josephine Shamash Department of Mathematics, Achva Academic College, Arugot, Israel

Department of Science Teaching, Weizmann Institute of Science, Rehovot, Israel

J. Philip Smith Department of Mathematics, Science and Technology, Teachers College, Columbia University, New York, NY, USA

Ashley L. Suominen Department of Liberal Arts, Savannah College of Art and Design, Atlanta, GA, USA

Peter D. Taylor Department of Mathematics and Statistics, Queen's University, Kingston, ON, Canada

Nicholas H. Wasserman Department of Mathematics, Science and Technology, Teachers College, Columbia University, New York, NY, USA

Keith Weber Graduate School of Education, Rutgers University, New Brunswick, NJ, USA

Diana White Department of Mathematical and Statistical Sciences, University of Colorado Denver, Denver, CO, USA

Rina Zazkis Faculty of Education, Simon Fraser University, Burnaby, BC, Canada

Rose Mary Zbiek Department of Curriculum and Instruction, The Pennsylvania State University, University Park, PA, USA

External Chapter Reviewers

Jonathan Bostic College of Education and Human Development, Bowling Green State University, Bowling Green, OH, USA

Shawn Broderick Department of Mathematics, Weber State University, Ogden, UT, USA

Paul Dawkins Department of Mathematical Sciences, Northern Illinois University, DeKalb, IL, USA

Benjamin Dickman Department of Mathematics, The Hewitt School, New York, NY, USA

Timothy Fukawa-Connelly College of Education, Temple University, Philadelphia, PA, USA

Steven Greenstein Department of Mathematical Sciences, Montclair State University, Montclair, NJ, USA

William McGuffey Department of Mathematics, Science and Technology, Teachers College, Columbia University, New York, NY, USA

Melissa Mills Department of Mathematics, Oklahoma State University, Stillwater, OK, USA

Teo Paoletti Department of Mathematical Sciences, Montclair State University, Montclair, NJ, USA

Jason Samuels Department of Mathematics, City University of New York-BMCC, New York, NY, USA

Julianna Stockton Woodrow Wilson Academy of Teaching and Learning, Cambridge, MA, USA

Megan Wawro Department of Mathematics, Virginia Polytechnic Institute and State University, Blacksburg, VA, USA

Ann Wheeler Department of Mathematics and Computer Science, Texas Woman's University, Denton, TX, USA

Sean Yee Department of Mathematics and Instruction and Teacher Education, University of South Carolina, Columbia, SC, USA

William Zahner Department of Mathematics and Statistics, San Diego State University, San Diego, CA, USA

About the Authors

James A. Mendoza Álvarez is professor of mathematics and distinguished teaching professor at the University of Texas at Arlington. He directs the master of arts program in mathematics for secondary teachers at UT Arlington. Currently, he also serves on the Mathematical Association of America Congress and on the Board of Directors of the Conference for the Advancement of Mathematics Teaching (CAMT). His scholarly interests include program and curriculum development, mathematical problem-solving, calculus learning, impact of mathematics-specific technology use on student learning and teacher preparation, and the mathematical education of teachers.

Erin E. Baldinger is an assistant professor of mathematics education at the University of Minnesota. Her research focuses on the preparation of secondary mathematics teachers. She is particularly interested in how preservice teachers develop mathematical knowledge for teaching and engage in mathematical practices. She also investigates how preservice teachers learn to enact core teaching practices such as leading whole-class discussions. She is a former middle school mathematics teacher.

Marita Barabash is an assistant professor of mathematics teaching in Achva Academic College and teaches in the Rothschild-Weizmann Program for excellence in mathematics and science teaching of the Weizmann Institute of Science. Her research focuses on mathematical subject matter knowledge as a didactic basis. Her mathematical background is PhD in applied mathematics. For the last 25 years, she educates mathematics teachers and is involved in research aiming at their ability to apply their mathematical knowledge to improve their teaching at school. She is also interested in geometry teaching and in issues related to educated usage of technology in teaching and learning mathematics.

Hyman Bass is the Samuel Eilenberg distinguished university professor of mathematics and mathematics education at the University of Michigan. His mathematical research has been in commutative homological algebra, algebraic K-theory, and geometric group theory. In the 1990s, he started collaborating with Deborah Ball to develop a practice-based theory of the mathematical understandings needed for teaching, with attention to equity. Currently he is also interested in mathematical practices, particularly learning how knowledge is generated and justified, starting in the early grades. He is also interested in novel kinds of problem-solving designs and in the roles of language in mathematics.

Anna Marie Bergman is a doctoral student at Portland State University, advised by Dr. Sean Larsen. Her research interests include undergraduate abstract algebra, particularly student's use of group theory within the context of chemistry.

John Caughman is a mathematics professor at Portland State University. His research expertise is focused on algebraic combinatorics and graph theory. He is also interested in mathematics education, particularly the teaching and learning of undergraduate combinatorics and abstract algebra.

John Paul Cook is an assistant professor of mathematics at Oklahoma State University. His research interests include undergraduate mathematics education, particularly investigating how students can come to understand the core concepts of abstract algebra by leveraging their own mathematical reasoning and intuition.

James Cummings is a professor in the Mathematical Sciences Department at Carnegie Mellon University. His research interests lie in set theory and the use of infinitary methods in finite combinatorics.

Al Cuoco taught high school mathematics to a wide range of students in the Woburn, Massachusetts, public schools from 1969 until 1993. He then moved to Education Development Center, where he works in curriculum and professional development. For the past three decades, he and Glenn Stevens have worked together on Boston University's PROMYS for Teachers, a professional development program for teachers based on an immersion experience in mathematics. A student of Ralph Greenberg, Al's research was in Iwasawa theory. Recent books include *Mathematical Connections: A Companion for Teachers and Others*, *Learning Modern Algebra* (with Joseph Rotman) and *Applications of Algebra and Geometry to the Work of Teaching* (with Bowen Kerins, Ben Sinwell, Darryl Yong, and Glenn Stevens), all published by MAA/AMS. But his favorite publication is a 1991 paper in the *American Mathematical Monthly*, described by his wife as an attempt to "explain a number system that no one understands with a picture that no one can see."

Allan Donsig is a mathematics professor at the University of Nebraska-Lincoln. His research interests are in operator algebras and in mathematics education, particularly improving teaching and learning in first-year mathematics courses. He is one of the leaders of a redesign of the first-year mathematics courses at UNL, which has significantly improved student success in precalculus. Together with

Kenneth R. Davidson, he has written an introductory analysis text, *Real Analysis and Applications: Theory and Practice* (Springer, 2010).

Ruhama Even is professor at the Weizmann Institute of Science and holds the Rudy Bruner Professorial Chair of Science Teaching. Her main research and development work is structured around three interrelated foci: (a) the professional education and development of mathematics teachers, (b) mathematics curriculum development and analysis, and (c) the interplay of factors involved in shaping students' opportunities to learn mathematics. Ruhama Even has been co-chair of ICMI Study 15 on the professional education and development of teachers of mathematics, and she serves as an editorial board member of the *Journal of Mathematics Teacher Education* (JMTE), *Mathematics Education Research Journal* (MERJ), and the *Encyclopedia of Mathematics Education* of which she is also a section editor.

Joshua Fagan is a PhD candidate in the final year of his dissertation at Texas State University. He has both a masters and bachelors in mathematics education. His chief area of research focuses on how students learn to construct, comprehend, and validate proofs and what/how they learn through proof as a curricular tool. He has also worked on research projects exploring students' cognitive processes in mathematical modeling tasks and the effects of professional development on teacher growth and student outcomes.

Patrick Galarza completed his MA degree in mathematics education at Teachers College, Columbia University in 2016 and remains at the institution, pursuing his PhD studies. His primary research interests are in the uses of mathematical games as pedagogical tools and problem-solving. He currently works as a private mathematics tutor and independent educational game designer.

Erin Glover is doctoral candidate in mathematics education at Oregon State University. She has extensive experience teaching mathematics to K-12 teachers. Her research focuses on the ways that instruction can support students in engaging in mathematical practices.

M. Kathleen Heid is distinguished professor of education (mathematics education) at the Pennsylvania State University, where she teaches undergraduate and graduate courses on the teaching and learning of mathematics. She served as editor of *Journal for Research in Mathematics Education*, member of the Board of Directors of National Council of Teachers of Mathematics, and member of the Board of Governors for Mathematical Association of America. Her research centers on the teaching and learning of mathematics in technological environments and on the mathematical understanding of secondary teachers. Her recent work has included the publication of *Mathematical Understanding for Secondary Teaching*.

Yvonne Lai is an assistant professor in the Department of Mathematics at the University of Nebraska-Lincoln. Her research focuses on mathematical knowledge for teaching and how this knowledge is promoted through educative curricula. She was a member of the writing team for the NCTM publication *Catalyzing Change* and was the founding chair of the Special Interest Group of the Mathematical Association of America on Mathematical Knowledge for Teaching (SIGMAA-MKT).

Sean Larsen is a professor in the Fariborz Maseeh Department of Mathematics and Statistics at Portland State University. His research interests include the teaching and learning of advanced mathematics. He was PI of the NSF-funded project Teaching Abstract Algebra for Understanding.

Younhee Lee is a lecturer in the Department of Mathematics at the Pennsylvania State University and finishing her PhD degree in mathematics education in the Department of Curriculum and Instruction. Prior to her graduate studies, she worked as a high school mathematics teacher in South Korea. Her research interests center on issues related to connecting school and collegiate mathematics. Those issues include how university students come to see school mathematics from an advanced viewpoint in their learning of collegiate mathematics and how one might teach collegiate mathematics in ways that help students make connections between school and collegiate mathematics.

Elise Lockwood is an assistant professor in the Department of Mathematics at Oregon State University. Her research focuses on undergraduate students' combinatorial reasoning, and she was recently awarded an NSF Career Grant to study students' combinatorial thinking in computational settings. She serves on the editorial board for the *International Journal of Research in Undergraduate Mathematics Education* and is affiliated with the Center for Computing in Science Education at the University of Oslo.

Ami M. Mamolo is an assistant professor in the Faculty of Education at University of Ontario Institute of Technology. Her research focuses on mathematical knowledge in and for teaching, including fostering mathematical reasoning and understanding via nonroutine problems and situations. She is a member of the Executive Committee for the Canadian Mathematics Education Study Group (Groupe Canadien d'Étude en Didactiques des Mathématiques), and she serves on the editorial board for the *Canadian Journal of Science, Mathematics, and Technology Education.*

Ofer Marmur is a postdoctoral fellow in the Faculty of Education at Simon Fraser University. His research interests include undergraduate mathematics education, teacher education, and specifically the interaction of affective and cognitive factors in student learning. He completed his doctoral studies in the Faculty of Education in Science and Technology at the Technion – Israel Institute of Technology in 2017.

During his studies, he was awarded the Jacobs Scholarship for Excellence in PhD Studies (2015), as well as the Kaplan Prize for Extraordinary Excellence in Research (2016).

William McCallum is a university distinguished professor of mathematics at the University of Arizona. Born in Sydney, Australia, in 1956, he received his PhD in mathematics from Harvard University in 1984, under the supervision of Barry Mazur. After spending 2 years at the University of California, Berkeley, and one at the Mathematical Sciences Research Institute in Berkeley, he joined the faculty at the University of Arizona in 1987. In 2005, he received the Director's Award for Distinguished Teaching Scholars from the National Science Foundation. In 2006, he founded the Institute for Mathematics and Education at the University of Arizona and is currently its director. In 2009–2010, he was one of the lead writers for the Common Core State Standards in Mathematics. His professional interests include arithmetical algebraic geometry and mathematics education. He has received grants and written articles, essays, and books in both areas.

Kathleen Melhuish is an assistant professor in the Department of Mathematics at Texas State University. Her primary research area is investigating student learning and understanding in abstract algebra. She developed and maintains the Group Theory Concept Assessment. Additionally, she studies how teachers and students conceptualize and engage in argumentation processes.

Eileen Murray is an assistant professor in the Department of Mathematical Sciences at Montclair State University. Her research centers around teacher learning along the professional continuum, from preservice teacher education through practicing teacher professional development. One aspect of this work focuses on the content preparation of secondary teachers through university mathematics courses, such as abstract algebra. Dr. Murray hopes to better understand the role such courses play in undergraduate mathematics teacher preparation. Specifically how courses that have natural content connections to secondary mathematics could impact instruction in middle and high school classrooms.

Josephine Shamash lectures at Achva Academic College and at the Weizmann Institute of Science. Her PhD from Yale University and postdoctoral fellowship at the Weizmann Institute were in pure mathematics. She has worked in mathematics education first at the Weizmann Institute Department of Science Education and then at Achva College, for the past 30 years, also teaching graduate algebra courses regularly at the Weizmann Institute, including in the Rothschild program in science education. Her research interests in mathematics education focus on the influence of abstract mathematical enrichment on the professional development of mathematics teachers and methods of teaching the evolution of mathematical ideas and concepts.

Philip Smith currently a visiting professor at Teachers College, Columbia, is a former mathematics department professor, chairman, dean of arts and sciences, vice president of academic affairs, and acting president at Southern Connecticut State University. His research interests lie in the areas of curriculum and problem-solving. In the late 1960s and early 1970s, he was a member of the Secondary School Mathematics Curriculum Improvement Study.

Ashley L. Suominen is a professor of mathematics at Savannah College of Art and Design in Atlanta. Her research interests include research in undergraduate mathematics education and curriculum. She has done extensive curriculum writing with Macmillan Publishers and CK-12 Foundation. She is also affiliated with the Advanced Placement Calculus Program.

Peter D. Taylor is a professor in the Department of Mathematics and Statistics at Queen's University, cross-appointed to the Department of Biology and the Faculty of Education. His areas of research are evolutionary game theory and secondary mathematics curriculum. He is a 3M fellow, a fellow of the Fields Institute, and past chair of the Education Committee of the Canadian Mathematical Society. He has done extensive curriculum writing with the Ontario Ministry of Education, and as preparation for this, he has taught two semesters in high school.

Nicholas H. Wasserman is an Assistant Professor of Mathematics Education at Teachers College, Columbia University. Previously, he taught mathematics for six years at the secondary level, in both a large public school in Austin and a private school in Manhattan. His scholarly interests focus on mathematics teachers' knowledge and development, particularly on how advanced content knowledge is relevant for teachers and influences secondary classroom teaching and practice. Dr. Wasserman's research has led him to examine how knowledge of Abstract Algebra becomes influential for the teaching of numerical and algebraic concepts in school mathematics. He is also collaborating with faculty from Rutgers University and Temple University to develop an instructional model for designing tasks in a real analysis course with secondary teachers in mind. His related interests include combinatorics education, both at the secondary and undergraduate levels, as well as how the use of dynamic technologies can influence mathematics teaching and learning.

Keith Weber is a professor of mathematics education at Rutgers University. His research interests include research in undergraduate mathematics education and the relationship between learning advanced mathematics and teaching secondary mathematics.

Diana White is an associate professor of mathematics and mathematics education in the Department of Mathematical and Statistical Sciences at the University of Colorado Denver. Her research and scholarly work has focused on commutative algebra, homological algebra, teacher education, history of mathematics, and mathematical outreach. She directs the National Association of Math Circles and the Rocky Mountain Math Circle Program.

Rina Zazkis is a professor of mathematics education at the Faculty of Education and associate member in the Department of Mathematics at the Simon Fraser University, Canada. Her research is in the area of undergraduate mathematics education, with a general focus on mathematical content knowledge of teachers and the ways in which this knowledge is acquired and modified. In 2016, Zazkis was appointed as Tier 1 Canada Research Chair, a prestigious recognition for excellence in research and research training.

Rose Mary Zbiek is a professor of mathematics education and department head for Curriculum and Instruction at the Pennsylvania State University. She is a former high school mathematics and computer science teacher. Her scholarly interests involve how teachers draw on their understandings of mathematics in their daily work of teaching. She is especially interested in teachers' work in contexts of algebra, mathematics technology, and mathematical modeling. She served as an associate editor for the *Journal for Research in Mathematics Education* and as the series editor for the 16 *Essential Understanding* books published by the National Council of Teachers Mathematics.

Chapter 1
Exploring Advanced Mathematics Courses and Content for Secondary Mathematics Teachers

Nicholas H. Wasserman

Introduction

Over the past century, mathematicians and mathematics educators have explored various ways in which abstract algebra is related to school mathematics. These have included Felix Klein's work as a mathematician—in which his synthesis of the study of geometry through abstract algebraic structures has proved influential on our approach to teaching secondary students geometry even today; his work as a mathematics teacher educator—famous for his observation of the "double-discontinuity" that secondary teachers face in their mathematical preparation; the provocative and controversial New Math curricular reforms in the 1960s in the USA, which reorganized and restructured the content of school mathematics to be more in accord with formal set theory and the study of algebraic structures; and various studies about, and investigations of, teachers' (advanced) mathematical knowledge in relation to their practices in the classroom and their student's outcomes. These efforts, and others, have considered the connection between school mathematics and the study of the abstract algebra structures they comprise.

The purpose of this edited volume is to continue to explore some of these connections, through a particular lens: secondary mathematics teacher preparation. That is, the sum of the chapters in this volume considers the role that studying abstract algebra can play in the preparation of teachers to teach secondary mathematics. Notably, the chapters do not consider this role as equivalent to only recognizing mathematical connections between two domains (i.e., advanced and secondary mathematics); they regard this role as also incorporating notions of student learning, disciplinary

N. H. Wasserman (✉)
Department of Mathematics, Science and Technology, Teachers College, Columbia University, New York, NY, USA
e-mail: wasserman@tc.columbia.edu

N. H. Wasserman (ed.), *Connecting Abstract Algebra to Secondary Mathematics, for Secondary Mathematics Teachers*, Research in Mathematics Education,
https://doi.org/10.1007/978-3-319-99214-3_1

knowledge, disciplinary practice, teacher practice, etc. Individual chapters probe these issues deeply, mining them for important insights. Collectively, the aim of the volume is to explore different facets of secondary teachers' study of abstract algebra concepts—as they might occur in an abstract algebra course or other types of content-focused teacher preparation or professional development courses. For abstract algebra instructors, mathematics teacher educators, researchers, secondary mathematics teachers, and others interested in these issues, I hope the content of these chapters proves to be both pertinent to, and stimulating for, your work.

In this introduction chapter, I highlight some of the core ideas that have shaped the volume's development and organizational structure as a whole. First, I highlight important mathematics education literature that provides some of the background and context for this volume. Second, I elaborate on four different kinds of connections to secondary teaching, especially as they help situate different aims of, and approaches to, teacher education. Indeed, these different kinds of connections are evident from the overall organization of the volume. Lastly, I provide an introduction to each section of the volume as well as the chapters within it.

Background for the Edited Volume

We consider the content of abstract algebra to be "advanced mathematics." That is, a first course in abstract algebra is relatively later on in one's course of mathematical study at the undergraduate level and it also represents a course in which very abstract mathematical ideas (e.g., algebraic structures) are the focus of study. Specifically with respect to the literature on teachers' mathematical knowledge, advanced mathematics has also been referred to as "horizon" (Ball, Thames, & Phelps, 2008) or "nonlocal" (Wasserman, 2018) content knowledge. Abstract algebra is also a proof-based course, with the presumption that students are engaged in proving theorems about the algebraic structures in question. Although some of the literature discussed relates specifically to abstract algebra, more generally, we discuss literature about advanced mathematics courses—of which, abstract algebra would be one.

Ostensible Benefits in Studying Advanced Mathematics

There appear to be two prominent arguments in the literature for secondary teachers studying advanced mathematics. The first is that the advanced ideas discussed in these courses are foundational for the mathematics studied at the secondary level (e.g., CBMS, 2012; Klein,1932). The second is that the study of advanced mathematics is productive for instilling broader mathematical norms, sensibilities, practices, etc., in teachers; for developing a sense of what "doing mathematics" entails.

In terms of the first argument—about connections to the content of school mathematics—Felix Klein (1932) was one of the first to advocate such an approach for teachers. Klein coined the phrase "elementary mathematics from an advanced standpoint," which has continued to be an influential perspective over the last century for preparing secondary mathematics teachers. His books are filled with numerous examples of how the study of advanced mathematics underpins more elementary mathematics. The *Mathematical Education of Teachers (I and II)*, more recent reports published by the Conference Board of the Mathematical Sciences (2001, 2012) that outline recommendations for mathematical content and courses to be included in teacher education programs, adopts a similar stance. They suggest, for example, that "[*i*]*t* would be quite useful for prospective teachers to see how $\mathbb{C}$ can be "built" as a quotient of $\mathbb{R}[x]$ and, more generally, how splitting fields for polynomials can be gotten in this way. The quadratic formula, Cardano's method, and the algorithm for solving quartics by radicals can all be developed from a structural perspective as a preview to Galois theory, bringing some coherence to the bag of tricks for factoring and completing the square that are traditional in high school algebra" (CBMS, 2012, pp. 59–60). Textbooks about mathematics for high school teachers (Bremigan, Bremigan, & Lorch, 2011; Sultan & Artzt, 2011; Usiskin, Peressini, Marchisotto, & Stanley, 2003) explore similar connections between the content of advanced mathematics and how it relates to the mathematics studied in secondary school. All of these suggest that advanced mathematics serves to deepen, and more rigorously confirm, the specific mathematical ideas secondary teachers will teach.

As for the second argument—about connections to the broader disciplinary practices of mathematics—the fact that secondary students are increasingly expected to engage in mathematical practices (e.g., NCTM's (2000) five process standards, CCSSM's (2010) eight mathematical practices) reinforces the importance of secondary teachers understanding them. Cuoco, Goldenberg, and Mark (1996) discussed these broader disciplinary practices as mathematical habits of mind. These included that students should be experimenters, describers, inventors, visualizers, conjecturers, etc.—attributes and habits of mind ascribed to mathematicians. The thinking is that, because the content of advanced mathematics courses is structured in a way that closely resembles such activities (i.e., beginning with descriptions of objects through precise definitions, then conjectures about those objects, then arguments for and proofs of those conjectures, etc.), such coursework can serve to help teachers develop these disciplinary habits. In elaborating on their provisional domain of "horizon content knowledge," Ball and Bass (2009) identified knowledge of major disciplinary ideas and structures, key mathematical practices, and core mathematical values and sensibilities as important components of this domain. That is, in terms of mathematics that relates to advanced mathematics courses, one aspect that is productive for teachers is the acquisition of broader disciplinary practices. Indeed, in designing the Rothschild–Weizmann master's program, Even (2011) expressed the justification for including mathematical studies at an advanced level: "The main purpose of the advanced mathematics component is to broaden secondary school teachers' perspective and understanding of and about the discipline of mathematics" (p. 944). All of these suggest that advanced mathematics can serve to reinforce and help teachers develop ways of thinking about and doing mathematics that align with disciplinary norms.

Are These Ostensible Benefits Realized by Secondary Teachers?

By and large, despite the strong arguments for how and why studying advanced mathematics might benefit secondary teachers, much of the research has found the opposite to be true: teachers and their students appear to gain little from a teacher's study of advanced mathematics.

Early on, Begle's (1972) quantitative study with 308 teachers found no significant correlation between teacher's understanding of modern algebra and their student's achievement in school algebra. Replication studies (e.g., Eisenberg, 1977) led to comparable conclusions. Two decades later, Monk (1994) investigated the relationship between the number of undergraduate mathematics courses taken by a teacher and that teacher's students' performance. This study had similar findings: courses beyond a fifth course taken by a teacher—i.e., an advanced mathematics course—appeared to have little to no effect on the learning outcomes of that teacher's students. Much of these early quantitative studies suggested that very little seemed to be gained by a teachers' study of advanced mathematics. However, it is difficult to create instruments that accurately measure knowledge in a valid and reliable way (e.g., Begle's approach), and counting mathematics courses is a coarse measure of "exposure" to content (e.g., Monk's approach), so perhaps such studies were not able to investigate the issue from a sufficiently fine-grained perspective. Even still, they are not encouraging when considering the role of a teacher's advanced study on their secondary students' performance.

Yet studies with teachers themselves seem to indicate more of the same. Zazkis and Leikin (2010) found from interviews with 52 practicing secondary teachers that they reported their knowledge of advanced mathematics as being rarely used and having little direct influence on their classroom practices. Other studies with teachers have reported similar results, indicating that the kinds of desirable connections between advanced and secondary content are not being made (e.g., Cofer, 2015; Goulding, Hatch, & Rodd, 2003; Rhoads, 2014; Wasserman, 2017). In interviews with 14 practicing secondary teachers, Wasserman, Weber, Villanueva, and Mejia-Ramos (2018) found that the majority of secondary teachers found explanations from real analysis about secondary mathematics topics as not useful for their teaching. Technical terminology and their perception of the capabilities of their students influenced the degree of relevance the teachers attributed to the explanations from advanced mathematics. The authors' use this to characterize teachers' perspectives as a "transport model"—where the perceived importance of an advanced mathematics explanation is dependent upon the teacher's ability to transport that explanation directly into their instruction in a secondary mathematics classroom. Consequently, their perceived value of such advanced mathematics coursework in their teacher preparation is inherently limited.

When looking at it from the perspective of what teachers *have* reported gaining from their advanced mathematical studies, one theme pertains to developing affective traits, such as empathy for students or confidence in their own mathematical ability. For example, Even (2011) reported that a major theme among teachers was of reminding them what learning mathematics feels like. Notably, although this can be a positive experience—e.g., when the instructor models good practices for learning mathematics—it may not always be. Many teachers had forgotten what it was like to experience difficulties when learning mathematics. The difficulties they faced in their own learning of new (advanced) ideas attuned them to students' "emotional difficulties." In other words, the struggle that the teachers themselves experienced while trying to learn difficult content opened their eyes to how a large number of students might feel every day in their classes (empathy); however, their ultimate ability to persevere through such study also provided some assurance that they could handle whatever they might encounter from students in the classroom (confidence). Zazkis and Leikin (2010) reported a similar theme from their interviews with secondary teachers about advanced mathematical knowledge. Now, both mathematical confidence and empathy for students are valuable traits for teachers to develop—ones that taking advanced mathematics seemingly provides. Yet developing affective traits would not be considered the main objective of requiring secondary teachers to take such advanced courses.

What Are Potential Explanations for Why Benefits Are Not Being Realized?

The explanations given in this section do not intend to be exhaustive in any sense of the word. There are many reasons why the ostensible benefits for studying advanced mathematics are not being realized. I elaborate on only a few.

First, advanced mathematics courses are hard. Numerous studies have revealed that students emerge from advanced mathematics courses with a limited understanding of the course material (e.g., Dubinsky, Dautermann, Leron, & Zazkis, 1994) and unable to engage in core disciplinary practices, such as proving (e.g., Weber, 2001). (See Rasmussen and Wawro (2017) for a recent review of this literature.) Second, although the mathematical connections between advanced and secondary mathematics are real, there are frequently differences in how the corresponding objects are represented in each domain (e.g., Moreira & David, 2008). Such differences add to the difficulty of establishing meaningful and productive connections between the two domains. Indeed, several studies (e.g., Rhoads, 2014; Wasserman, 2017) suggest that prospective teachers' understanding about secondary content potentially inhibits them from making meaningful and rich connections to the content of advanced mathematics. Third, the goals that teachers have in taking advanced mathematics courses may not align with the goals of mathematics educators. Ticknor (2012) argued that many teachers' goals are simply to pass such courses; thus, expecting teachers to reflect on how the material in abstract algebra might relate to secondary mathematics is not realistic given their goals. Lastly, even if prospective teachers learned the advanced mathematical content and could relate what they learned to the secondary mathematical content, this would not necessarily alter their pedagogical practice. That is, among other scholars, many of whom have contributed chapters to this volume, Wasserman (2016, 2018) and Wasserman, Fukawa-Connelly, Villanueva, Mejia-Ramos, & Weber (2017) and Wasserman, Weber, & McGuffey (2017) have argued that there is a difference between improved knowledge of secondary mathematics and improved knowledge for the teaching of secondary mathematics.

The point in summarizing some of this literature is not to "cut us off at the knees" before we begin (i.e., Why read further? There seems to be little value gained by secondary teachers in studying advanced mathematics anyway). Rather, it makes a different point. Based on extant literature, the study of advanced mathematics does not appear to be *inherently* beneficial to secondary teachers. Any exposure to or instruction about, say, abstract algebra content, is not innately relevant for secondary teachers, nor perceived by them as relevant for their future teaching. But that is not the same as claiming that it *cannot* be beneficial. Indeed, recent research (e.g., Even, 2011; Wasserman, Weber, et al., 2017) indicates that teachers can benefit from such study. With this in mind, the aim of this edited volume is to share potential approaches for overcoming this disconnect for secondary teachers with respect to the study of abstract algebra. Because it is not inherently beneficial, we need to be much more explicit about potentially productive ways of engaging secondary teachers with ideas in advanced mathematics. This includes being explicit about the pedagogical aims of such approaches to teacher education.

Points of Connection to Secondary Teaching

In what follows, I elaborate on different approaches to advanced mathematics instruction in secondary teacher education, based on four different points of connections between the study of advanced mathematics and secondary teaching: (1) content connections; (2) disciplinary practice connections; (3) classroom teaching connections; and (4) modeled instruction connections. The first two relate to connections discussed previously. The third relates to the idea that studies in advanced mathematics can be applied to the work of teaching; that is, situations in teaching can be used as a site for a kind of applied mathematics. The fourth is based on the notion that we teach how we were taught; that is, modeling particular kinds of instruction in advanced mathematics can shape a teacher's ideas about the mathematical learning process. Broadly, these four represent *how* an instructor of an advanced mathematics course might attempt to make their

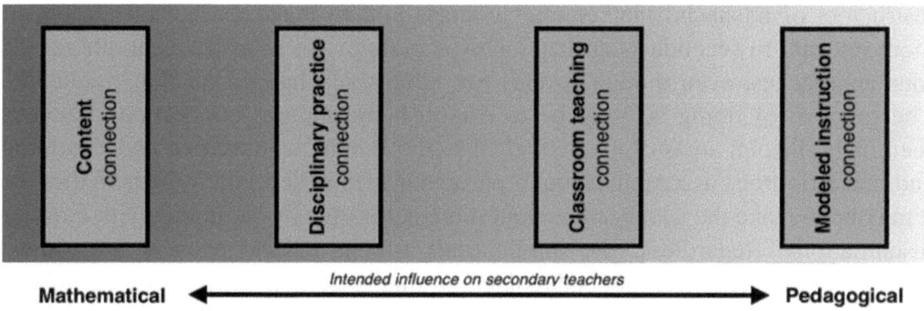

Fig. 1.1 Four points of connection to secondary teaching in advanced mathematics instruction, along a spectrum of intended implications

instruction more meaningful to future secondary mathematics teachers. I am not claiming this list to be exhaustive; however, listing also serves as a way of becoming more explicit, with the presumption that by being more explicit here, instruction in advanced mathematics courses can communicate more clearly the various points of connections being made to teaching.

In addition, I align these four points of connections on a spectrum with respect to the intended aim of such connections—ranging from "content" to "pedagogical" implications. (See Fig. 1.1.) In essence, their alignment along the spectrum indicates *what kind* of influence these sorts of connections and instructional approaches might aim for with respect to secondary teachers' future instruction—especially, in relation to either more mathematical or more pedagogical aspects of their teaching. Content connections are mathematical, with the intent of influencing the mathematical aspects of a teacher's teaching; on the other end, modeled instruction aims at the pedagogical aspects of a teacher's instruction. Disciplinary practices are primarily mathematical in nature, but they also are oriented toward a particular kind of pedagogical practice—one that engages learners in the doing of mathematics; classroom teaching connections are primarily pedagogical since they address specific situations in teaching, but they also draw on an application of some mathematical ideas in the process, which makes their utility at least partly about mathematical aspects of teaching.

Content Connections

In an advanced mathematics course, by a *content connection* being the point of connection to secondary teaching, I am meaning a connection between the content of advanced mathematics and the content of school mathematics. That is, there is a meaningful relationship between a concept discussed in the advanced course (e.g., a group in abstract algebra) and some secondary content (e.g., invertible functions). Explicitly discussing such mathematical connections is one way that instructors of advanced mathematics courses aim to make the advanced content more relevant to secondary teachers. This is perhaps the most frequent among the four approaches, even though it does not necessarily happen all that frequently. The purpose of doing so may be to establish meaningful connections between seemingly disjoint areas of mathematical study; it may be to deepen understanding and make rigorous assertions about a particular idea in secondary mathematics; or it may be to make the advanced content more accessible by beginning with familiar instantiations. Regardless, the implications for secondary teachers are almost universally mathematical (not pedagogical)—that is, in their teaching, teachers have improved mathematical knowledge, they better understand particular mathematical concepts, they can identify mathematical connections, they are less prone to making mathematical errors, etc. There is no pedagogical imperative; the primary purpose is to increase the mathematical expertise of secondary teachers with regard to the mathematical content they will be responsible for teaching.

Disciplinary Practice Connections

By a *disciplinary practice connection* being the point of connection to secondary teaching, I am meaning that the same kind of disciplinary practice that one engages in while studying advanced mathematics is also engaged in while studying secondary mathematics. That is, the processes that one engages in while "doing" advanced mathematics are related to some of the important mathematical practices that have been identified by mathematicians and mathematics educators—e.g., National Council of Teachers of Mathematics (NCTM) (2000) process standards, Common Core State Standards in Mathematics (CCSSM) (2010) mathematical practice standards, Cuoco et al.'s (1996) mathematical habits of mind, etc. In a sense, these kinds of connections serve a dual purpose. First, they serve a mathematical purpose. By becoming better "doers" of mathematics, secondary teachers have a better grasp on mathematics itself—i.e., the epistemological nature of mathematics, mathematical norms and sensibilities, etc. Second, though, these connections also serve a pedagogical purpose. That is, by learning more about the discipline of mathematics, there is a hope that secondary teacher's pedagogical choices will, in fact, engage their own students in these forms of thinking and doing. Thus, while these may be primarily about an improved mathematical sensibility (more on the mathematical end of the spectrum) there is also an embedded pedagogical implication (at least partially toward the pedagogical end of the spectrum).

Classroom Teaching Connections

In terms of a *classroom teaching connection* being the point of connection to secondary teaching, I am meaning some connection regarding the content of advanced mathematics, but as applicable to a specific secondary teaching situation. That is, the advanced mathematics is serving as a means to motivate particular and specific kinds of pedagogical actions in the classroom. Cuoco (2000), for example, considered the mathematical work that is involved for teachers in designing problems with "nice" solutions, making connections to specific algebraic structures. As another example, Wasserman and Weber (2017) explore how the study of proofs of the algebraic limit theorems, in a real analysis course, can be applied to situations when secondary teachers interact with secondary students about rounding and operating on rounded values. Classroom teaching connections are epitomized by leveraging "classroom teaching situations" as a means to connect to advanced mathematics topics. The primary implication here is about shaping a teacher's pedagogical response to a specific teaching situation—which may be about designing problems with particular characteristics, about responding to students, about sequencing activities, etc. However, such situations are also mathematical, in the sense that the intended point to exploring the teaching situation also includes applying and incorporating mathematical (and not strictly pedagogical) ideas.

Modeled Instruction Connections

Finally, by a *modeled instruction connection* being the point of connection to secondary teaching, I am indicating that another way that instructors of advanced mathematics may attempt to make pertinent connections to secondary teaching is by intentionally modeling particular kinds of instruction in mathematics. Essentially, having teachers *learn* mathematics in particular ways can shape their own approaches to *teaching*. Advocates of educational reform hope that, for example, experiencing inquiry-oriented approaches in their own learning will help teachers see the value in using these approaches with their own students. Modeling instruction can take many forms: use of technology, group work, inquiry learning, etc. In each of these cases, although mathematics is being learned, the intended implication for secondary teachers is about their pedagogy. That is, completely aside from the mathematical content, the implicit intent behind modeling instruction in mathematics courses for secondary teachers is that particular kinds of pedagogical choices or pedagogical models would be incorporated into their teaching.

A Few Remarks

The purpose of discussing these four points of connection is to be explicit about *some* of the ways that instructors of advanced mathematics have attempted (or might attempt) to make the study of course content more relevant for secondary teachers. These four were selected because of their prevalence—some more recent, some less so. Additionally, some of my recent work has indicated that each of these types of connections can have influence on, and be evident from, secondary teachers' observed classroom practice. Their alignment along the spectrum, spanning mathematical and pedagogical implications on teaching, further situates these in relation to each other. Primarily, it allows us—as instructors of mathematics courses and mathematics teacher education courses—to be more explicit about how the particular points of connection we might utilize are related to teaching. These two facets, mathematics and pedagogy, are sensible in that they are both important to mathematics teaching. However, there are other kinds of implications for teaching that may stem from studying advanced mathematics—such as affective implications. From the study of advanced mathematics, for example, a teacher might change their beliefs about the nature of mathematics, or increase their sense of mathematical confidence, or, by struggling themselves to learn difficult content, become more attuned to the struggles that their own students would have in learning mathematics. In addition, it might be the case that the descriptors of mathematics and pedagogy are too broad; further refinement of these ideas may help situate these different points of connection more clearly. Nonetheless, it is at least a first step in trying to organize some of the more-common approaches and points of connection as they relate to teaching.

Introduction to Sections and Chapters in the Edited Volume

Notably, the sections and chapters of this volume have been coordinated with, and organized by, some of the ideas already mentioned. In particular, this volume has four sections.

The first section explores the learning of abstract algebra concepts; challenges in learning were one of the potential reasons why the ostensible benefits from studying abstract algebra were not necessarily realized. Chapters in this section use cognitive analyses to provide insight into why, and how, concepts studied in abstract algebra pose challenges to learners, particularly as those abstract algebra concepts are related to secondary mathematics content. Insights from these chapters point out ways to attend to students' conceptual development, which, for instructors, can be useful for improving, and modeling, their own instruction—a possible point of connection to teaching. The second section explores in more detail the mathematical connections between abstract algebra and secondary mathematics. Chapters in this section primarily leverage mathematical analyses to explore specific kinds of content connections; the audience for the mathematical connections in these chapters ranges from secondary students, to secondary teachers, to secondary teacher educators. The third and fourth sections relate more explicitly to teacher education. That is, they comprise design research and other ideas that explore ways in which teacher education can connect to the work of secondary teaching. In particular, while all of the chapters include specific content connections (which was also the explicit focus of the second section), these two sections look especially at approaches to secondary teacher education that make use of disciplinary practice connections and classroom teaching connections. Respectively, the third section considers the use of disciplinary practice connections in secondary teacher education; and the fourth section considers using classroom teaching connections. Each of the four sections ends with a commentary chapter, coauthored to bring in perspectives of both mathematics educators and mathematicians; a reflection to the field on the broader theme of the section, as well as the contribution of the chapters within the section. In sum, the volume as a whole spans a range of mathematical and pedagogical implications that stem from various connections to, and approaches to instruction in, abstract algebra for secondary teacher education.

Exploring Students' Learning of Abstract Algebra Concepts

The three chapters in this section (two chapters and a commentary chapter) explore and summarize some of the cognitive issues that can arise in the learning of abstract algebra. Leveraging both larger quantitative studies and smaller qualitative studies, the authors of these chapters consider some of the challenges that arise in learning about abstract algebraic structures, such as groups and rings. In particular, they focus their analyses on concepts that play a fundamental role in both abstract algebra and secondary mathematics, including functions, binary operations, and the zero product property. These chapters provide insight into the ways that students' reasoning about these concepts—which are familiar from secondary mathematics—can impede or promote learning in abstract algebra. These cognitive analyses help depict the kinds of rich mathematical understandings that need to be developed by secondary mathematics students, and help inform how the study of abstract algebra might enrich secondary teachers' mathematical knowledge. The commentary chapter picks up these themes and explores more explicitly how these understandings might be leveraged to develop mathematical knowledge for teaching. Notably, although the chapters do not go into great detail about implications for instruction in abstract algebra, overall the authors promote instruction that develops deep conceptual understanding and that builds on students' reasoning and sense-making. They discuss both guided reinvention[1] and engineering backward transfer opportunities as potentially useful instructional approaches. That is, the work of these authors can promote improving the quality of instruction in abstract algebra. For secondary teachers, having such instruction modeled in their own mathematical learning might serve as a point of connection to their future teaching.

[1] Although not the focus of the chapters in this section, for those interested in how an abstract algebra course might draw on guided reinvention as a means to model inquiry-oriented instruction in undergraduate mathematics education, see also: Larsen (2009); Larsen, Johnson, and Weber (2013); and Weber and Larsen (2008).

Exploring Content Connections to Secondary Mathematics

The five chapters in this section (four chapters and a commentary chapter) consider some of the mathematical connections between the content of abstract algebra and the content of secondary mathematics. The most basic content connections include the fact that the objects of study in secondary mathematics are frequently examples of the algebraic structures studied in abstract algebra—of groups, of rings, and of fields. But these chapters also explore other connections, and from different perspectives. Two chapters look at connections to the secondary curriculum. One provides a look at the Secondary School Mathematics Curriculum Improvement Study (SSMCIS) from the 1960s, which brought ideas from modern algebra into the design of curricular materials for secondary students, and fleshes out the various examples and points of connection to abstract algebra in those curricular materials. The other provides a look at how modern curricular standards, and, specifically the Common Core State Standards in Mathematics, have used the higher standpoint of abstract algebra to inform their specific emphases and architecture (albeit not necessarily a direct incorporation of those ideas in the curriculum); this chapter gives a historical perspective about quadratic equations as an example. In the remaining chapters, one provides an approach to secondary teacher education that uses some unique content connections (e.g., division with remainder), and two types of problem-solving activities, as its point of connection to teaching, and another provides an analysis about how mathematicians and mathematics educators describe connections between abstract algebra and school mathematics. The commentary chapter reminds readers that to make abstract algebra courses rich opportunities for connecting to secondary mathematics, more work is needed in terms of developing curricular materials, conducting research, and, in particular, designing professional development opportunities for faculty who might teach such courses.

Exploring Approaches to Secondary Teacher Education via Disciplinary Practice Connections

The six chapters in this section (five chapters and a commentary chapter) comprise various approaches to teacher education that explicitly emphasize disciplinary practices. Some chapters report data from studies, others present and exemplify ideas. But each chapter uses disciplinary practices as the primary point of connection between abstract algebra and teaching, and considers these in the context of secondary teacher education. The commentary chapter serves as a reminder to consider some of the potentially unique contributions in this regard that studying abstract algebra (as opposed to other content areas) might provide. As a whole, the authors of these chapters consider different kinds of disciplinary practices and activities, including mathematical noticing, reasoning,creating, generalizing, defining, extending, and unifying. These disciplinary practices cut across various content examples, which span from equations and algebraic structures, to geometry and transformations, to historical developments and modern-day theorems. The authors use these examples to showcase the ways in which particular activities in an abstract algebra setting might be used to develop secondary teachers' sense of what "doing mathematics" entails—such as noticing symbolic form or axiomatizing transformational geometry. Although each of the chapters elaborates on connections to content, their approaches to teacher education are captured by their use of, and emphasis on, disciplinary practices as a central point of connection to teaching. That is, the authors exemplify that connections to teaching and learning in mathematics are not just about the results (e.g., content), but are also about the activities that produce those results. The commentary chapter picks up these themes and raises some important questions for the field about how we might tighten up theoretical perspectives and clarify goals as a means to improve the quality and scope of research in this nascent area.

Exploring Approaches to Secondary Teacher Education via Classroom Teaching Connections

The five chapters in this section (four chapters and a commentary chapter) explore various ways in which ideas from abstract algebra are germane to situations in secondary teaching—all of which are relevant for thinking about teacher education. Some chapters report data from studies; others present and exemplify ideas. But each chapter highlights ways in which knowledge of abstract algebra might influence the work of teaching. The commentary chapter identified four aspects: planning, task design, norm enactment, and in-the-moment responses. In relation to planning instruction, for example, some authors discuss how ideas in abstract algebra might be applied to how teachers design problems for students—especially problems that have particular pedagogical features; others discuss how it might be applied to expanding the example space used in teaching secondary students about the concept of a function, or to emphasizing mathematical properties in the process of solving equations. In other chapters, it is in relation to situations of contingency—when a teacher has to respond, in real time, to a student's question or solution approach. For example, some authors discuss how knowledge of abstract algebra might influence what a teacher notices in a student's question or solution approach—that is, they might "hear" or "see" classroom interactions differently; others discuss how it might influence the way in which a teacher goes about investigating and responding to a student's ideas. The commentary chapter explores the nature of these connections in further detail and comments on approaches that might support teachers' mathematical (and pedagogical) development as well as some of the challenges that they raise. Regardless of the nature of the teaching situation, each of the chapters in this section explores how, and why, attention to structural insights from abstract algebra can be productive for the work of teaching, and how these classroom teaching connections can be leveraged in secondary teacher education.

References

Ball, D. L., & Bass, H. (2009). *With an eye on the mathematical horizon: Knowing mathematics for teaching to learners' mathematical futures*. Paper presented at the 43rd Jahrestagung der Gesellschaft fur Didaktik der Mathematik, Oldenburg.

Ball, D. L., Thames, M. H., & Phelps, G. (2008). Content knowledge for teaching: What makes it special? *Journal of Teacher Education, 59*(5), 389–407.

Begle, E. G. (1972). *Teacher knowledge and pupil achievement in algebra (NLSMA technical report number 9)*. Palo Alto, CA: Stanford University, School Mathematics Study Group.

Bremigan, E. G., Bremigan, R. J., & Lorch, J. D. (2011). *Mathematics for secondary school teachers*. Washington, DC: Mathematical Association of America (MAA).

Cofer, T. (2015). Mathematical explanatory strategies employed by prospective secondary teachers. *International Journal of Research in Undergraduate Mathematics Education, 1*(1), 63–90.

Common Core State Standards in Mathematics (CCSSM). (2010). Retrieved from: http://www.corestandards.org/the-standards/mathematics

Conference Board of Mathematical Sciences (CBMS). (2001). *The mathematical education of teachers*. Retrieved from: https://www.cbmsweb.org/archive/MET_Document/index.htm

Conference Board of Mathematical Sciences (CBMS). (2012). *The mathematical education of teachers II*. Providence, RI: American Mathematical Society and Mathematical Association of America.

Cuoco, A. (2000). Meta-problems in mathematics. *The College Mathematics Journal, 31*(5), 373–378.

Cuoco, A., Goldenberg, E. P., & Mark, J. (1996). Habits of mind: An organizing principle for mathematics curricula. *Journal of Mathematical Behavior, 15*(4), 375–402.

Dubinsky, E., Dautermann, J., Leron, U., & Zazkis, R. (1994). On learning fundamental concepts of group theory. *Educational Studies in Mathematics, 27*(3), 267–305.

Eisenberg, T. A. (1977). Begle revisited: Teacher knowledge and student achievement in algebra. *Journal for Research in Mathematics Education, 8*(3), 216–222.

Even, R. (2011). The relevance of advanced mathematics studies to expertise in secondary school mathematics teaching: Practitioner's views. *ZDM, 43*(6–7), 941–950.

Goulding, M., Hatch, G., & Rodd, M. (2003). Undergraduate mathematics experience: Its significance in secondary mathematics teacher preparation. *Journal of Mathematics Teacher Education, 6*(4), 361–393.

Klein, F. (1932). *Elementary mathematics from an advanced standpoint: Arithmetic, algebra, analysis (trans. Hedrick, E. R. & Noble, C. A.)*. Mineola, NY: Macmillan.

Larsen, S. (2009). Reinventing the concepts of groups and isomorphisms: The case of Jessica and Sandra. *Journal of Mathematical Behavior, 28*(2–3), 119–137.

Larsen, S., Johnson, E., & Weber, K. (Eds.). (2013). The teaching abstract algebra for understanding project: Designing and scaling up a curriculum innovation. *Journal of Mathematical Behavior (Special Issue), 32*(4), 691–790.

Monk, D. H. (1994). Subject area preparation of secondary mathematics and science teachers and student achievement. *Economics of Education Review, 13*(2), 125–145.

Moreira, P. C., & David, M. M. (2008). Academic mathematics and mathematical knowledge needed in school teaching practice: Some conflicting elements. *Journal of Mathematics Teacher Education, 11*(1), 23–40.

National Council of Teachers of Mathematics (NCTM). (2000). *Principles and standards for school mathematics*. Reston, VA: Author.

Rasmussen, C., & Wawro, M. (2017). Post-calculus research in undergraduate mathematics education. In J. Cai (Ed.), *Compendium for research in mathematics education* (pp. 551–579). Reston, VA: National Council of Teachers of Mathematics (NCTM).

Rhoads, K. (2014). High school mathematics teachers' use of beliefs and knowledge in high quality instruction. *Unpublished doctoral dissertation*.

Sultan, A., & Artzt, A. F. (2011). *The mathematics that every secondary school math teacher needs to know*. New York, NY: Routledge.

Ticknor, C. S. (2012). Situated learning in an abstract algebra classroom. *Educational Studies in Mathematics, 81*(3), 307–323.

Usiskin, Z., Peressini, A., Marchisotto, E. A., & Stanley, D. (2003). *Mathematics for high school teachers: An advanced perspective*. Upper Saddle River, NJ: Pearson.

Wasserman, N. (2016). Abstract algebra for algebra teaching: Influencing school mathematics instruction. *Canadian Journal of Science Mathematics and Technology Education, 16*(1), 28–47.

Wasserman, N. (2017). Making sense of abstract algebra: Exploring secondary teachers' understanding of inverse functions in relation to its group structure. *Mathematical Thinking and Learning, 19*(3), 181–201.

Wasserman, N. (2018). Nonlocal mathematical knowledge for teaching. *Journal of Mathematical Behavior, 49*(1), 116–128.

Wasserman, N., Fukawa-Connelly, T., Villanueva, M., Mejia-Ramos, J. P., & Weber, K. (2017). Making real analysis relevant to secondary teachers: Building up from and stepping down to practice. *PRIMUS, 27*(6), 559–578.

Wasserman, N., & Weber, K. (2017). Pedagogical applications from real analysis for secondary mathematics teachers. *For the Learning of Mathematics, 37*(3), 14–18.

Wasserman, N., Weber, K., & McGuffey, W. (2017). Leveraging real analysis to foster pedagogical practices. In A. Weinberg, C. Rasmussen, J. Rabin, M. Wawro, & S. Brown (Eds.), *Proceedings of the 20th annual conference on research in undergraduate mathematics education (RUME)* (pp. 1–15). San Diego, CA: RUME.

Wasserman, N., Weber, K., Villanueva, M., & Mejia-Ramos, J. P. (2018). Mathematics teachers' views about the limited utility of real analysis: A transport model hypothesis. *Journal of Mathematical Behavior, 50*(1), 74–89.

Weber, K. (2001). Student difficulty in constructing proofs: The need for strategic knowledge. *Educational Studies in Mathematics, 48*(1), 101–119.

Weber, K., & Larsen, S. (2008). Teaching and learning group theory. In M. Carlson & C. Rasmussen (Eds.), *Making the connection: Research and teaching in undergraduate mathematics* (pp. 137–149). Washington, DC: Mathematical Association of America (MAA).

Zazkis, R., & Leikin, R. (2010). Advanced mathematical knowledge in teaching practice: Perceptions of secondary mathematics teachers. *Mathematical Thinking and Learning, 12*(4), 263–281.

Part I
Exploring Students' Learning of Abstract Algebra Concepts Closely Connected to Secondary Mathematics

Chapter 2
Connecting the Group Theory Concept Assessment to Core Concepts at the Secondary Level

Kathleen Melhuish and Joshua Fagan

Introduction

Abstract algebra is a fundamental course required of the majority of preservice secondary teachers in the USA (Blair, Kirkman, & Maxwell, 2013). It is in this course that students experience generalized structures that represent the number systems embedded in the K-12 curriculum. However, within this context, students may struggle to develop conceptual understanding. It is here that students often reason for the first time with "difficult notions of mathematical abstraction and formal proof" (Weber & Larsen, 2008, p. 139). In the past, researchers have stated that students find the content of abstract algebra to be "one of the most troublesome undergraduate subjects" (Dubinsky, Dautermann, Leron, & Zazkis, 1994, p. 268). The research on student understanding in abstract algebra plays out this fact, particularly in the context of introductory group theory (Dubinsky et al., 1994; Hazzan, 1999; Leron, Hazzan, & Zazkis, 1995). Furthermore, research focused on secondary preservice teachers suggests they may not be making connections between the content of such courses and the content at the secondary level (Ticknor, 2012; Wasserman, 2017). We posit that building rich connections between group theory and secondary content requires a deep level of conceptual understanding. Without such a precursor, students do not have the foundation necessary to build connections.

The focus of our research is on students' conceptual understanding of introductory group theory topics, and, particularly, connections to the core concepts of function and binary operation. In this chapter, we focus explicitly on connections between the Group Theory Concept Assessment (GTCA) (Melhuish, 2015) and the

K. Melhuish (✉) · J. Fagan
Department of Mathematics, Texas State University, San Marcos, TX, USA
e-mail: melhuish@txstate.edu

N. H. Wasserman (ed.), *Connecting Abstract Algebra to Secondary Mathematics, for Secondary Mathematics Teachers*, Research in Mathematics Education,
https://doi.org/10.1007/978-3-319-99214-3_2

secondary level, in terms of both the population of preservice secondary teachers and the mathematical content. We structure our discussion around three core connections: (1) establishing the degree to which preservice secondary teachers are building conceptual understanding of group theory (as measured by the GTCA); (2) exploring how understanding of key secondary topics (binary operation and function) can inform or hinder success on GTCA tasks; and (3) leveraging GTCA tasks to theorize how engagement in complex group theory tasks can strengthen student understanding of binary operations and functions via backward transfer.

Literature and Theoretical Perspective

Underlying our theoretical framing are three key constructs: conceptual understanding, abstract structures, and core concepts. We begin by outlining what we mean by conceptual understanding and, particularly, how conceptual understanding interplays with abstract structures in introductory group theory. This framing situates the larger GTCA instrument. We then leverage the notion of core concepts to identify and analyze two essential concepts that transcend mathematical curricula: binary operation and function. We conclude this section by situating conceptual components of binary operations and functions as found in the literature base.

Conceptual Understanding in Group Theory

We treat learning as a cognitive activity in which students construct knowledge through building on and adapting prior knowledge during new experiences (cf., von Glasersfeld, 1995). We then conceptualize knowledge on a continuum: on one end, being rich in connections and internally cohesive, and on the other end, disconnected and potentially contradictory (to the observing expert). Strong conceptual understanding reflects knowledge rich in connections (Hiebert & Lefevre, 1986). Furthermore, we treat conceptual understanding as reflecting internally coherent and consistent knowledge. Tall and Vinner (1981) provided a fundamental framework for probing this coherence: *concept image*. A concept image moves beyond just a concept definition or even a personal definition (the words an individual uses to define a concept) to include all of the individual's surrounding cognitive structures. A robust understanding of a particular concept requires significantly more than an accurate definition. Concept images contain exemplars, metaphors, representations, and any number of cognitive structures that are activated when engaging in tasks related to a particular concept. When engaging in a task, students draw upon various components of their concept images in ways that can be supportive, or problematic (if their images lack important components or vital connections). When considering the construct of conceptual understanding in the group theory context, it becomes necessary to consider what makes concepts in this setting different from many of the other settings in which we have explored conceptual understanding. We note two significant differences: proof as the primary mechanism for activity and the abstract nature of concepts in the course. In regard to the former, we distinguish proof activity (i.e., producing logical, deductive arguments) from the understanding of the concepts themselves. Although we recognize that conceptual understanding can inform proof production, we do not aim to directly examine proof production as part of conceptual understanding. This leaves the abstract, stipulated nature of concepts as a necessary aspect on which to make sense of conceptual understanding in group theory.

We borrow from Hazzan's (1999) treatment of abstraction, in which abstract captures notions of familiarity, generality, and process–object duality. Within group theory, students must work at an appropriate abstraction level to accurately engage with tasks. In order to probe conceptual understanding, students must reflect accurate conceptual connections to engage in tasks with *unfamiliar* structures (such as modular groups or unfamiliar binary operations), and *general* classes of *objects* (such as all groups rather than a specific example). In this way, tasks that probe a particular, familiar group (such as the integers under addition) are less likely to unearth incoherence within a concept image. Rather, the unfamiliar and general provide impetus for students to navigate abstraction in a way that reflects their conceptual understanding. Although the contexts are new, by the nature of group theory, they still evoke aspects of core concepts from familiar settings that may or may not be a cohesive part of the new cognitive structures developed around the abstract versions of these core concepts. In the next section, we then unpack two core concepts: binary operation and function.

Core Concepts: Binary Operation and Function

We use core concepts (cf., Akkoç & Tall, 2002) to refer to concepts that are threaded throughout the mathematical curriculum from course to course. As such, core concepts occurring in abstract algebra have rich potential to connect to the secondary level. We focus primarily on two concepts that underlie much of the group theory curriculum: function and binary operation. While we acknowledge that binary operations are a type of function, we see their treatment as largely independent in curricula (e.g., the treatment of function versus arithmetic operations in Common Core State Standards Initiative, 2010), research (e.g., Wasserman, 2017), and in our experience within students' concept images (Melhuish & Fagan, 2017). For this reason, we treat binary operation and function as different core concepts, rather than situating binary operation as a subclass of function. In this section, we organize literature on student concept image components related to binary operation and function in terms of process–object duality, metaphors, representations, and properties. We note that the literature is quite robust related to function, but much more limited with respect to binary operation.

Function Concept Image Components

The literature on function can roughly be divided into several categorizations: process–object duality focused, representation-focused, metaphor-focused, property-focused, and covariation-focused. For our purposes, we restrict our overview to the first four categories as covariational reasoning superimposes an ordinal system on function, which does not underlie many of the discrete structures in abstract algebra.

From a process–object duality (cf., Sfard, 1991) standpoint, a function can be thought of as something that acts on elements from a domain or, alternately, as an object itself that can be operated on (such as composing two functions). Breidenbach, Dubinsky, Hawks, and Nichols (1992) further decompose function conceptions into an action conception (operating on individual elements one at a time) and a process conception (seeing a function as a process in totality). Much of the work in this realm has established that students struggle to move beyond action conceptions of function to process conceptions and from process conceptions to object conceptions (e.g., Breidenbach et al., 1992; Oehrtman, Carlson, & Thompson, 2008). Students' concept images of function may reflect varying degrees of action, process, or object conceptions.

Functions also lend themselves to a number of representations including symbolic, graphic, diagrammatic, and tabular. One class of studies has addressed which representations of function students accept and prefer. For example, Knuth (2000) found that students prefer symbolic forms of function rather than graphic, even when graphic forms would be beneficial in problem-solving. Another class has explored how students interact with certain representation types. For example, Clement (2001) found students do not always see tabular-defined functions as functions. In general, students struggle to move between various representations (Akkoç & Tall, 2002; Schwarz, Dreyfus, & Bruckheimer, 1990) and often see different representations as *distinct* functions rather than as different representations of the same function (Elia, Panaoura, Eracleous, & Gagatsis, 2007). Across most studies, students have a clear preference for symbolic representations. As noted by Thompson (1994) when describing undergraduates' notions of function, "[a] predominant image evoked in students by the word 'function' is of two written expressions separated by an equal sign" (p. 5). The dominance of this particular symbolic representation exists across the secondary and post-secondary level.

Conceptual metaphors (cf., Lakoff & Núñez, 2000) also play a powerful role in concept images. In addition to students' personal definitions for functions, they often draw on informal metaphors to provide meaning in tasks. One of the most powerful metaphors related to functions is that of an input–output machine (Tall, McGowen, & DeMarois, 2000). In the undergraduate population (linear algebra students), Zandieh, Ellis, and Rasmussen (2017) identified a range of metaphors in addition to a machine, including input/output, traveling, morphing, and mapping. Such metaphors can be supportive when engaging in tasks, so long as they remain mathematically accurate.

The last cognitive aspect we consider is that of properties. A given concept has a set of required properties and properties that may be varied. A property-focused view represents an alternative route toward abstraction (rather than focusing from process to object reification) where students extrapolate properties from examples. Slavit (1997) provides a property-based decomposition of function with a focus on both the required properties and the varying properties. Particularly relevant to group theory are the varying function properties of injective and surjective, that distinguish homomorphisms from isomorphisms. On one end, students may overgeneralize these properties and, for example, have a strong preference for all functions to be injective (e.g., Leinhardt, Zaslavsky, & Stein, 1990). On the other end, students may not identify all required properties in their concept images and, for example, miss the requirement of well-definedness. Even (1993) documented that this issue exists among secondary preservice teachers.

Binary Operation Concept Image Components

In contrast to function, which is studied as a generally defined construct (even though it is frequently limited to common exemplar functions situated in familiar numeric systems (Tall et al., 2000)), the majority of research on binary operation remains specific to different arithmetic operations. In this way, the literature about general metaphors and properties is scarce. We lightly treat the same cognitive structures as functions: process–object duality, representations, metaphors, and properties.

The general notion of a binary operation has been decomposed into Action, Process, Object, and Schema by Brown, DeVries, Dubinsky, and Thomas (1997) and, more recently, Wasserman (2017). An action conception equates to being able to evaluate a single instance of a binary operation, such as evaluating $2 + 3$, and may rely on a physical representation. A process conception of binary operation reflects viewing a binary operation as a process of providing two inputs and arriving at an output without needing to reference the explicit elements. An object conception reflects treating a binary operation as an object itself to be operated on (such as composing two binary operations). As illustrated by Brown et al. (1997), having only an action or process conception of binary operation may limit the tasks in which students can engage successfully.

As in the case of functions, binary operations also have different representations. In a recent exploration of abstract algebra textbooks, the first author identified several, including: element-wise defined, Cayley tables, verbal descriptions, and symbolic representations (Melhuish, 2015). In a recent study focused on binary operation, we found that students preferred symbolic representations (Melhuish & Fagan, 2017). As noted by Thompson (1994), students often have a predominant image of function. For binary operation, we found this image to be element–operator–element (E-O-E). This is an idea we elaborate on in the discussion of students' performance on the GTCA.

The literature on conceptual metaphors for binary operation is lacking. Generally, research presents metaphors for specific arithmetic operations. For example, Lakoff and Núñez (2000) elaborate on a collection metaphor as underlying the standard arithmetic operations. In this metaphor, addition is the joining of two collections. As far as more general metaphors for binary operation, we have found some power in adapting the function metaphors (Melhuish & Fagan, 2017). For example, we found that students using an input–output metaphor when engaged in tasks prompted them to identify examples of binary operations.

Properties of binary operation are treated somewhat more extensively in the literature. Analyses focused on "structure sense," position binary operations as a central feature (Novotná & Hoch, 2008) that ties together structural properties. Important facets of structure sense include making sense of properties, their quantification, and the relations between properties. A number of studies have specifically focused on the properties of commutativity and associativity. Some outcomes include: students struggle to differentiate these two properties (Larsen, 2010), students may see commutativity as implying associativity (Tirosh, Hadass, & Movshovitz-Hadar, 1991), and students' example space of binary operation may not be robust enough for them to identify cases of binary operations that are not associative, but are commutative (Zaslavsky & Peled, 1996). Commutativity and associativity are varying properties of a binary operation and, therefore, may or may not hold for a given binary operation. On the other hand, a binary operation does have some essential properties, including being a function, having a binary input (i.e., two inputs, or, alternately, one input that is an ordered pair), and maintaining closure. In our recent work, we have found that the closure property (and not the binary or function properties) dominates students' concept images (Melhuish & Fagan, 2017). See Table 2.1 for an overview of the concept image structures discussed.

Function and Binary Operation from the Secondary Level to Abstract Algebra

Function and binary operation play essential roles in many abstract algebra topics. This is reflected in their existence across numerous works providing genetic decompositions of group theory concepts (Brown et al., 1997; Dubinsky et al., 1994). Furthermore, binary operation also plays a fundamental role in developing structure sense levels that connect high school algebra and abstract algebra (Novotná & Hoch, 2008).

The literature almost universally documents that students struggle to develop rich conceptions of abstract algebra concepts that implicitly or explicitly rely on functions and binary operations (Dubinsky et al., 1994; Hazzan, 1999; Leron et al., 1995). Much of this difficulty can be understood in terms of the increasingly abstract and general structures found in abstract algebra. The literature richly documents that

Table 2.1 Cognitive aspects of binary operation and function

	Binary operation	Function
Metaphors	Inputs/outputs Combining (arithmetic metaphors: collection-driven)	Input/output Traveling Morphing Mapping Function machines
Process–object duality	A *process* from combining any two elements An *object* that can be operated on	A *process* for mapping from inputs to outputs An *object* that can be operated on
Representations	Element-defined Tabular E-O-E symbolic	Element-defined Tabular Function-diagram $y = f(x)$ symbolic Graphical
Properties	Required: Binary Closed Function Variable: Associative Commutative	Required: Well-defined Everywhere defined Domain set, range set Variable: Injective Surjective

when students move from arithmetic to the more generalized high school algebra, conceptual limitations around arithmetic are unearthed (Slavit, 1998; Warren, 2003). In a similar way, we may expect that the shift from high school algebra to general structures in abstract algebra would unearth limited conceptions of underlying topics, including function and binary operation.

In secondary settings, functions are almost universally defined on the set of real numbers. Although a function is defined generally, its applications in practice are specific. In terms of binary operation, this leap is more pronounced because most students have never encountered a general definition of a binary operation. Rather, operations in secondary mathematics are specific to numeric and algebraic contexts. However, even high school students are placed in a position to generalize when they encounter properties (such as associativity) that span across operations. We argue that rich conceptions of the general constructs of binary operation and function can enhance understandings of the specific examples of these topics at the secondary level. We also argue that students need flexible and rich conceptions of binary operation and function to be successful in abstract algebra. Without them, they may incorrectly revert to familiar numerical settings (e.g., Hazzan, 1999), or lack the ability to leverage important metaphors or representations while engaging in abstract algebra tasks. In this chapter, we document that students struggle in these ways when engaged with abstract algebra tasks, and provide insights into the understandings of function and binary operation that can be productive for helping students successfully complete such tasks.

Instrument and Methods

The data presented in this chapter comes from data collected during the development and validation of the GTCA (Melhuish, 2015). The GTCA was created to probe students' conceptual understanding, as defined by knowledge of concepts rich in relationships (Hiebert & Lefevre, 1986). Particularly, tasks were created such that students had to engage at a level of abstraction that was generally higher than that of previous courses, such as dealing with new structures (like modular groups) or general structures. The GTCA covers the following topics: (1) binary operation (including identities, inverses, and associative and commutative properties), (2) isomorphisms and homomorphisms (including the Fundamental Homomorphism Theorem and kernels), (3) Lagrange's theorem, (4) cyclic groups, (5) groups, subgroups, and normal subgroups, (6) cosets and quotient (factor) groups, and (7) order of elements. Tasks were created based on results from a textbook analysis, expert consensus on topic coverage, expert review of items, and literature review. During the initial development, 387 students (27 in pilot round; 350 in open-ended round) spanning 24 undergraduate institutions in the USA responded to a set of 18 open-ended tasks in either an early pilot or open-ended round. The first author then interviewed a subset of 15 students with representative responses. She prompted the students to share their thinking about the questions, their responses, and their understanding of various concepts. The most common responses during the open-ended round became the options for the multiple-choice version given to an additional 462 students (87 in round 1; 375 in round 2) spanning 41 institutions. Similar validation interviews were conducted with another 15 students. Throughout this process, questions were refined to connect to student concept images and to optimize test characteristics. See Table 2.2 for a breakdown of institution types and students for each round.

This chapter focuses on two parts of this project: overall performance on four GTCA tasks differentiated by major, and qualitative analysis (stemming primarily from the interview data). We analyzed our open-ended responses and interview data to identify different ways that knowledge of binary operation and function may play a supportive or inhibitive role in group theory task performance.

Research Questions

We take the view that every task in group theory relies on fundamental ideas from prior education. A group itself is no more than a set with a binary operation following certain rules (i.e., closure, associativity, and the existence of an identity and inverse elements). Furthermore, particular functions provide the mechanism for larger ideas, such as isomorphic groups, quotient groups, and other fundamental theorems. In this sense, preservice teacher performance on the GTCA as a whole is immediately important. In terms of overall GTCA scores, mathematics education majors performed slightly worse than other students ($t(161.067) = -3.175$,

Table 2.2 Sample institution characteristics for GTCA development rounds

	Least selective (>75% admitted)	Mid-level selective (50 – 75% admitted)	More selective (25 – 50% admitted)	Most selective (<25% admitted)	Not classified	Total
Open-ended round	13 Classes (138 students)	12 Classes (108 students)	4 Classes (47 students)	1 Class (57 students)	0 Classes	20 Classes (350 students)
Closed-form round 1	2 Classes (17 students)	0 Classes	3 Classes (26 students)	4 Classes (44 students)	0 Classes	9 Classes (87 students)
Closed-form round 2	13 Classes (131 students)	10 Classes (127 students)	6 Classes (87 students)	1 Class (14 students)	2 Classes (16 students)	32 Classes (375 students)
Total	28 Classes (286 students)	22 Classes (236 students)	13 Classes (157 students)	6 Classes (119 students)	2 Classes (19 students)	61 Classes (812 students)

$p = 0.002$). This difference remained consistent even when controlling for gender, institution-type, and students nested in classrooms. The overall performance of mathematics education majors reflects that many did not demonstrate conceptual understanding on a number of tasks. We share this information not for the purpose of comparison, but rather for the purpose of motivation. Secondary teachers play an important role in developing secondary students' conceptual understanding of function and binary operation. It is imperative to have insight into the kinds of understandings that might be particularly productive to develop about these core concepts.

We use the GTCA data for this purpose answering questions:

1. How are mathematics education majors[1] performing on the GTCA group theory tasks?
2. How may students' knowledge of functions and binary operation aid or hinder their ability to correctly answer a subset of GTCA prompts?
3. How might GTCA tasks be leveraged to promote backward transfer to enrich knowledge of function and operation?

We explore the first two questions empirically, leveraging data from the GTCA development study. We address the third question theoretically in a discussion session.

Task Performance Results

We use meaning and metaphors, properties, and representations as the guiding foci to identify the relationship between task performance and concept images of the core concepts of function and binary operation. We begin each section by presenting two related tasks from the GTCA and providing a breakdown of student responses by frequency. We then provide exemplars of students using meanings, metaphors, representations, and properties when approaching the tasks. We note that the tasks are not exhaustively treated; rather, we focused on a subset of responses connected to supportive or problematic areas related to the core concepts.

Functions

We begin our exploration with functions. In group theory, common functions include homomorphisms and isomorphisms. For the scope of this chapter, we focus on two prompts related to homomorphisms:

---

[1]Note: The GTCA survey collected data on major. However, this distinction may not cover all preservice secondary teachers at institutions where preservice teachers obtain a mathematics major.

Table 2.3 Student responses to function prompt 1

Response	Other majors	Mathematics education majors
(a) {1}	25.7 % ($n = 75$)	36.1 % ($n = 30$)
(b) {4}	9.6 % ($n = 28$)	10.8 % ($n = 9$)
(c) **Multiples of 4**	51.0 % ($\mathbf{n} = 149$)	48.2 % ($\mathbf{n} = 40$)
(d) Empty set	13.7 % ($n = 40$)	4.8 % ($n = 4$)

Table 2.4 Student responses to function prompt 2

Response	Other majors	Mathematics education majors
(a) **1, 3, 12**	18.8 % ($\mathbf{n} = 55$)	22.9 % ($\mathbf{n} = 19$)
(b) 1, 4, 12	30.8 % ($n = 90$)	27.7 % ($n = 23$)
(c) 1, 2, 3, 4, 6, 12	28.4 % ($n = 83$)	22.9 % ($n = 19$)
(d) 4, 8, 12	21.9 % ($n = 64$)	26.5 % ($n = 22$)

1. Let $i = \sqrt{-1}$. Consider the homomorphism $\phi(n) = i^n$, that maps $\mathbb{Z}$ under addition to the set $H = \{1, -1, i, -i\}$ (a subgroup of $\mathbb{C}$ under multiplication). What is the kernel of this homomorphism?

This prompt requires students to deal with an unfamiliar function (with an imaginary component) to an unfamiliar group, the subgroup of the complex numbers: $\{1, -1, i, -i\}$. Of particular note, this function does not have the injective property that students often prefer. In order to identify the kernel, a student must realize that many elements will map to the identity in the subgroup H. Table 2.3 presents the multiple-choice options and the frequency they were selected in the large-scale round. The correct response is bolded.

2. Let ϕ be a homomorphism from group G to group H. Let G have order 12. Suppose G has only one normal (nontrivial and proper) subgroup K and K has order 4. What are the possible orders for $\phi(G)$?

This prompt connects to what is often the pinnacle theorem in a 10-week introduction to group theory: The Fundamental Homomorphism Theorem (FHT). In order for a student to engage meaningfully with this prompt, he/she must recognize the role of various functions in the theorem. A homomorphism collapses to an isomorphism via collecting elements that are in the same preimage. These collections can be structured by identifying the kernel (which must be a normal subgroup). In this case, the options for the order of the image of the homomorphism are 1, 3, or 12, where the kernel is the entire group G, the normal subgroup of order 4, or the trivial subgroup, respectively. Although we acknowledge that this prompt requires a great deal of knowledge, in addition to understanding functions, functions provide the foundation for making sense of the theorem. Table 2.4 presents the multiple-choice options and the frequency they were selected in the large-scale round. On the surface, these distributions may appear to reflect random guessing (25% each); however, the validation interviews illustrated that students had sensible reasoning behind each answer selection. The correct response is bolded.

Meanings and Metaphors of Function: Machines and Collapsing

A function is a map (or relation) from a domain to range, such that each domain element maps to exactly one element in the range. In contrast to binary operation, meanings and metaphors of function have been extensively treated in the literature. Most of this work stems from the secondary level, with a recent expansion to the undergraduate linear algebra level (Zandieh et al., 2017). Metaphors, such as input–output machines or morphing from domain element to range element, are prevalent.

Students leveraged similar metaphors in the GTCA tasks. Using a metaphor was often supportive in these situations. For example, in order to find the kernel of ϕ from the first prompt, students must recognize a particular preimage: the preimage of the identity. Metaphors that allow for functions to have multiple inputs to single output, and for imagining a preimage (necessitating at minimum a process conception of function) can support an accurate exploration of such a prompt. One successful student explained the kernel as:

> [a]ll the elements in the integers that would fit through the function and go to the identity on the codomain. Then start plugging things into that function. Plugging integers into that function ... I just knew from experience that [the multiples of four] has a pattern and generalized.

This student is using language that reflects a function as something that elements "fit through" or that elements are "plugg[ed]" into. The student attends not just to a single element that "goes to" the identity, but *all* elements that "go" to it. This metaphor parallels a function machine where inputs go through and become outputs.

Another powerful student metaphor for these two tasks is that of "collapsing." In some ways, this is analogous to a morphing metaphor where elements are altered to arrive at their image. However, we regard collapsing language as particularly powerful because it connects directly to a function having preimages with multiple elements. Each preimage set is collapsed into the same image element.

Properties of Function: Injective

The collapsing metaphor helps to address one notable issue related to properties: the treatment of all functions as if they were injective. In the kernel task, this type of overgeneralizing was found amongst students who identified the kernel as containing only one element: 4. In some sense, presuming injectivity represents a simpler treatment of function (despite having an additional property). If a function is injective and surjective, then the codomain and image of a function are not distinct entities, and there are only single elements in a given element's preimage. We found, even when students defined kernels in a general manner (e.g.: "kernels are what map to the identity element of your second group"), they often reverted back to treating functions as injective when engaging in the tasks. The same student from the parenthetical quote went on to explain, "What is *the* [emphasis added] x such that i to the x equals 1?" They knew that the identity in the codomain was a singular element, {1}; thus, if they falsely assumed that the mapping was a bijection, the associated kernel could only have a singular element as well. This may reflect an action conception of function where only one element is treated at a time. Students may then rely on the first element they identify in the preimage. This treatment may be a subconscious, rather than an explicit, assumption about maintaining the injective property. However, such reasoning highlights that students may not be attending directly to injectivity as a variable aspect of function.

Representation of Function: Symbolic and Function Diagrams

In terms of representations, symbolic formulas were ubiquitous across tasks. In the open-ended rounds, students infrequently provided any alternate representation. One of the interviewed students provided a notable exception to this trend. This student navigated the complexities of the FHT prompt using the function diagram in Fig. 2.1, explaining it as follows:

> So we have some group, we'll call that H and G we can partition into things that go to specific parts of H.... Suppose we have this ϕ, and so everything G_1, G_2, G_3, let's say G_1. Everything in G_1 is going to map to the *identity* and everything in G_2 is going to map to a. In order to have a quotient group, we have to have some kernel K...

At this point, the student continued walking through the kernel possibilities. This robust function diagram reflects the advanced visual imagery associated with the theorem in Nardi's (2000) work. This student successfully partitions G into cosets of elements with the same image in H. Making sense of the FHT requires a significant amount of coordination, including understanding the role of kernels, the nature of homomorphisms, and the relationship between the quotient groups formed and the image of the homomorphism. This function diagram tracks many of these complexities and supported this student in unpacking the task.

A majority of students either did not have access to this function representation or otherwise did not engage in leveraging the representation. Although one could argue that the complexity of the FHT may be a mitigating factor in the use of diagrams, we have evidence from our other research and other tasks that function diagrams are infrequently used even in simpler contexts within group theory (see Melhuish, Larsen, & Cook, 2018).

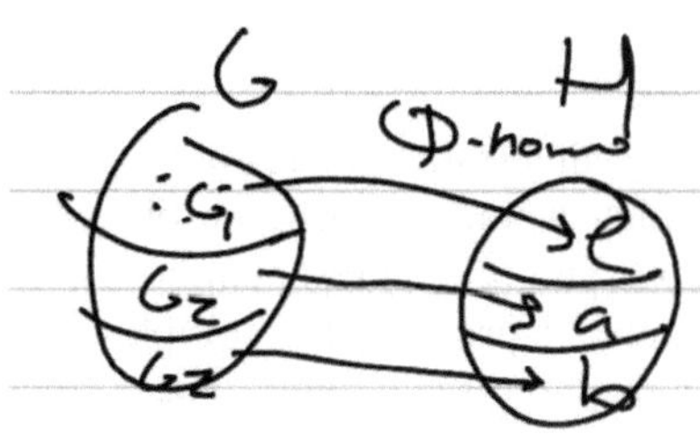

Fig. 2.1 Function diagram representation of the fundamental homomorphism theorem

Conclusion

This pair of tasks from the GTCA manifested a number of student conceptions and activities related to functions. Students' work on the GTCA function tasks supported the following conclusions: (1) Students' existing metaphors for functions (input–output machines or morphing) can be supportive in more abstract mathematics settings involving functions; (2) Students may overgeneralize properties (such as injective) despite having accurate language for surrounding concepts (e.g., kernel); (3) Students tend to prefer symbolic manipulation and representation even when an alternative representation may be more supportive.

Binary Operation

We start this section with reflection on what it means for an operation to be a *binary* operation on a set. Binary operations on a set can be thought of as a way to combine two elements from a set resulting in a third element from that set. Formally, a binary operation on set S is a function from $S \times S$ to S. The required properties are an ability to combine any *two* elements (binary) and produce an element remaining in the set (closure). Furthermore, binary operations are functions and therefore lend themselves to metaphorical meanings such as taking (two) *inputs* and producing an *output*. In this section, we focus on two tasks related to the concept of a binary operation from the GTCA:

1. Consider the binary operation of averaging $\diamond$ on the set of real numbers defined below. Is this operation associative?

   $a \diamond b = \frac{1}{2}(a + b)$

   The binary operation in this prompt is composed of two familiar operations (addition and multiplication) to produce the third: averaging. Although averaging is likely familiar to students, it is unlikely they have thought of averaging as a binary operation. Students may struggle to make sense of exactly what this operation is and how to compose the operation in order to test associativity. Table 2.5 presents the multiple-choice options and the frequency they were selected in the large-scale round. The correct response is bolded.

2. Consider the set: $S = \{1, 2, 4\}$. Can an operation be defined such that S forms a group?

   In this task, students are asked to define a binary operation on a set, $\{1, 2, 4\}$,[2] which would not have an obvious symbolic form. This necessitates the use of a non-symbolic representation to define an operation that meets group requirements. Table 2.6 presents the multiple-choice options and the frequency they were selected in the large-scale round. The correct response is bolded.

[2] In pilot rounds the set was $\{1, 2, 3\}$.

Table 2.5 Student responses to binary operation prompt 1

Response	Other majors	Mathematics education majors
(a) Yes, addition (and multiplication) are associative, so $\diamond$ is also	27.1 % ($n = 79$)	33.7 % ($n = 28$)
(b) Yes, because $\frac{1}{2}(a + b) = \frac{1}{2}(b + a)$	20.5 % ($n = 60$)	24.1 % ($n = 20$)
(c) No, because $\left(\frac{1}{2}a\right) + b \neq \frac{1}{2}(a + b)$	17.5 % ($n = 51$)	16.9 % ($n = 14$)
(d) **No, because** $\frac{a}{2} + \frac{b}{4} + \frac{c}{4} \neq \frac{a}{4} + \frac{b}{4} + \frac{c}{2}$	34.9 % ($\boldsymbol{n} = 102$)	25.3 % ($\boldsymbol{n} = 21$)

Table 2.6 Student responses to binary operation prompt 2

Response	Other majors	Mathematics education majors
(a) **Yes, because an operation can be defined on any three element set to form a group**	19.2 % ($\boldsymbol{n} = 56$)	15.7 % ($\boldsymbol{n} = 13$)
(b) Yes, multiplication mod 6	40.8 % ($n = 119$)	36.1 % ($n = 30$)
(c) No, the set will not be closed under any operation	18.5 % ($n = 54$)	16.9 % ($n = 14$)
(d) No, the identity element 0 would be needed	8.6 % ($n = 25$)	14.5 % ($n = 12$)
(e) None of the above reasoning is valid[a]	13 % ($n = 35$)	16.9 % ($n = 14$)

[a]Note: All prompts began with a "none of the above" option to allow for other reasoning to emerge. This was one of the few tasks where this option did not become obsolete via eliminating it when less than 5% of students answered this way. The students selecting this answer did so for inconsistent reasons

Meaning and Metaphors of Binary Operation: Operator Symbol, Input/Output, and Weighting

Students tended towards two images for binary operation: inputs and outputs (much like that of function), and elements combined through a specific E-O-E structure: that is, the operation is the action between two elements. Students with the latter image struggled to grapple with binary operations that did not have the "*a operator b*" structure. This thinking came up in a number of places in follow-up interviews. When asked if $a \diamond b = \frac{1}{2}(a + b)$ is associative, students with an E-O-E image of binary operation often responded in the negative, stating that $\frac{1}{2}(a + b) \neq \left(\frac{1}{2}a\right) + b$. In interviews, students who reverted to moving around the parentheses without attention to which aspects were input variables, treated the operation as varying sets of E-O-E structures: $\frac{1}{2} \cdot a$ and $a + b$. In this way, the students were unable to identify *the* operation as there were seemingly two embedded in the expression. One student explicated this thought when prompted, "I don't know what is the binary rule." (For a more detailed treatment of this idea, see Melhuish & Fagan, 2017) However, when students had a concept image that allowed for alternate input–output structures, they often found success.

Many of the students who answered the prompt correctly treated the binary operation by using an input and output metaphor. For example, one student walked through his process of making sense of "($a \diamond b$) $\diamond$ c" as "now, I'm doing whatever that output [$a \diamond b$] is diamond with c." He flexibly thought of the resulting $a \diamond b$ as an output that could then be input into the complete expression. In this way, the entire expression $\frac{1}{2}$ $(a + b)$ can be treated holistically. From there, he unpacked the binary operation and symbolically tested the associativity requirement.

Other metaphors that emerged were specific to handling a particular operation (analogous to the collection metaphor for arithmetic operations). The averaging operation can be thought of using a "weighting" metaphor, where expressions are weighted by 1/2. One student leveraged the weighting metaphor in relation to determining whether the diamond operator was associative, stating:

> Then I looked at them as one and two. Then I asked the question, "is [sic] one and two equivalent?" In operation 1, c is given the same weight as the average of a and b, right? And computed over average 1 and 2. In 2, a is given the same weight as [inaudible]. Not all the elements, um, are weighed equally when computing the average." [Gives a counterexample: 4, 6, 8.] I used my intuition for averages first, then went through the steps of showing it.

This metaphor allowed the student to focus on an important aspect of binary operation, of treating two elements at a time. This metaphor was not strictly input–output, but a morphing metaphor where elements are changed together in pairs. Such metaphors place focus on the binary structure of the operation.

Properties of Binary Operation: Binary, Commutative, and Associative

In this section, we briefly focus on both a required property—binary—and on some varying properties—commutativity and associativity. We found that students often omitted a standard piece of the meaning of binary operation: the requirement for two input elements. For instance, in the associativity prompt, students who did not see the importance of having two input elements may see this expression as an operation of three elements: $\frac{1}{2}$, a, and b. This may have resulted in the selection of choice c as discussed in the prior section.

This prompt also highlighted problems with varying properties. As with functions, students often over-attributed properties (e.g., "all binary operations are associative"), conflated properties (e.g., commutativity and associativity), or inserted inappropriate implications (e.g., commutativity implies associativity). Although this may appear to be an issue remembering terms, further investigation showed that these properties were frequently tangled up in students' concept images. Both properties are fundamentally about order and therefore may seem more similar than they are. Students selecting choice b (in prompt 1) may not be differentiating between the role of commutativity and associativity. See Larsen (2010) for a detailed treatment of this conflation amongst abstract algebra students.

We also saw many students overgeneralize the idea of inheriting associativity. For example, in an interview one student stated of the diamond operator that, "This operation is associative because it is comprised of two operations that are associative on the reals." That is, for $\frac{1}{2}(a+b)$, this student recognized the familiar operations of multiplication and addition and assumed that combining these operations would preserve associativity. This may reflect a lack of object understanding of binary operations. That is, a new operation is not treated as an independent object with its own properties.

Representation of Binary Operation: Symbolic Forms and Tables

Finally, we consider representations of binary operation. As discussed previously, students often preferred symbolic forms, and particularly the E-O-E symbolic form. We expand this discussion by identifying two places for flexibility: internal to symbolic forms (moving between an E-O-E representation and a non-E-O-E format) and using representations other than symbolic representations.

We illustrate the flexibility of symbolic representation through exploring two different students' treatment of the first prompt about the associative property. Most students we interviewed could produce a general form of the associative property, often relying on a star to represent a generic operation: $a \times (b \times c) = (a \times b) \times c$. This formulation contains the E-O-E format. Moving from the general statement to the specific case of averaging can be challenging if the format drives the exploration. When prompted what the operation was (i.e., what the star represented in the averaging binary operation), one set of students responded that it was the "addition" or "the plus sign" from the expression. In contrast, students with more a flexible understanding treated the star symbol quite differently. One student explained:

> I did the test for associativity using the operation. I just kind of picked the star to represent the operation. So I was saying a combined with b times c. And I did it a times parenthesis b times c. So I did it both ways and I got different results that it's not associative because if it was associative they'd be equal.

This student used a star notation as well, but in this case attributed the star as being a placeholder for the full operation, $\frac{1}{2}(a+b)$. The student went on to successfully explain why the property did not hold for this operation. The first case, where the student identified star as the plus sign, is emblematic of students who reasoned using E-O-E formatting: they focused on combining two elements based on a symbol representing the operator. Such focus often led to the problematic treatment of $\frac{1}{2} \cdot (a+b)$ as E-O-E-O-E without attention to the differing operators. The question of associativity then becomes equality or non-equality of: $\frac{1}{2} \cdot (a+b)$ and $\left(\frac{1}{2} \cdot a\right) + b$. The second case, where the star was a placeholder, illustrates a flexible understanding of binary operation that can move between E-O-E formats and other symbolic representations of binary operation.

The desire for a binary operation to be of the form E-O-E is also related to acceptable representations. The E-O-E form reflects a specific representation—symbolic—and a particular format of that symbolic representation. The desire for a symbolic rule was most evident in the second prompt. Students struggled to define

Fig. 2.2 Re-creation of a student's Cayley table

o	1	2	3
1	1	2	3
2	2	3	1
3	3	1	2

a binary operation on the set {1,2,3}[3] or $\{1,2,4\}$ in the respective rounds. These students tested operations with known forms rather than creating a binary operation using a Cayley table or an element-wise defined operation. One student explained, "When we usually test it, we test with addition and multiplication, and that didn't work." This student even discussed a Cayley table, but did not see it as something that could be created without a symbolic representation first. In contrast, students who created a table first found success with this prompt. Consider the follow student explanation from the open-ended round prompting about the set $\{1,2,3\}$. A re-creation of their table is in Fig. 2.2.

> [Filling out the first row and column] "1,2,3 1,2,3. You pick one to be the identity. So that has to turn back into [motions to the table] right there. And so now closure. This right here, I have to have one and three. You already have three one there so one there and one there. And that's all that's left. So there's one three element group.

In order to probe whether the student was treating the numbers as arbitrary labels, the interviewer followed up by asking, "So the first step in here, you said to use one as the identity?" The student quickly responded, "You can start with anything." In this excerpt, the student was able to look beyond known binary operations to create his own binary operation. The student used the table to *define* the operation, rather than the table being presented as an alternate representation of a known, symbolically represented operation. This flexibility is seen when the student explains that any number can serve as the identity. He recognized that "1" was not the only possible identity for the set, but was in fact dependent upon however he had chosen to set up the binary operation. From here, the student successfully argued that his operation fulfilled all the properties of the group.

Conclusion

These two tasks illustrated many relevant understandings around binary operation: (1) Students do not always attend to essential properties (inputs are two elements) and overgeneralize and conflate nonessential properties (associativity and commutativity); (2) Students see the E-O-E format as an essential aspect of binary operations; (3) Students often do not use non-symbolic representations, such as Cayley tables, flexibly; however, students who could move between representations engaged more productively in the tasks.

[3]This change occurred because a minority of students did create a correct symbolic binary operation for the $\{1,2,3\}$, and the intention was to remove this option.

Discussion of Opportunities for Backward Transfer

The opportunity for backward transfer is one of the most notable implications for secondary teachers regarding the prominent role of binary operation and function in abstract algebra. Hohensee (2014) described backward transfer as, "when learning about new concepts influences learners' ways of reasoning about previously encountered concepts" (p. 135). As a field, we typically treat transfer as bringing old knowledge into new tasks. Backward transfer upends this directionality to consider how new tasks may impact old knowledge. That is, a student's schemata around a particular concept may be reorganized as a result of engaging in a task about a new, related concept. Rather than asking what about the old concept transferred to the new situation, the question becomes what impact did the new situation have on the old knowledge? When students engage in complex tasks in abstract algebra, they not only draw on their prior knowledge of function and operation, but they also may enrich their prior understanding of these concepts. Although we often conceptualize connections between abstract algebra and the secondary level in terms of explicit connections about structures and properties, it also may be beneficial to think of the implicit connections that can be developed. What can preservice secondary teachers learn about functions from engaging in tasks about isomorphism or homomorphism in abstract algebra contexts? What can preservice secondary teachers learn about number operations or function operations through encountering general and unfamiliar operations in abstract algebra contexts? We break down several of the GTCA tasks using this lens and identify five areas of potential backward transfer. For each area, we unpack a relevant conceptual hurdle related to function or binary operation. We then unpack the group theory tasks to hypothesize how productive engagement in these tasks may help students strengthen their knowledge about secondary content.

Moving Students Beyond Surface Level Form of Binary Operation

In our work, we have documented that many students (including preservice secondary teachers) demonstrate a strong preference for the E-O-E formatting of binary operations. In many ways, this parallels students' preferences for the symbolic rule form of functions (e.g., Thompson, 1994). We illustrated this preference in the associativity task. Many students identified $\frac{1}{2}(a+b)$ as having the operation "addition" by focusing only on the portion that is in E-O-E formatting. Alternately, students may treat the entire expression as a series of five entities (E-O-E-O-E), where ½ is treated as one of the elements (despite not being a variable input) and multiplication and addition are the operators. Students with this preference may see a binary operation as a string of entities, rather than treating the operation holistically as a single entity.

If preservice secondary teachers are asked to engage in determining whether the associative property holds for the averaging operation, $\frac{1}{2}(a + b)$, they are put in a situation where they might reorganize their understanding of binary operation. If they hold the above conceptions, they are likely to either make claims that this operation is associative due to addition being associative, or that this operation is not associative because the expressions $\frac{1}{2} \cdot (a + b)$ and $\left(\frac{1}{2} \cdot a\right) + b$ are not equivalent. If preservice secondary teachers are put in a position to confront these invalid types of reasoning, they may develop a more robust understanding of binary operation as a single entity. It is more productive to think of this expression as the "averaging operation on two quantities." This takes the focus away from addition (and multiplication) as the operation in between two elements from the string of entities. In this way, a common operation, such as addition, can similarly be conceptualized as the process of "summing on two quantities," rather than a string of three entities of the form: "$x + y$." From the lens of backward transfer, this shift can support students in seeing operations as holistic processes. Further, such reorganizing has the potential to support students in recognizing and treating other operations that are not always symbolized in an E-O-E format, such as function composition, which is denoted as $f(g(x))$.

Beyond the specifics of formatting, students also demonstrated a general preference for a symbolic rule representation of an operation. This preference was most notable in the task asking students to define an operation on the set $\{1, 2, 4\}$. In this task, building and testing symbolic rules did not lead students to successfully define a binary operation on the set. Students exhausted their symbolic rules and often concluded there was no possible binary operation. This likely reflects an assumption that *all* binary operations can be written in symbolic shorthand.

If preservice secondary teachers engage in this task, instructors can promote them to engage in productive disequilibrium. Rather than leave preservice secondary teachers to test possible operations and conclude there is no binary operation, the task could be altered to have them "define" the binary operation. The most productive way to engage in this task is via defining an operation explicitly through a table or other element-wise approach. Undoubtedly, some students will approach the prompt in this manner, which can lead to an explicit conversation about the acceptability of alternate representations. Notably, this treatment may also promote a shift from an operation *as* the symbolic rule to an operation *as* the way we combine elements. Detaching from the symbol can be productive for addressing operations that look the same, but behave differently. At the secondary level, a notable example is multiplication of numbers versus multiplication of matrices. A shift from a focus on the multiplication symbol to how individual elements are combined can allow for the conceptualization that one "multiplication" is commutative and one is not.

Developing a Conceptual Understanding of the Associative Property

The associative property is quite complex. To deal with this cognitive complexity, students may attend to only a subset of relevant attributes of associativity, such as the movement of parentheses or the general relationship to ordering. In our data, we found students frequently addressed the less complex order property of commutativity by determining if the expressions $\frac{1}{2}(a+b)$ and $\frac{1}{2}(b+a)$ are equivalent, or alternately, moving parentheses in a superficial way and determining if the expressions $\frac{1}{2}\cdot(a+b)$ and $\left(\frac{1}{2}\cdot a\right)+b$ are equivalent. In the case of conflating properties, this is an issue that has been previously documented amongst abstract algebra students (e.g., Larsen, 2010). In order to see commutativity and associativity as fundamentally different, there must be a shift in attention from the elements to the operators as the objects being ordered. Seeing the operator as an object, rather than a process, is an increase in conceptual sophistication. In the case of focusing on parentheses, students may conceptualize associativity as moving parentheses around, regardless of the operation, an overgeneralization that has been documented amongst teachers (Zaslavsky & Peled, 1996). This procedure is quite successful until a binary operation is complex enough that it contains parentheses for the purposes of indicating order of operation, like the averaging operation.

Determining if averaging is associative provides opportunities to enrich pre-service secondary teachers' understanding of the associative property. As such, this prompt provides productive grounds for backward transfer to the associative property of addition and multiplication. First, averaging is an example of an operation that is commutative, but not associative. Such an observation can motivate the need to disentangle the meaning behind the two properties. In most secondary contexts, associativity and commutativity both existed or did not exist for a given operation (such as the four standard arithmetic operations). This task puts students in a position to notice two things: the ordering for associativity is based on operator not elements, and the operator in the statement of associativity property is the same. This can lead to a restructuring of knowledge around the associative property of multiplication and addition to incorporate this increased understanding of the general associative property, where the focus can be about the meaning of regrouping, rather than the movement of parentheses.

This also provides a case for challenging assumptions about how associativity and commutativity are established. Literature on statistics has shown that students assume that averaging is commutative and associative, perhaps because averages are built from operations that have those properties (Mevarech, 1983). We have similarly documented that abstract algebra students make the same inference. Binary operations are more than the sum of their parts. When we frequently use language like "associativity is inherited," we may be failing to distinguish between situations where this occurs (such as subgroups) and where it does not (such as with a different operation). It can be powerful for preservice secondary teachers' understanding to see a binary operation that is composed of other operations, but which does not have the same properties as the parent structures. To connect this to a secondary context, when finding the mean of a set of data, a lack of associativity prevents us from averaging two unequal size sets independently and then averaging those together to arrive at a mean.

Distinguishing Between Constants and Variables

We briefly touch on one other area that can be impacted via engagement in these types of tasks: constants and variables. Binary operations, as special cases of functions, have two varying inputs. Such operations can have any number of constants as part of their expressions. We know that students struggle with variables in terms of their multiple roles, such as parameters versus varying quantities (Philipp, 1992). We were surprised to find abstract algebra students may still struggle to make delineations between constants and varying quantities. We saw students do so on the associativity task, treating $\frac{1}{2}$, a, and b as variables.

The associativity task can provide opportunity for discussion on the role of varying quantities within operations (or even in the larger context of functions). Binary operations can be created through any combination of constants with the consistent inclusion of two inputs that serve as variables. By engaging in the task around $\frac{1}{2}(a + b)$, preservice secondary teachers are placed in a position where they have to make sense of what elements are inputs. While students may see a and b as inputs when confronting the binary operation as is, the change in setting to testing the associative property can lead many students to start treating ½ as a variable when unpacking: $(a \diamond b) \diamond c$. Opportunity for backward transfer occurs when students reflect on a statement with three variables compared to a statement with two variables and a constant. Such an exploration can be used to emphasize the different role of constants and variables in an expression. This can aid in reorganization of the role of variable and constant in a preservice teachers' knowledge related to a number of secondary contexts.

Understanding the Implications and the Nonnecessity of Functions Being Injective

We know that students have a strong preference for injective functions (e.g., Leinhardt et al., 1990). Dealing with injective functions is less cognitively complex: a single element (rather than some set) maps to a given element of the codomain. It may be that students overgeneralize from a relatively populated class of injective functions from high school algebra. It has also been documented that students may conflate the definition which requires a function be well-defined with the injective property (e.g., Vinner, 1983). That is not to say that abstract algebra students could not produce examples of functions that are not injective (such as $y = x^2$ on the real numbers), but that functions may still be treated *as if* injective in many settings. In our research, we documented this was the case for many students when approaching an unfamiliar function in the kernel prompt. We conjecture that this may also be the case in other abstract settings such as when a function is defined generally.

The kernel task and FHT task provide opportunities for preservice secondary teachers to engage richly with functions that are not injective. In the kernel task, a student can arrive at an incorrect conclusion by assuming the mapping must be injective (identifying a kernel with one element). In the high school setting, it is in the context of inverse functions that students most frequently deal with preimages. Only injective functions have inverses. Thus, the opportunity for backward transfer is twofold. First, engaging with a concrete function that is not injective (and for which the horizontal line test is not meaningful) can lead to restructuring knowledge around injective functions that is not reliant on a particular representation. Second, dealing with multiple member sets in a preimage can aid in restructuring knowledge around inverse functions and the particular role the injective property plays in this context.

The FHT prompt provides a different sort of setting: working with a general function rather than one defined on a particular set of elements. In such a setting, traditional methods of attending to injectivity fail (such as looking at a graph or table.) Without such aids, preservice teachers may assume injectivity on the homomorphism in FHT. However, with such an assumption, the theorem becomes trivial. In order to address the GTCA task, preservice teachers must recognize that the FHT is fundamentally about building subsets of elements (which may have more than one element) based on their images. With supported engagement in tasks around the FHT, preservice teachers can be placed in a position to notice implicit assumptions of injectivity. In these tasks, a different function metaphor may support this noticing: that of collapsing. This type of metaphor can prevent the conflation of injectivity and well-definedness as the directionality from domain to codomain is built within the imagery. As such, preservice teachers may restructure their knowledge of injective functions at the secondary level to incorporate the powerful notion of collapsing.

Developing Flexibility in Moving Between and Leveraging Alternate Function Representations: Function Diagrams

Our final consideration is related to representations of functions. We know students have a strong preference for symbolic, algebraic rule representations (e.g., Thompson, 1994). This can be quite limiting in terms of problem solving (e.g., Knuth, 2000). In group theory tasks, functions are often described abstractly and, as such, a symbolic form is not particularly powerful in conveying information. The FHT relies on a complex set of functions to move from a homomorphism to an isomorphism. In our research, when engaging with the FHT, students reached for a variety of inapplicable tools as they struggled to make sense of the essence of the map between preimages of a homomorphism and the codomain. Despite the importance of functions in the FHT, students rarely attempted to use visual function representations to make sense of the theorem.

The FHT task provides a setting where preservice teachers can be motivated to leverage a visual representation of function (function diagrams). They are not given a particular homomorphism, but asked to identify possible orders of an undefined homomorphism's image (on a given domain). This removes the availability of a symbolic form. If the preservice teacher robustly understands the FHT, the application is immediate. However, this understanding requires a coordination of many components and a particularly rich usage of functions. Function diagrams provide one representation that does not require a specific function, but that still captures powerful intuition about functions. By placing preservice secondary teachers in this complex situation, we can necessitate more robust types of function representations. As such, there is a great opportunity for backward transfer to general function knowledge at the secondary level. Students may reorganize their knowledge to more robustly incorporate function diagrams as a representation linked closely with even the most complex or abstract functions. Function diagrams in curricular materials may be associated with early definitions, but these kinds of diagrams, ultimately, seem to move to the background, with graphs becoming the dominant visual representation. However, function diagrams can be a powerful way to make sense of a function's behavior that directly connects to the meaning of various properties (such as injectivity).

Backward Transfer Conclusion

Group theory provides a context for productive backward transfer. In the tasks described above, students could engage robustly to develop a deeper understanding of binary operation and function. For example, dealing with a nonroutine operation, such as averaging, provides an opportunity to dig into the meaning of associativity and what, exactly, we mean by an operation. Dealing with complicated functions (like those in homomorphisms) provides an opportunity where students may find more access via leveraging alternative representations. The abstract group concept (where an operation need not be symbolically defined) can provide a situation where students need to leave a standard symbolic rule to approach binary operations or functions. Some argue that group theory is too disconnected from high school topics to provide a fertile ground for building connections and deep understanding of relevant topics. We counter that more abstract and complex problems can actually lead to more robust understandings by necessitating alternative treatments of core concepts.

Conclusion

In this chapter, we share data and theoretical exploration to illustrate that (1) future secondary mathematics teachers are not always developing conceptual understanding in group theory; (2) conceptual understanding in groups is tied to understanding core concepts of function and binary operation; (3) if students are put in positions where they can engage in conceptual tasks, there may be rich opportunities for backward transfer to deepen their conceptions around function and binary operation.

In response to point 1, we can speculate on the cause of limited conceptual understanding. One possible explanation is that the course focuses fundamentally on proof and formal representation systems. Students may not engage with metaphor, meaning, examples, or nonformal representations, which can build important connections and flexibility. A second possible explanation is that the concepts in group theory are inherently difficult because of the jump in abstraction, the complexity of various concepts and theorems, and the unfamiliar settings. Regardless of the reason, it is not surprising that preservice secondary teachers are not always making connections between abstract algebra and what they will eventually teach (e.g., Wasserman, 2017). If they are not developing a deep conceptual understanding, what will they build connections between?

We contend that attention to the role of binary operation and function can help combat this issue. If our goal is for preservice teachers to deepen their understanding related to the content they teach, then backward transfer provides a way for us to conceive of these relations. It is not only that they are being exposed to abstract structures and properties that guide our number system, but they are also being re-exposed to core concepts from the K-12 curriculum. Nontraditional and conceptually focused tasks position students to develop their understanding of the core concepts that were previously treated. The abstract nature and complexity of topics in abstract algebra necessitate deeper understanding and more-developed usage of binary operation and function. As abstract algebra instructors, if we use tasks to intentionally develop students' ability to make sense of topics, then we can provide powerful opportunities for learning related to the content preservice teachers' will eventually teach.

References

Akkoç, H., & Tall, D. (2002). The simplicity, complexity and complication of the function concept. In A. D. Cockburn & E. Nardi (Eds.), *Proceedings of the 26th conference of the International Group for the Psychology of Mathematics Education* (Vol. 2, pp. 25–32). Norwich: PME.

Blair, R., Kirkman, E. E., & Maxwell, J. W. (2013). *Statistical abstract of undergraduate programs in the mathematical sciences in the United States: Fall 2010 CBMS survey*. Providence, RI: American Mathematical Society.

Breidenbach, D., Dubinsky, E., Hawks, J., & Nichols, D. (1992). Development of the process conception of function. *Educational Studies in Mathematics, 23*(3), 247–285.

Brown, A., DeVries, D. J., Dubinsky, E., & Thomas, K. (1997). Learning binary operations, groups, and subgroups. *The Journal of Mathematical Behavior, 16*(3), 187–239.

Clement, L. L. (2001). What do students really know about functions? *The Mathematics Teacher, 94*(9), 745.

Common Core State Standards Initiative. (2010). *Common core state standards for mathematics*. Washington, DC: National Governors Association Center for Best Practices and the Council of Chief State School Officers Retrieved from http://www.corestandards.org

Dubinsky, E., Dautermann, J., Leron, U., & Zazkis, R. (1994). On learning fundamental concepts of group theory. *Educational Studies in Mathematics, 27*(3), 267–305.

Elia, I., Panaoura, A., Eracleous, A., & Gagatsis, A. (2007). Relations between secondary pupils' conceptions about functions and problem solving in different representations. *International Journal of Science and Mathematics Education, 5*(3), 533–556.

Even, R. (1993). Subject-matter knowledge and pedagogical content knowledge: Prospective secondary teachers and the function concept. *Journal for Research in Mathematics Education, 24*, 94–116.

Hazzan, O. (1999). Reducing abstraction level when learning abstract algebra concepts. *Educational Studies in Mathematics, 40*(1), 71–90.

Hohensee, C. (2014). Backward transfer: An investigation of the influence of quadratic functions instruction on students' prior ways of reasoning about linear functions. *Mathematical Thinking and Learning, 16*(2), 135–174.

Hiebert, J., & Lefevre, P. (1986). Conceptual and procedural knowledge in mathematics: An introductory analysis. In J. Hiebert (Ed.), *Conceptual and procedural knowledge: The case of mathematics* (pp. 1–27). Hillsdale, NJ: Lawrence Erlbaum Associates.

Knuth, E. J. (2000). Student understanding of the Cartesian connection: An exploratory study. *Journal for Research in Mathematics Education, 31*, 500–507.

Lakoff, G., & Núñez, R. E. (2000). *Where mathematics comes from: How the embodied mind brings mathematics into being*. New York, NY: Basic Books.

Larsen, S. (2010). Struggling to disentangle the associative and commutative properties. *For the Learning of Mathematics, 30*(1), 37–42.

Leinhardt, G., Zaslavsky, O., & Stein, M. K. (1990). Functions, graphs, and graphing: Tasks, learning, and teaching. *Review of Educational Research, 60*(1), 1–64.

Leron, U., Hazzan, O., & Zazkis, R. (1995). Learning group isomorphism: A crossroads of many concepts. *Educational Studies in Mathematics, 29*(2), 153–174.

Melhuish, K. (2015). The design and validation of a group theory concept inventory (Doctoral dissertation). *Dissertations and theses*. Retrieved from: http://pdxscholar.library.pdx.edu/open_access_etds/2490

Melhuish, K. M., & Fagan, J. (2017). Exploring student conceptions of binary operation. In A. Weinberg, C. Rasmussen, J. Rabin, M. Wawro, & S. Brown (Eds.), *Proceedings of the 20th annual conference on research in undergraduate mathematics education* (pp. 166–180). San Diego, CA: SIGMAA.

Melhuish, K., Larsen, S., & Cook, S. (2018). When students prove a theorem without explicitly using a necessary condition: Digging into a subtle problem from practice. Manuscript under review.

Mevarech, Z. R. (1983). A deep structure model of students' statistical misconceptions. *Educational Studies in Mathematics, 14*(4), 415–429.

Nardi, E. (2000). Mathematics undergraduates' responses to semantic abbreviations, 'geometric' images and multi-level abstractions in group theory. *Educational Studies in Mathematics, 43*(2), 169–189.

Novotná, J., & Hoch, M. (2008). How structure sense for algebraic expressions or equations is related to structure sense for abstract algebra. *Mathematics Education Research Journal, 20*(2), 93–104.

Oehrtman, M., Carlson, M., & Thompson, P. W. (2008). Foundational reasoning abilities that promote coherence in students' function understanding. In M. Carlson & C. Rasmussen (Eds.), *Making the connection: Research and teaching in undergraduate mathematics education* (pp. 27–42). Washington, DC: Mathematical Association of America.

Philipp, R. A. (1992). The many uses of algebraic variables. *The Mathematics Teacher, 85*(7), 557–561.

Schwarz, B., Dreyfus, T., & Bruckheimer, M. (1990). A model of the function concept in a three-fold representation. *Computers & Education, 14*(3), 249–262.

Sfard, A. (1991). On the dual nature of mathematical conceptions: Reflections on processes and objects as different sides of the same coin. *Educational Studies in Mathematics, 22*(1), 1–36.

Slavit, D. (1997). An alternate route to the reification of function. *Educational Studies in Mathematics, 33*(3), 259–281.

Slavit, D. (1998). The role of operation sense in transitions from arithmetic to algebraic thought. *Educational Studies in Mathematics, 37*(3), 251–274.

Tall, D., & Vinner, S. (1981). Concept image and concept definition in mathematics with particular reference to limits and continuity. *Educational Studies in Mathematics, 12*(2), 151–169.

Tall, D., McGowen, M., & DeMarois, P. (2000). The function machine as a cognitive root for the function concept. In M. L. Fernandez (Ed.), *Proceedings of the annual meeting of the North American Chapter of the International Group for the Psychology of Mathematics Education* (Vol. 1, pp. 255–261). Tucson, AZ: PME-NA.

Thompson, P. W. (1994). Students, functions, and the undergraduate curriculum. *Research in Collegiate Mathematics Education, 1*, 21–44.

Ticknor, C. S. (2012). Situated learning in an abstract algebra classroom. *Educational Studies in Mathematics, 81*(3), 307–323.

Tirosh, D., Hadass, R., & Movshovitz-Hadar, N. (1991). Overcoming overgeneralizations: The case of commutativity and associativity. In F. Furinghetti (Ed.), *Proceedings of the fifteenth annual conference of the International Group for the Psychology of Mathematics Education* (Vol. 3, pp. 310–315). Assisi: International Group for the Psychology of Mathematics Education.

Vinner, S. (1983). Concept definition, concept image and the notion of function. *International Journal of Mathematical Education in Science and Technology, 14*(3), 293–305.

von Glasersfeld, E. (1995). *Radical constructivism: A way of knowing and learning*. London: Falmer Press.

Wasserman, N. H. (2017). Making sense of abstract algebra: Exploring secondary teachers' understandings of inverse functions in relation to its group structure. *Mathematical Thinking and Learning, 19*(3), 181–201.

Weber, K., & Larsen, S. (2008). Teaching and learning abstract algebra. In M. Carlson & C. Rasmussen (Eds.), *Making the connection: Research and teaching in undergraduate mathematics*. Washington, DC: Mathematical Association of America.

Warren, E. (2003). The role of arithmetic structure in the transition from arithmetic to algebra. *Mathematics Education Research Journal, 15*(2), 122–137.

Zandieh, M., Ellis, J., & Rasmussen, C. (2017). A characterization of a unified notion of mathematical function: The case of high school function and linear transformation. *Educational Studies in Mathematics, 95*(1), 21–38.

Zaslavsky, O., & Peled, I. (1996). Inhibiting factors in generating examples by mathematics teachers and student teachers: The case of binary operation. *Journal for Research in Mathematics Education, 27*, 67–78.

Chapter 3
Monster-Barring as a Catalyst for Bridging Secondary Algebra to Abstract Algebra

John Paul Cook

Introduction

Abstract algebra is seen as an important course in the mathematical preparation of secondary teachers, largely because of its potential to enable students to view the familiar content of secondary algebra through a more advanced lens. For example, it is recommended that prospective teachers come to regard the secondary algebra that they will be teaching as "the algebra of rings and fields" (CBMS, 2010, p. 59). While this seems sensible—the ring and field properties feature prominently in primary and secondary algebra—there are reports that adopting such an advanced perspective is a challenging endeavor for prospective teachers, who see very little relevance between advanced mathematics and secondary mathematics (e.g., Wasserman, 2016; Wasserman, Weber, Villanueva, & Mejia-Ramos, 2018; Zazkis & Leikin, 2010). Wasserman et al. (2018) proposed that prospective teachers might not see the relevance of their advanced mathematics courses because they struggle to understand the advanced content and the connections to teaching in these courses. This is especially true for abstract algebra, which is a notoriously difficult course for students (e.g., Dubinsky, Dautermann, Leron, & Zazkis, 1994; Hazzan, 1999).

A productive avenue of insight is to investigate student thinking about the algebraic properties that characterize such fundamental structures as rings, integral domains, and fields. Interestingly, even though these properties appear frequently throughout primary and secondary algebra, it can be challenging for students to employ them effectively in abstract algebra settings (Larsen, 2010). One possible explanation is that students enter abstract algebra with context-specific understandings of algebraic properties that are not readily generalizable to other

J. P. Cook (✉)
Department of Mathematical Sciences, Oklahoma State University, Stillwater, OK, USA
e-mail: cookjp@okstate.edu

N. H. Wasserman (ed.), *Connecting Abstract Algebra to Secondary Mathematics, for Secondary Mathematics Teachers*, Research in Mathematics Education,
https://doi.org/10.1007/978-3-319-99214-3_3

structures (Cook, 2018; Hazzan, 1999; Wasserman, 2017). This raises the overarching question that motivated the current study: how might beginning abstract algebra students be able to adapt their existing understandings of an algebraic property to be effective in abstract algebra?

To answer this question, I conducted a teaching experiment (Steffe & Thompson, 2000) with a pair of prospective teachers preparing to take an introductory course in abstract algebra. The purpose of the teaching experiment was to investigate how prospective teachers might "assimilate their understanding of secondary mathematics with advanced mathematics" (Wasserman, 2017, p. 199) by focusing on (1) student thinking related to the zero-product property (ZPP), a tool for solving equations in secondary algebra and the definitive characteristic of integral domains in abstract algebra, and (2) how such thinking might be leveraged to enable students to develop an effective understanding of the ZPP in abstract algebra.

This chapter makes two primary contributions. First, I document a learning trajectory that details how a student's outright rejection of counterexamples to the ZPP—an activity called *monster-barring* (Lakatos, 1976; Larsen & Zandieh, 2008)—can be repurposed to develop an effective understanding of the ZPP in abstract algebra. That is, a student's refusal to accept a key component of the concept to be learned can, if carefully and intentionally repurposed, serve a productive pedagogical purpose and support the student's development of an effective understanding of that concept. Second, I discuss two implications for teacher preparation that stem from this work: the first about developing deeper mathematical understandings with respect to secondary mathematics, and the second about modeling mathematics instruction in ways that build on students' thinking, even when their ideas initially appear counterproductive.

Literature and Theoretical Framing

Investigating student thinking about zero-divisors and the ZPP is a useful means to gain insight into how students might adapt their thinking about an algebraic property to be effective in abstract algebra because of (1) the ZPP's prominence in both secondary algebra (as a tool for solving equations) and abstract algebra (as the defining characteristic of integral domains), and (2) the fact that counterexamples to the ZPP (zero-divisors) do not typically appear in the mathematics curriculum in a meaningful way until abstract algebra. With respect to my research question, I employed Thompson's (2008) tools for *conceptual analysis* in order to describe an effective understanding of the ZPP in abstract algebra. As part of the conceptual analysis, I also hypothesize a trajectory by which students might be able to develop the aforementioned effective understanding in abstract algebra.

Effective Ways of Understanding

A *way of understanding* is a meaning or conception that a student has for a particular mathematical concept (Harel, 1998). A way of understanding might include a system of strategies, analogies, informal descriptions, and examples and non-examples. For example, a student's initial, implicit way of understanding what it means for an element to be a zero-divisor might include some notion of elements that cause repetition in the respective rows and columns of that structure's multiplication table (Cook, 2014), or elements that cause issues with the multiplicative cancellation law (Simpson & Stehlikova, 2006). I adopt the perspective that such intuitive, informal ways of understanding can play an important role in the development of formal mathematical concepts (Freudenthal, 1973; Gravemeijer, 1998).

Harel (1998) proposed that a student holds an *effective* way of understanding a mathematical idea if, in addition to retaining that way of understanding over time, she is able to:

- Criterion I: reformulate and articulate it in her own words.
- Criterion II: think about it in a general way.
- Criterion III: coordinate it with her ways of understanding other ideas.

These criteria provide an observable way to determine if a student holds an effective way of understanding, but it remains unclear exactly what these criteria mean for zero-divisors and the ZPP in an abstract algebra setting. While criterion I—the student's ability to formulate the concept in her own words—is relatively straightforward, in order to operationalize Harel's criteria it is necessary to specify what it means for a student to think about zero-divisors and the ZPP in a general way (criterion II), and also to incorporate her thinking about other concepts (criterion III).

In order to operationalize criterion II—what it means to think about a concept in a general way—I adopted Alcock and Simpson's (2011) perspective that classification of examples is a fundamental mathematical task. Indeed, a fundamental task for introductory abstract algebra students is to determine if a new example structure is an integral domain, which essentially amounts to determining whether the structure contains zero-divisors. Though the ability to consistently classify examples is rarely the final objective, it can be a useful opportunity for students to gain some initial experience with the underlying concept (e.g., Ross & Makin, 1999). Particularly, students with a way of understanding that is not fully developed will probably be unable to use it to consistently classify examples (e.g., Davis & Vinner, 1986). Thus, I used the ability to consistently classify algebraic structures on the basis of a particular property as evidence that a student was thinking about that property in a general way.

Regarding criterion III, although there are several other mathematical ideas with which students can coordinate to understand an algebraic property, including set and binary operation (Brown, DeVries, Dubinsky, & Thomas, 1997; Dubinsky et al., 1994; Simpson & Stehlikova, 2006), here I will focus on the idea of an additive identity because of its specific relevance to zero-divisors and the ZPP. Particularly, in order to recognize instances of zero-divisors and the ZPP in unfamiliar algebraic structures, students must also have a way of understanding the notion of identity that is general enough to recognize instances of identity elements which are not 0 or 1 (or do not consist solely of 0 and 1 s). Specifically, they must conceive of the identity as an element that when combined with any other element via the relevant binary operation yields that same element.

A Hypothesis for How Students Might Achieve Such a Way of Understanding

I adopted the instructional design principles of Realistic Mathematical Education (RME) to guide my design of instructional tasks. RME is a perspective pioneered by Freudenthal (1973) in which the central tenet is that mathematics should be taught in a way that reflects how people actually create it. Instead of immediately considering formally stated mathematical definitions and theorems—a presentation that, Freudenthal argued, hides the process behind which the underlying ideas were conceived—students should be given opportunities to reinvent such mathematical ideas for themselves. RME operationalizes these views in the form of principles for instructional design to explore how students might, through continual reflection on their own informal and intuitive mathematical activity, gradually develop effective ways of understanding sophisticated mathematical concepts.

Freudenthal (1973) himself provided design principles that characterize the means by which students might overcome the challenges of abstraction in order to develop their intuitive, informal notions of an algebraic property. Specifically, he conjectured that students might be able to develop an effective way of understanding an algebraic property by engaging in the following hierarchical levels of activity:

- Level 1: An algebraic property emerges implicitly in the student's intuitive activity with an example structure.
- Level 2: The property appears explicitly in the student's general description of her activity with the structure.
- Level 3: The student repurposes the property as a lens to classify other structures.

This characterization is underpinned by the assumption that "the intuitive activity on the lower level becomes a means of organization on a higher level" (Freudenthal, 1973, p. 123). In order to use a property to investigate a new and unfamiliar context, then, the property should emerge first in a student's reasoning within another example structure. Specifically, I conjectured that:

- Level 1: the ZPP would emerge implicitly in the students' reasoning as they solved equations[1] in $(\mathbb{R}, +, \cdot)$.
- Level 2: the students would, in response to my prompting, explicitly formulate their use of the ZPP.
- Level 3: the students would then operationalize their explicit formulation of the ZPP as a lens through which to investigate $(\mathbb{Z}_{12}, +_{12}, \cdot_{12})$, a finite ring containing zero-divisors. I anticipated that the presence of multiple solutions to multiplicative linear equations (e.g., Cook, 2014; Simpson & Stehlikova, 2006) would eventually lead them to revisit their assumptions about the ZPP and engage in productive conversations about possible counterexamples (i.e., zero-divisors).

By iterating this process through several examples structures, I hypothesized that the students would be able to refine their way(s) of understanding the ZPP so that they would be able to describe the property in their own words (criterion I); be able to use it as a means to classify algebraic structures (criterion II); and be able to connect it to other concepts like the additive identity (criterion III). I provide more detail about the other example structures I used in my description of the study.

Methods

I adopted the teaching experiment methodology (Steffe & Thompson, 2000) as a means of exploring and refining the conceptual analysis—that is, the characterization of an effective way of understanding the ZPP and my hypothesis about how students might come to achieve such a way of understanding. The teaching experiment methodology was appropriate for this study for two reasons. First, it characterizes mathematics as "the product of a functioning human intelligence" (Steffe & Thompson, 2000) and thus is compatible with the RME principle that mathematics is a human activity (Freudenthal, 1973). An important consequence here is that the researcher assumes that a student's mathematical behavior is rational—that is, that there are plausible explanations for how the student is thinking that provide a reasonable context for her mathematical activity. This leads to the construction of *models of student thinking*. The conceptual analysis is a key part of this process because it serves as a framework with which to propose such models (Thompson, 2008). Second, refining a conceptual analysis via a teaching experiment results in a concrete example of how a student might achieve an effective way of understanding a particular topic. Examples of such learning trajectories can serve as "important examples of how mathematically substantive, conceptually-grounded conversations can be held with students" (Thompson, 2008, p. 60). Thus, the act of refining a conceptual analysis via the teaching experiment methodology

[1]Informally, an equation *has a solution in the ring* $(R, +, \cdot)$ *if (1) its coefficients are all elements of the set* R, *and (2)* x *is an element of* R. For example, $4(x-5)=0$ has a unique solution ($x=5$) in $(\mathbb{R}, +, \cdot)$ but four solutions ($x = 2, 5, 8, 11$) in $(\mathbb{Z}_{12}, +_{12}, \cdot_{12})$.

has tangible benefits for teachers (who obtain examples of how students might overcome cognitive hurdles as they develop effective ways of understanding) as well as educational researchers (who obtain a robust analytical framework for characterizing students' thinking and learning).

Participants

Two undergraduate students participated in this study. Brian and Julie (both pseudonyms) were both beginning the first semester of their junior years at a small, public liberal arts college as mathematics education majors and prospective secondary mathematics teachers. Both had completed a course in linear algebra (both earning B's) but had not yet taken an introduction to proof course. This was typical for mathematics education majors at this particular institution, who instead were required to take an "abstract algebra for future secondary teachers" course that focused more on the relevance of abstract algebra to secondary algebra than on the rigors of proof (both Brian and Julie were preparing to begin this course when they participated in this study). I selected them for participation in this study for two reasons. First, their responses to an initial recruitment survey indicated that their current ways of understanding zero-divisors and the ZPP were suitable to answer this study's research question. Specifically, they (1) successfully employed the ZPP to solve linear and quadratic equations in $(\mathbb{R}, +, \cdot)$, yet, in a subsequent response, overgeneralized the property and asserted that the ZPP held in the ring of 2×2 matrices under matrix addition and multiplication $(M_2(\mathbb{R}), [+], [\cdot])$, a structure containing zero-divisors with which they were familiar from linear algebra. Second, they appeared to be students who could be easily articulate their thinking and express their mathematical thoughts without reservation.

Data Collection and Analysis

The teaching experiment consisted of 4 sessions lasting between 75 and 90 min each; I served as the teacher-researcher for all sessions. Each session was recorded with LiveScribe pen technology, which records the students' pen strokes with synchronized audio (called a pencast). I constructed models of Brian and Julie's ways of understanding using *ongoing* and *retrospective* analysis techniques (Steffe & Thompson, 2000). A central tenet of the teaching experiment methodology is that researchers do not have actual access to a student's ways of understanding. Thus, researchers can only formulate conjectures about what students' ways of understanding might be, and support or reject these conjectures using their utterances and written work—in this case, interactions with the students during the sessions as well as their pencasts—as evidence. These conjectures about students' ways of understanding, paired with the evidence (in the form of observable behaviors) that supports them, are the *models* to which Steffe and Thompson (2000) refer. Briefly, the purpose of *ongoing analysis* is to (1) construct and refine models of students' ways of understanding during and between teaching experiment sessions and (2) decide how to use these models to advance the students' learning. At the conclusion of the teaching experiment, I engaged in *retrospective* analysis (Steffe & Thompson, 2000) in order to identify common themes and changes in the models of students' ways of understanding over the course of the teaching experiment.[2]

Instructional Tasks

The instructional tasks center on solving equations, a mathematical activity that is familiar to students from school algebra that can serve as a useful means of gaining insight into the algebraic structures—like groups (e.g., Wasserman, 2014) and rings (e.g., Cook, 2014)—that form the foundation of abstract algebra. A key benefit of equation solving—especially with respect to the goals of this chapter—is that it allows students to start with a familiar activity from secondary algebra (and the ways of understanding they associate with this activity) and potentially leverage it to develop an effective way of understanding in abstract algebra. The structures that Brian and Julie explored included: the real numbers $(\mathbb{R}, +, \cdot)$, the integers modulo 12 $(\mathbb{Z}_{12}, +_{12}, \cdot_{12})$, 2×2 matrices over the real numbers $(M_2(\mathbb{R}), [+], [\cdot])$, the integers modulo 5 $(\mathbb{Z}_5, +_5, \cdot_5)$, and abstract, context-independent renderings of the finite field of order 4 $(\mathbb{F}_4)$ and a noncommutative ring of order 4 that contains zero-divisors (which I refer to as N_4) in which all binary operations are explicitly given in operation tables.[3]

Results

In this section, I focus on Brian's learning trajectory as he developed an effective way of understanding the ZPP in abstract algebra. Although Julie's learning trajectory was not the focus of this analysis—she adapted to the presence of zero-divisors rather easily—I include examples and analysis of her mathematical activity as needed to create a backdrop against which to contrast Brian's activity. The following subsections are organized according to the key developments in Brian's activity that emerged during my retrospective analysis of the data, with two analytical subsections interjected to describe key insights that shaped these developments. This organization, and these developments, mirror the learning trajectory developed from

[2] Although a comprehensive account of these analysis techniques is beyond the scope of this chapter, the interested reader may consult Cook (2018) for a detailed account from a very similar teaching experiment.

[3] Henceforth, for brevity, I suppress the notation for a ring's binary operations and simply denote the set.

the study which is summarized in the discussion section. Throughout these results, in addition to describing and analyzing Brian's mathematical activity, I also include my own reflections and conjectures from ongoing analysis in order to illustrate the significant role that I played as the teacher-researcher.

Emergence of the ZPP in Brian's Equation Solving Activity in $\mathbb{R}$

I first noticed the ZPP in Brian and Julie's activity early in the teaching experiment when they were solving a quadratic equation in the real numbers. While they had already solved linear equations like $4x = 0$ and $4(x - 5) = 0$ in $\mathbb{R}$, there was no observable use of the ZPP in their activity. Rather than directly question the students about their solutions strategies for these (admittedly simple) equations, I designed and administered additional tasks involving pre-factored quadratic equations. Julie's response to a task asking them to solve $(x + 2)(x + 3) = 0$ in $\mathbb{R}$ appeared to implicitly invoke the ZPP (level 1 of Freudenthal's framework). Accordingly, I asked her to explain her reasoning so that she might start to explicitly formulate any rules they implicitly used (level 2). Julie replied that "we only have two factors that multiply to equal zero, so the only way to solve this is that one of the factors has to equal zero." Brian added, "there's no other way to get a zero." When I asked them to describe their justification in the form of a rule, Brian wrote "anytime you have an equation where two factors multiply and equal 0, then you can tell that one of them is zero." I concluded that both students held a reasonably functional understanding of the ZPP in $\mathbb{R}$, the predominant context of secondary algebra.

Brian Operationalizes His ZPP Rule to Solve Equations in $\mathbb{Z}_{12}$

Shortly thereafter, I turned Brian and Julie's attention to $(\mathbb{Z}_{12}, +_{12}, \cdot_{12})$, an accessible yet initially unfamiliar structure that contains zero-divisors. Their first task, in order to ensure computational fluency with the binary operations (addition and multiplication modulo 12), was to construct addition and multiplication tables for the set $\{0, 1, 2, \ldots, 11\}$ using clock arithmetic as an experientially real starting point.[4] Both Brian and Julie noticed that certain rows and columns of the multiplication table displayed repetitive patterns. From my perspective, this was a potentially valuable way to single out and attend to the zero-divisors in $\mathbb{Z}_{12}$ (e.g., Cook, 2014). I questioned them about what such a pattern might mean:

[4]The interested reader can consult Cook (2014) for more information about how $\mathbb{Z}_{12}$ can be introduced in an experientially real way.

Excerpt 1

Brian:	Some of these on the table have patterns, which made them easier. I just recognized the pattern. And we could easily fill it in after that.
Researcher:	Okay. So give me an example. Like which one? Can you give me . . . ?
Brian:	For example, row four so the four. Four times zero. 0, 4, 8. That pattern 0, 4, 8, 0, 4, 8.
Researcher:	Okay. So it goes 0, 4, 8 all the way across.
Brian:	Yes sir.
Researcher:	Okay. So this zero right here, can you ... what does that zero mean? Why is that zero there?
Brian:	Four times three is zero.
Researcher:	Okay. So this zero in the pattern is here because four times three is zero.
Brian:	Mm-hmm.
Researcher:	What about over here? What about this one?
Brian:	Four times six would also be zero.
Researcher:	And right here?
Brian:	Nine times four, or four times nine, sorry, is also zero.
Researcher:	Four times nine is also zero.
Brian:	Mm-hmm.

Brian's response indicated to me that he had some awareness that two nonzero elements could be multiplied together to produce zero, though neither he nor Julie exhibited signs that they regarded this as a peculiarity, or viewed such instances as counterexamples to their rule. On one hand, I had initially conjectured that both students would indeed have reason to single out calculations like $4 \cdot 3 = 0$ because their explicit formulation of their rule seemed at the time to be robust enough to enable them to detect $4 \cdot 3 = 0$ as a counterexample. On the other hand, as Freudenthal (1973) noted, it is natural for students who are faced with a new, unfamiliar algebraic structure to proceed at first as if their existing rules hold. Thus, rather than question the students about the apparent discrepancy between such calculations and their rule, I decided to problematize it by introducing linear equations with multiple solutions.

Hoping to provide sharp contrast against which the students could compare the multiplicity of solutions to certain linear equations in $\mathbb{Z}_{12}$, I encouraged Brian and Julie to solve some of the same equations in $\mathbb{Z}_{12}$ that they had already solved in $\mathbb{R}$. I was curious to observe their activity as they solved the equations $4x = 0$ and $4(x - 5) = 0$ in $\mathbb{Z}_{12}$ because Brian had correctly calculated and articulated that $4 \cdot 3 = 0$, $4 \cdot 6 = 0$, and $4 \cdot 9 = 0$ (Excerpt 1). But, just as in $\mathbb{R}$, they both asserted that $x = 0$ is the only solution to $4x = 0$ and $x = 5$ is the only solution to $4(x - 5) = 0$ in $\mathbb{Z}_{12}$, with Brian specifically mentioning that "the only way for 4 times a number to equal 0 is by multiplying by 0." Similarly, Julie's solution to solving $(x + 2)(x + 3) = 0$ in $\mathbb{Z}_{12}$ employed what appeared to be the ZPP and proceeded almost identically to her response to the same equation in $\mathbb{R}$, the only difference being that her solutions were $x = 9$ and $x = 10$ (instead of $x = -2$ and

$x = -3$). Brian's response made it clear that he also did not detect any differences between $\mathbb{R}$ and $\mathbb{Z}_{12}$:

Excerpt 2

Brian:	Uh ... what was the point of that?
Researcher:	What was the point of what?
Brian:	That is literally the exact same as normal math.
Researcher:	OK, so ... [laughs]. OK, so I want to break this down. What is, what is that? What are you ... what is the same as normal math?
Brian:	The way she solved it with $\mathbb{Z}_{12}$ is the exact same way you solve that in normal factoring.

Brian Monster-Bars Zero-Divisors in $\mathbb{Z}_{12}$ and $M_2(\mathbb{R})$

Simpson and Stehlikova (2006) proposed that, in cases in which students struggle to identify critical aspects of an algebraic structure for themselves, instructors should "explicitly guide attention to, first, those aspects of the structure which will be the basis of later abstraction" (p. 368). As my efforts to guide their attention to zero-divisors implicitly via task design were unsuccessful, I decided to heed these recommendations and explicitly point out an instance of zero-divisors. Specifically, referring to the task in which Brian and Julie had proposed that $x = 5$ was the only solution to $4(x - 5) = 0$ in $\mathbb{Z}_{12}$, I asked them about the possibility that $x - 5 = 3$ (Excerpt 3) so that they might recognize that $4(x - 5) = 4 \cdot 3 = 0$. I phrased my inquiry somewhat unconventionally in terms of the element $x - 5$ (as opposed to simply offering $x = 8$ as an additional solution) because I wanted to maintain focus on the equation's product structure and, potentially, the ZPP. Julie immediately realized (and accepted) that they had overlooked such cases, remarking that she had stopped looking for solutions after identifying $x = 5$ because she had only expected one solution. Brian, on the other hand, rejected the possibility of additional solutions:

Excerpt 3

Researcher:	What do you think, Brian, you don't look, you've got a skeptical look on your face.
Brian:	I still think that this [motions to $4 \cdot 0 = 0$] is 0, right, but this ....
Researcher:	So, can you say what you're pointing to right now?
Brian:	The 4, um, as long as $x = 5$, then that's 0, and I think that's the only way to 0. This is some type of convoluted plan or a scheme you've come up with. There's no way that this is a 0.

Brian's outright rejection of zero-divisors surprised me—I had predicted that he would react like Julie and reluctantly concede that he had overlooked several solutions (which would then have been an opportunity to encourage them to revisit their rule and check whether or not it holds in $\mathbb{Z}_{12}$). Instead, however, I decided to explore Brian's reasons for rejecting (the additional solutions created by) zero-divisors. My first conjecture was that perhaps the clock arithmetic metaphor from the initial task that introduced $\mathbb{Z}_{12}$ was influencing Brian's thinking. Perhaps, for example, he viewed $4 \cdot 3$ as 12, and, as a result, did not identify 12 with 0.

To test this conjecture, I shifted to another example structure, thinking that, if Brian raised no objection to zero-divisors in the new context, then I could conclude that the nature of his previous objection was context-specific to $\mathbb{Z}_{12}$. If, however, he maintained his objections, this would indicate that he was potentially objecting to idea of zero-divisors altogether. I chose $M_2(\mathbb{R})$ as the new example structure because it contains zero-divisors, and it would have been familiar to Brian from linear algebra, thus leaving him with fewer reasons to doubt its legitimacy.[5] I asked if their rule held in $M_2(\mathbb{R})$, and Julie, who seemed relatively unperturbed by the presence and effects of zero-divisors in $\mathbb{Z}_{12}$, drew an analogy with $\mathbb{Z}_{12}$ and seemed to accept the possibility of such elements in $M_2(\mathbb{R})$ (though she was unable to identify any at first), remarking that "when I look back ... there are some other ways to get 0 without multiplying by 0, so I think that maybe there could be a way to multiply two matrices so that you can get the zero matrix." Brian, on the other hand, remained steadfast in his apparent belief that the ZPP was universally inviolable, and responded before even trying to produce a counterexample that "in order to get a zero matrix, you have to multiply by 0." I responded by presenting them with a pair of zero-divisors—specifically, $\begin{bmatrix} 1 & 0 \\ 0 & 0 \end{bmatrix}$ and $\begin{bmatrix} 0 & 0 \\ 0 & 1 \end{bmatrix}$. Brian, after multiplying the two matrices together to obtain the zero matrix, again stood by his original assertion:

Excerpt 4

Brian:	I don't understand how this example ... can count. [sighs].
Researcher:	So why, why wouldn't it count?
Brian:	Because you're still ... you still have zeros here. Like you literally just added a 1 somewhere, and said, here you go! It works!
Researcher:	OK, um, when you said 'zeros here,' can ... unfortunately, the Livescribe pen can't, uh, can't tell us which ones you're pointing to.
Brian:	OK ... these ones [motions to and marks the zeros in $\begin{bmatrix} 1 & 0 \\ 0 & 0 \end{bmatrix}$ and $\begin{bmatrix} 0 & 0 \\ 0 & 1 \end{bmatrix}$]. So there are zeros involved.
Researcher:	There are zeros involved.
Brian:	Yes, so I don't think this should, this should count as an example that we can use. I, I just don't believe that, that this is OK.

[5]It is not completely inconceivable that Brian viewed $\mathbb{Z}_{12}$ as a contrivance that I created purely for the purposes of this teaching experiment. There would be no such concerns with $M_2(\mathbb{R})$.

Because the nature of Brian's objection in this case was that "there are still zeros involved," I responded by presenting him with a zero-divisor pair that did not involve any 0 entries: $\begin{bmatrix} 1 & 2 \\ 3 & 6 \end{bmatrix}$ and $\begin{bmatrix} 2 & 2 \\ -1 & -1 \end{bmatrix}$. This time, after verifying for himself that the product of these two matrices was indeed the zero matrix, he maintained his skepticism, this time on the grounds that 0 was *not* involved:

Excerpt 5

Brian:	Um, I'm still skeptical because I still think you need zeros to get zeros, and ... you're not multiplying *A* times, er ... you're not multiplying *A* and *B* together to get 0, um, because *A* and *B* have to be 0.

Brian's refusal to accept zero-divisors in both $\mathbb{Z}_{12}$ and $M_2(\mathbb{R})$ suggests that his reasons for doing so were not context-specific and that he was indeed objecting to counterexamples to the ZPP in a more general way.

Analysis, Part I: Monster-Barring and Its Pedagogical Potential

Brian's rejection of zero-divisors across algebraic contexts is an example of *monster-barring*. In his seminal text *Proofs and Refutations*, Lakatos (1976) defined monster-barring as the outright rejection of a counterexample on the grounds that it is "a pathological case" (p. 14). Similarly, Larsen and Zandieh (2008), who repurposed Lakatos's methods for mathematical discovery as design heuristics for RME, characterized monster-barring as "any response in which the counterexample is rejected on the grounds that it is not a true instance of the relevant concept" (p. 208). This includes cases in which students summarily reject a counterexample without an apparent reason. Indeed, several of Brian's comments support the assertion that he viewed zero-divisors as pathological (see, for example, Excerpts 3 and 4) and, as a result, he refused to consider them as counterexamples to the ZPP.

Although monster-barring might, at first, seem to be counterproductive and in need of correction via direct instruction, Lakatos (1976) suggested there was potential for such activity to be productively repurposed, commenting that mathematical ideas "are frequently proposed and argued about when counterexamples emerge" (p. 16). Accordingly, Larsen and Zandieh (2008) proposed that having students consider and render judgments about the validity of proposed counterexamples and underlying definitions is a form of informal mathematical thinking that can be leveraged to support the development of more formal mathematical concepts.

Zazkis and Chernoff (2008) cautioned that the pedagogical potential in a given counterexample is learner- and situation-dependent; that is, counterexamples fail to provoke a student to reflect on their assumptions and ways of understanding when the student does not view them as meaningful. An effective instructional response might often involve simply being more strategic about the choice of a counterexample for the student to consider. But regarding examples of zero-divisors that Brian might consider meaningful, students are exposed to few instances of zero-divisors prior to abstract algebra. The examples that typically appear prior to abstract algebra involve modular and matrix arithmetic, structures in which Brian had already rejected zero-divisors. Thus, presenting Brian with zero-divisors from another algebraic context at this time seemed unproductive.

Instead of attempting to find a more suitable example of zero-divisors for Brian to consider, I altered course to determine if there was a way in which he might come to regard the examples he had already seen as meaningful. From an RME perspective, a mathematical idea is meaningful if it is experientially real to the learner—that is, if the idea emerges in his own intuitive thinking about a mathematical context or situation (Freudenthal, 1973). This distinction might explain why Brian did not initially consider the zero-divisors in $\mathbb{Z}_{12}$ and $M_2(\mathbb{R})$ to be meaningful, as they were not the products of his engagement with the instructional tasks; rather, they were mine. But it also suggested that I should focus on encouraging Brian to expound upon his reasons for monster-barring zero-divisors to try to find some aspect of his activity that might enable him to view zero-divisors from a different perspective. Thus, I changed tactics from trying to convince Brian that zero-divisors are indeed legitimate counterexamples to trying to understand his intuitive thinking about these elements and how I might productively incorporate his thinking into my instructional design.

Leveraging Monster-Barring to Characterize Zero-Divisors as "Awkward"

During this new line of inquiry, I asked Brian to identify exactly which products he objected to in the multiplication table for $\mathbb{Z}_{12}$. He and Julie responded by turning to their multiplication table and circling entries.

Excerpt 6

Brian:	So 6 times 2, 6 times 4 ...
Julie:	6 times 6, 6 times 8, 6 times 10.
Researcher:	So you're just going down ...
Brian:	We're just finding the places that ... it doesn't look like a 0 needs to be there. Like it's awkward, like it shouldn't be on the multiplication table. So, numbers that multiply ... don't look like they multiply together would equal 0, we'll find they do.
Julie:	6 times 4, there's a 0.
Researcher:	OK.
Julie:	So, like, the same thing with, like, 8 times 3.
Researcher:	And that's, so, Brian, that's what you're calling an awkward ...
Brian:	Yes.
Researcher:	Like zero showing up in an awkward place?

Brian:	Yes.
Researcher:	Where, where does ... what are the non-awkward appearances of 0?
Brian:	The places where 0, the top row and the first column in the table show that every one of those numbers is multiplied by 0 to get 0. Those are the normal ways ... to get zero.
Researcher:	Are there, so are there any normal ways that are not in the first row or the first column?
Brian:	No.

This was an important exchange for several reasons. First, Brian used the phrase "awkward ways to make zero" to refer to combinations of elements in which "it doesn't look like a zero needs to be there ... numbers that ... don't look like they multiply together would equal 0." Similarly, "normal ways to get zero" are those involving multiplication by 0. This mirrors the distinction between the ZPP (which is equivalent to the absence of zero-divisors in a ring) and its converse (which always holds in a ring). Second, Julie, who was relatively unperturbed by zero-divisors, was able to quickly operationalize Brian's distinction, as evidenced by her immediate engagement in the task. I interpreted this as a sign that Brian's criteria could be a meaningful way for Brian (and even Julie) to engage with zero-divisors and use them to make distinctions between algebraic structures. This hypothesis shaped my instructional decisions and analysis in the remaining sessions of the teaching experiment.

Using Zero-Divisors as "Awkward" to Classify $\mathbb{Z}_{12}$, $M_2(\mathbb{R})$, and $\mathbb{Z}_5$

To further elicit Brian and Julie's thinking about awkward and normal ways to make zero, I designed classification tasks that prompted them to decide if a given structure behaved more like $\mathbb{R}$ or more like $\mathbb{Z}_{12}$ (as they had already concluded that $\mathbb{R}$ contains no awkward ways to make zero, unlike $\mathbb{Z}_{12}$). The first structures they considered were $\mathbb{Z}_{12}$ and $M_2(\mathbb{R})$, both of which they had worked with earlier in the teaching experiment. Brian immediately responded that $M_2(\mathbb{R})$ should be classified as "more like $\mathbb{Z}_{12}$."

Excerpt 7

Brian:	Definitely $\mathbb{Z}_{12}$.
Researcher:	Why? What makes you so sure?
Brian:	Well, earlier we discussed that $\mathbb{Z}_{12}$ has some awkward ways to make zero and we also talked earlier that the matrices have awkward ways to make zero. Real numbers don't have awkward ways to make zero. So they share that comparison.
Julie:	That does make a little bit more sense because I guess in $\mathbb{Z}_{12}$ three times four is zero. So that would be an awkward way to make zero. You would have to multiply by zero in [the] real [numbers].

Brian's classification of $M_2(\mathbb{R})$ as "more like $\mathbb{Z}_{12}$" suggested that this adaptation to his way of understanding the ZPP might also be generalizable to other contexts. Brian's statements that "$\mathbb{Z}_{12}$ has some awkward ways to make zero" and "the real numbers don't have awkward ways to make zero" are comparable to the more conventional "$\mathbb{Z}_{12}$ contains zero-divisors" and "$\mathbb{R}$ does not contain zero-divisors." Notably, it is not difficult to find superficial similarities between $M_2(\mathbb{R})$ and $\mathbb{R}$: both are uncountably infinite and, moreover, $M_2(\mathbb{R})$ can be viewed as having been constructed from $\mathbb{R}$. The use of Brian's characterization of zero-divisors seemed to supersede such considerations.

Up to this point, Brian had only applied this way of understanding to $\mathbb{Z}_{12}$ and $M_2(\mathbb{R})$, the contexts from which it had emerged in his reasoning, both of which contain zero-divisors. Subsequently, I asked Brian and Julie to classify $(\mathbb{Z}_5, +_5, \cdot_5)$, a structure that, based upon purely superficial characteristics, might be classified as more similar to $\mathbb{Z}_{12}$. However, $\mathbb{Z}_5$ contains no zero-divisors and is thus more similar in this regard to $\mathbb{R}$. Initially, both Brian and Julie hypothesized that $\mathbb{Z}_5$ was more similar to $\mathbb{Z}_{12}$ and $M_2(\mathbb{R})$ because, Brian predicted, "they're [probably] awkward ways to make 0 for $\mathbb{Z}_5$ as well." As they attempted to justify this conjecture by constructing the operation tables, however, they changed their minds:

Excerpt 8

Julie:	That is more like the real numbers, actually. The only way we ended up getting zero is multiplying by zero. And so that would be more like the real numbers, because in $\mathbb{Z}_{12}$ we could do awkward ways like three times four and get zero. But in the real numbers we have to multiply by zero, and $\mathbb{Z}_5$ also, to get zero.
Researcher:	Do you agree, Brian?
Brian:	I would say it's like the real numbers, yes, after drawing the table out.
Researcher:	And what about the table changed your mind?
Brian:	Looking over, there are no other zeros where other numbers should be, except for where zero is multiplied by another number.
Researcher:	Yeah. I was gonna ask you about that. So I see zeros in the first row and the first column here. Are those not awkward?
Brian:	No. Those are normal ways to get zero. Multiply by zero.

In the above exchange, both students indicated awareness that the "normal" ways to get zero are the only such ways—for example, Julie mentioned that "we *have* to multiply by zero ... to get zero" and Brian noticed that "there are no other zeros where other numbers should be." This is notable because it demonstrates that both Brian and Julie were able to operationalize the awkward/normal distinction to identify a structure without zero-divisors.

Extending Understanding of Zero-Divisors and the ZPP by Incorporating Additive Identity

Although Brian had implicitly used (level 1 of Freudenthal's framework), explicitly formulated (level 2), and subsequently used his way of understanding the ZPP to classify additional algebraic structures (level 3), it was unclear how it might generalize to structures in which the additive identity does not involve 0. This is related to two key components of Harel's (1998) criteria for a way of understanding to be effective: the ability to think about it in a general way (by using it to classify a varied collection of example structures), and the ability to coordinate it with ways of understanding other mathematical ideas.

In order to observe the generalizability of Brian's[6] way of understanding, I presented him with two abstractly defined rings with explicitly defined binary operations to see how he might determine what "awkward way to make zero" means in such contexts. The first context was a finite noncommutative ring with four elements that contains zero-divisors, N_4; for contrast, the second was the finite field of order 4, $\mathbb{F}_4$. In each case, the elements were given as letters ($\{X, Y, Z, W\}$ and $\{A, B, C, D\}$, respectively) to ensure that Brian's conclusion would be attributable only to his current thinking about these contexts (and not facets that he might remember from previous courses, for example). The operation tables for these structures are given in Fig. 3.1.

I anticipated that, because the idea of zero seemed to feature so prominently in his thinking, Brian would struggle to classify the structure N_4 (in terms of being more similar to $\mathbb{R}$ or $\mathbb{Z}_{12}$), at least initially. But he immediately asserted that N_4 was "more like $\mathbb{Z}_{12}$ and matrices." Of course, there are a number of reasons other than the presence of zero-divisors that might explain such a conclusion. For example, this structure is noncommutative, which might have led Brian to identify

N_4

+	*x*	*y*	*z*	*w*
x	*w*	*z*	*y*	*x*
y	*z*	*w*	*x*	*y*
z	*y*	*x*	*w*	*z*
w	*x*	*y*	*z*	*w*

·	*x*	*y*	*z*	*w*
x	*x*	*y*	*z*	*w*
y	*w*	*w*	*w*	*w*
z	*x*	*y*	*z*	*w*
w	*w*	*w*	*w*	*w*

$\mathbb{F}_4$

+	*A*	*B*	*C*	*D*
A	*C*	*D*	*A*	*B*
B	*D*	*C*	*B*	*A*
C	*A*	*B*	*C*	*D*
D	*B*	*A*	*D*	*C*

·	*A*	*B*	*C*	*D*
A	*D*	*A*	*C*	*B*
B	*A*	*B*	*C*	*D*
C	*C*	*C*	*C*	*C*
D	*B*	*D*	*C*	*A*

Fig. 3.1 Operation tables for abstractly defined finite rings with (top, N_4) and without (bottom, $\mathbb{F}_4$) zero-divisors

[6]Unfortunately, Julie was unable to complete the final session of the teaching experiment.

it with matrices. It is also finite like $\mathbb{Z}_{12}$ and has repetitive rows and columns in its multiplication table. When I asked Brian to explain his thinking, however, the identification of the additive identity emerged as a critical component of his thinking:

Excerpt 9

Brian:	All these ones right here [gestures to the middle of the multiplication table] are awkward ways to make zero.
Researcher:	And what makes those awkward?
Brian:	Well, I'm assuming that the *W* is a zero because in the multiplication table, excuse me, in the addition table right here, I can add zero to anything and still get to any of these other variables, and still get the variable I added it to.
Researcher:	Okay. So give me an example.
Brian:	So *X* plus the *W* will still give me an *X*. The *Y* plus the *W* will still give me a *Y*.
Researcher:	Okay.
Brian:	So this *W* plus the *W* will still give me a *W*, which makes me assume that the *W* is a zero.
Researcher:	Okay. So what makes the ones that you circled awkward and not normal ways to make zero?
Brian:	Because I'm not using the *W* to make another *W*.
Researcher:	Okay.
Brian:	I'm not. In these cases here, you're not multiplying by *W* to get a *W*.

Brian correctly identified *W* as the additive identity element by relying on a context-independent way of understanding the identity. This was apparent in his remarks about specific calculations ("*X* plus the *W* will still give me an *X*") and his articulation of the general pattern underpinning these calculations ("I can add zero to anything ... and still get the variable I added it to"). A student without such a way of understanding the identity—who, for example, holds a way of understanding the identity element strongly linked to the real numbers—might have had trouble identifying the additive identity in such a situation. Indeed, there are several indications in the literature that holding only an $\mathbb{R}$-centric way of understanding the identity can inhibit the development of ways of understanding other ideas (e.g., Bagley, Rasmussen, & Zandieh, 2015; Zazkis & Kontorovich, 2016). An inability to determine the identity would significantly compromise a student's ability to determine if zero-divisors appear in the structure. In Brian's case, however, having such a flexible way of understanding the identity enabled him to quickly identify instances of zero-divisors. His remark that certain appearances of *W* are awkward because "I'm not using the *W* to make another *W*" harmonizes with remarks from earlier in the teaching experiment—for example, Brian commented on the importance of looking for "zeros where other numbers should be, except for when 0 is multiplied by another number" (Excerpt 8). I propose, therefore, that this task, rather than representing a significant upheaval in Brian's thinking, reflected instead a simple application of his way of understanding. That is, Brian's way of understanding the appearance of zero-divisors by characterizing them as awkward products was stable enough that he was able to coordinate it with his way of understanding the identity quite easily. Predictably, he quickly classified the finite field of order 4 ($\mathbb{F}_4$) in a similar way, correctly identifying C as the additive identity, and then making a convincing argument that there were no zero-divisors because "anything times C in the multiplication table part will equal C, and it's the only way to get a C," citing the absence of C's outside of the row and column for C as his justification.

Analysis, Part II: Assessing Brian's Way of Understanding the ZPP Using Harel's Criteria

I conclude this section by returning to Harel's criteria and claiming that, indeed, Brian's way of understanding zero-divisors and the ZPP was effective for the purposes of an introductory unit in abstract algebra. First, it is clear that Brian was able to articulate this way of understanding in his own words, as evidenced by the continual identification and explanation of the distinction between awkward and normal ways to make 0 (both students, for example, explicitly articulated this distinction in their own words and successfully employed it in Excerpts 7 and 8). Second, Brian successfully used this understanding to classify a variety of structures with varying elements and binary operations. I claim that this collection of six structures—all of which Brian correctly classified—is sufficiently general because it contains structures that are finite rings ($\mathbb{Z}_{12}$, $\mathbb{Z}_5$, $\mathbb{F}_4$, N_4), infinite rings ($\mathbb{R}$, $M_2(\mathbb{R})$), integral domains ($\mathbb{R}$, $\mathbb{Z}_5$, $\mathbb{F}_4$), rings that are not integral domains ($\mathbb{Z}_{12}$, $M_2(\mathbb{R})$, N_4), rings in which the additive identity involves 0 ($\mathbb{R}$, $\mathbb{Z}_{12}$, $\mathbb{Z}_5$, $M_2(\mathbb{R})$), rings in which the additive identity does not involve[7] 0 ($\mathbb{F}_4$, N_4), commutative rings ($\mathbb{R}$, $\mathbb{Z}_{12}$, $\mathbb{Z}_5$, $\mathbb{F}_4$), noncommutative rings ($M_2(\mathbb{R})$, N_4), rings whose binary operations and elements were familiar to the students ($\mathbb{R}$, $M_2(\mathbb{R})$), and rings whose binary operations were initially unfamiliar to the students ($\mathbb{Z}_{12}$, $\mathbb{Z}_5$, $\mathbb{F}_4$, N_4). Third, Brian demonstrated the ability to leverage a way of understanding another mathematical idea—particularly, additive identity—to apply his way of understanding to unfamiliar, abstractly defined structures. Specifically, Brian's general way of understanding the identity supported generalization to structures mentioned above in which the additive identity did not explicitly involve the number 0.

[7] I did not include an example in which the additive identity could be interpreted as a nonzero *number* (for example, viewing $\mathbb{Z}_{12}$ as the set $\{1, \ldots 12\}$). Exploring how such a representation of the additive identity might shape a student's conceptions of zero-divisors is an interesting question that I leave for future research.

Discussion

The primary objective of this chapter was to address the issue that prospective teachers do not see the relevance of their abstract algebra coursework to the secondary mathematics they will be teaching, in part because they struggle to understand fundamental concepts in abstract algebra (e.g., Cook, 2018; Wasserman, 2017). In response, guided by the tools of *conceptual analysis* (Thompson, 2008), I conducted a teaching experiment (Steffe & Thompson, 2000) that investigated how students might be able to adapt their ways of understanding familiar properties from secondary algebra to be effective in abstract algebra. Focusing specifically on the zero-product property (ZPP), my primary research question was: How might beginning abstract algebra students be able to adapt their existing understandings of the ZPP to be effective in abstract algebra?

The primary contribution of this chapter—and my answer to this research question—is the learning trajectory that I derived and generalized from Brian's mathematical activity in this teaching experiment. The second contribution of this chapter is that I discuss two implications for teacher education, which are based on reframing the student's monster-barring activity productively. In this section, I explicate these contributions along with the associated implications.

A Learning Trajectory for Developing an Effective Way of Understanding the ZPP in Abstract Algebra

The learning trajectory below explicates one possible way in which students might develop an effective way of understanding the ZPP using their prior knowledge of the ZPP as a starting point. The first three steps of the learning trajectory are independent of how students ultimately respond to zero-divisors:

1. Rules equivalent to the ZPP appears implicitly in students' equation solving activity in a familiar algebraic structure that does not contain zero-divisors.
2. After explicitly describing their use of the ZPP, the students are given an opportunity to apply their rules to a familiar or easily accessible structure that does contain zero-divisors.
3. Students are given an opportunity—either implicitly through task design or explicitly through direct intervention by a classmate or the instructor—to consider zero-divisors and their relationship to the ZPP.

The remaining steps cater specifically to those students who reject or otherwise disregard the significance or legitimacy of zero-divisors as counterexamples to the ZPP. However, as Julie's ability to make sense of and operationalize Brian's way of understanding in this teaching experiment suggests, these steps are not altogether irrelevant to, and perhaps can even enhance the understanding of, students (such as Julie) who might initially accept zero-divisors.

4. Using features from the students' own mathematical activities as inspiration (if possible), negotiate a "re-characterization" of zero-divisors in a way that acknowledges the students' skepticism yet also allows for them to engage in productive mathematical activity with zero-divisors.
5. The students apply their new characterization of zero-divisors as a means to classify the structures with which they have already worked (which should involve, at the very least, one structure with zero-divisors and one without). Details about how to employ this characterization and its relationship to the ZPP emerge and are discussed.
6. The students apply their new characterization of zero-divisors as a means to classify a varied set of additional algebraic structures. How "varied" should be interpreted depends upon each individual classroom, but, at the very least, the set of structures should include structures with zero-divisors and structures without zero-divisors.
7. The students extend their new characterization of zero-divisors as a means to classify examples of algebraic structures with different types of elements and binary operations, particularly structures in which the additive identity does not involve the number 0.

Although this learning trajectory is empirically derived and can indeed lead to an effective way of understanding the ZPP in abstract algebra—this chapter serves as an existence proof of this claim—I do not claim that students *should* learn about or come to understand the ZPP in this way, nor do I claim that this is the only such route to an effective way of understanding. Instead, this learning trajectory and the associated analysis offer one such example of how learning might proceed. Additionally, the theoretical analysis of Brian and Julie's thinking in this chapter is generalizable insofar as it can be used to make sense of and organize how other students might reason in similar mathematical situations. However, the chapter does suggest that a student's monster-barring of instances of zero-divisors can be repurposed in a way that still leverages the students' way of thinking and that, simultaneously, can be productive for learning abstract algebra concepts.

Implications for Secondary Teacher Preparation

In addition to the contributions discussed above, I believe this study offers two insights for instructors of teacher education courses in advanced mathematics.

First, it illustrates one way to approach the relationship between abstract algebra and secondary algebra through the point of view of a student who is coming to understand that same relevance for himself. The learning trajectory itself could be a useful way to engage prospective teachers in discussions about the relationship between secondary mathematics and advanced mathematics and the ways of understanding and thinking that might be involved in adapting flexibly from one to the other. More so than that, however, what we see from the learning trajectory is Brian's deepening mathematical understanding, not only of the notion of a zero-divisor, but also of the meaning of the ZPP. But unlike zero-divisors, which are more strictly in the realm of abstract algebra, the ZPP is studied throughout elementary and secondary mathematics. I see this deepening in two ways. First, the activities forced Brian to disentangle a statement and its converse. Namely, in all structures, the implication, $x = 0 \vee a = 0 \Rightarrow ax = 0$, is true; however, its converse, the ZPP, $ax = 0 \Rightarrow a = 0 \vee x = 0$, is not. Although Brian seemed to be able to differentiate the meaning of these two statements in the activities, his initial responses indicated that these two should always go together—they should be a single biconditional (and not two implications). Second, and perhaps more importantly, Brian was exposed to structures that included "awkward" ways to make zero. This exposure to examples and non-examples was productive, namely, in that it at least started Brian down the path of distinguishing "universal mathematical law" from "mathematical property." That is, the ZPP was viewed as an inherent mathematical truth to Brian—his monster-barring activity was a product of this belief. But classifying different structures as more-similar to $\mathbb{R}$ or $\mathbb{Z}_{12}$ exposed Brian to structures which adhered to this law and others which did not. This activity can be regarded as the beginning of his transition to "apprehending" (Simpson & Stehlikova, 2006) the ZPP—of recognizing the ZPP as a property of a particular system and not an innate mathematical truth. I see the descriptive nature of such activities (i.e., describing properties of systems) as then being leveraged to recognize the prescriptive nature of such properties. Freudenthal's (1973) distinction between *descriptive* and *constructive* views of mathematical definitions is illuminating here. A *descriptive* view of defining involves singling out a few characteristic properties to outline a well-known object, whereas a *constructive* view involves the classification of new mathematical objects. From this perspective, then, this teaching experiment demonstrates the potential value in leveraging one's *descriptive* views of an algebraic property (e.g., using the ZPP to describe familiar structures like the real numbers) in order to recognize their *constructive* potential (e.g., viewing the ZPP as a means to characterize a new kind of structure unlike the real numbers in which the ZPP does not hold). That is, the moment a structure is recognized as maintaining a particular property, then the structure also maintains all consequences of this property—i.e., a descriptive property is also constructive in some sense. Although I only reported on the ZPP in this chapter, other algebraic properties discussed in secondary mathematics (e.g., commutativity) could be viewed in a similar way. What this study indicates is that such properties might be viewed by students (and secondary teachers) as innate mathematical truths; exposure to, and classification of, examples and non-examples in an abstract algebra course is one activity that might help deepen teachers' understanding of this secondary content by having them view these properties both descriptively and constructively.

Second, and somewhat independent of the subject matter, I believe that this study has some implications for thinking about pedagogy in mathematics teaching more broadly. Namely, it provides an example for how students' experiences, even if they seem counterproductive and irrelevant at first, can be leveraged effectively to advance their mathematical thinking in productive ways. I see this as a more specific case of a broader phenomenon—an approach to teaching that builds on students' thinking. Much of the mathematics education literature advocates for such an approach. In fact, these findings were brought to light by applying Steffe and Thompson's (2000) methodological principle that researchers—and, indeed, teachers—should assume that students' behavior is rational and that there is great value in attempting to understand and build upon it. This study adhered to this principle by using Brian's thinking as he engaged with the notion of a zero-divisor. However, even more so, this study indicates that such an approach is possible *even when a students' thinking initially appears to be counterproductive*. That is, when a student's thinking is mathematically productive, it is relatively easy to build on it; however, when a student outright rejects an idea or appears to be firmly convinced of something that is untrue, our instincts (my own included in this teaching experiment) suggest we should directly correct them. Although this certainly might be the appropriate response in some teaching situations, the findings from this study suggest that it is possible to build on their thinking even when their ideas appear to be unproductive. This suggests two things to me about instruction in abstract algebra for an audience of secondary teachers. First, abstract algebra instruction can model good pedagogical practices. As was done in this study, using student thinking to develop abstract algebra ideas models good pedagogy. For secondary teachers, learning mathematics in ways that mirror good teaching contributes to their development as teachers. Second, not only can we model good pedagogical practices as abstract algebra instructors, we can also be explicit about this modeling. That is, as instructors, we can draw attention to the ways that we are building on students' thinking in our own classrooms. And, as evident from this study, building on student thinking is possible even in extreme cases, when their ideas appear to be unproductive.

Conclusion

Thompson (2008) argued that conceptual analysis—particularly, a characterization of what conceptual understanding is and a learning trajectory describing how a student might attain it—provides a key example of "how mathematically substantive, conceptually-grounded conversations can be held with students ... having positive examples of such conversations will be very important for mathematics teacher education" (p. 60). Along these lines, as courses in advanced mathematics present many ideas for which there may be no direct analog in secondary mathematics, fluency with techniques to reframe unfamiliar ideas and examples in terms of student thinking could be a useful, important technique for teacher education courses in advanced mathematics.

References

Alcock, L., & Simpson, A. (2011). Classification and concept consistency. *Canadian Journal of Science Mathematics and Technology Education, 11*(2), 91–106.

Bagley, S., Rasmussen, C., & Zandieh, M. (2015). Inverse, composition, and identity: The case of function and linear transformation. *Journal of Mathematical Behavior, 37*, 36–47.

Brown, A., DeVries, D. J., Dubinsky, E., & Thomas, K. (1997). Learning binary operations, groups, and subgroups. *Journal of Mathematical Behavior, 16*(3), 187–239.

Conference Board of the Mathematical Sciences. (2010). *The mathematical education of teachers II* (Vol. 17). Providence, RI: American Mathematical Society.

Cook, J. P. (2018). An investigation of an undergraduate student's reasoning with zero-divisors and the zero-product property. *Journal of Mathematical Behavior, 49*, 95–115.

Cook, J. P. (2014). The emergence of algebraic structure: Students come to understand units and zero-divisors. *International Journal of Mathematical Education in Science and Technology, 45*(3), 349–359.

Davis, R. B., & Vinner, S. (1986). The notion of limit: Some seemingly unavoidable misconception stages. *Journal of Mathematical Behavior, 5*(3), 281–303.

Dubinsky, E., Dautermann, J., Leron, U., & Zazkis, R. (1994). On learning fundamental concepts of group theory. *Educational Studies in Mathematics, 27*(3), 267–305.

Freudenthal, H. (1973). What groups mean in mathematics and what they should mean in mathematical education. In A. G. Howson (Ed.), *Developments in mathematical education, proceedings of ICME-2* (pp. 101–114). Cambridge: Cambridge University Press.

Gravemeijer, K. (1998). Developmental research as a research method. In A. Sierpinska & J. Kilpatrick (Eds.), *Mathematics education as a research domain: A search for identity* (pp. 277–296). Dordrecht: Kluwer.

Harel, G. (1998). Two dual assertions: The first on learning and the second on teaching (or vice versa). *The American Mathematical Monthly, 105*(6), 497–507.

Hazzan, O. (1999). Reducing abstraction level when learning abstract algebra concepts. *Educational Studies in Mathematics, 40*(1), 71–90.

Lakatos, I. (1976). *Proofs and refutations*. Cambridge: Cambridge University Press.

Larsen, S. (2010). Struggling to disentangle the associative and commutative properties. *For the learning of Mathematics, 30*(1), 37–42.

Larsen, S., & Zandieh, M. (2008). Proofs and refutations in the undergraduate mathematics classroom. *Educational Studies in Mathematics, 67*(3), 205–216.

Ross, B. H., & Makin, V. S. (1999). Prototype versus exemplar models in cognition. In R. J. Sternberg (Ed.), *The nature of cognition* (pp. 205–241). Cambridge, MA: MIT Press.

Simpson, S., & Stehlikova, N. (2006). Apprehending mathematical structure: A case study of coming to understand a commutative ring. *Educational Studies in Mathematics, 61*(3), 347–371.

Steffe, L., & Thompson, P. (2000). Teaching experiment methodology: Underlying principles and essential elements. In A. Kelly & R. Lesh (Eds.), *Handbook of research design in mathematics and science education* (pp. 267–306). Mahwah, NJ: Erlbaum.

Thompson, P. W. (2008). Conceptual analysis of mathematical ideas: Some spadework at the foundations of mathematics education. *Proceedings of the annual meeting of the International Group for the Psychology of Mathematics Education, 1*, 31–49.

Wasserman, N. H. (2014). Introducing algebraic structures through solving equations: Vertical content knowledge for K-12 mathematics teachers. *PRIMUS, 24*(3), 191–214.

Wasserman, N. H. (2016). Abstract algebra for algebra teaching: Influencing school mathematics instruction. *Canadian Journal of Science Mathematics and Technology Education, 16*(1), 28–47.

Wasserman, N. H. (2017). Making sense of abstract algebra: Exploring secondary teachers' understanding of inverse functions in relation to its group structure. *Mathematical Thinking and Learning, 19*(3), 181–201.

Wasserman, N., Weber, K., Villanueva, M., & Mejia-Ramos, J. P. (2018). Mathematics teachers' views about the limited utility of real analysis: A transport model hypothesis. *Journal of Mathematical Behavior, 50*, 74–89.

Zazkis, R., & Chernoff, E. J. (2008). What makes a counterexample exemplary? *Educational Studies in Mathematics, 68*(3), 195–208.

Zazkis, R., & Kontorovich, I. (2016). A curious case of superscript (−1): Prospective secondary mathematics teachers explain. *Journal of Mathematical Behavior, 43*, 98–110.

Zazkis, R., & Leikin, R. (2010). Advanced mathematical knowledge in teaching practice: Perceptions of secondary mathematics teachers. *Mathematical Thinking and Learning, 12*(4), 263–281.

Chapter 4
What Kind of Opportunities Do Abstract Algebra Courses Provide for Strengthening Future Teachers' Mathematical Knowledge for Teaching?

Sean Larsen, Erin Glover, Anna Marie Bergman, and John Caughman

Introduction

This section of the volume is focused on the cognitive complexity of abstract algebra. The chapters highlight specific cognitive challenges around concepts that are closely connected to important concepts in secondary mathematics. Drawing on these connections, the authors leverage their research to develop insights about preservice teacher education. In order to frame our commentary, we begin by briefly reviewing the literature around students' understanding of abstract algebra. We situate the studies described in the two previous chapters within this literature. Then, to support our efforts to examine the proposed implications for the mathematical preparation for teachers, we introduce a framework (Ball, Thames, & Phelps, 2008) that we will use to characterize and critically analyze the connections the authors make between their research into students' understanding of abstract algebra and the mathematical preparation of teachers.

Students' Understanding of Abstract Algebra

The early research on the teaching and learning of abstract algebra was often focused on students' difficulties and misunderstandings. Much of this work relied

S. Larsen (✉) · A. M. Bergman · J. Caughman
Fariborz Maseeh Department of Mathematics and Statistics, Portland State University, Portland, OR, USA
e-mail: slarsen@pdx.edu

E. Glover
College of Education, Oregon State University, Corvallis, OR, USA

N. H. Wasserman (ed.), *Connecting Abstract Algebra to Secondary Mathematics, for Secondary Mathematics Teachers*, Research in Mathematics Education,
https://doi.org/10.1007/978-3-319-99214-3_4

on the APOS (Action, Process, Object Schema) framework (Dubinsky, Dautermann, Leron, & Zazkis, 1994; Dubinsky & Leron, 1994) to develop and refine proposed cognitive models, called genetic decompositions, which were then used to investigate students' understanding of topics, such as groups (Dubinsky et al., 1994), binary operations (Brown, DeVries, Dubinsky, & Thomas, 1997), coset multiplication (Asiala, Dubinsky, Mathews, Morics, & Oktaç, 1997), and permutations (Asiala, Kleiman, Brown, & Mathews, 1998). For the most part, these studies consistently found that many students possessed a very limited understanding of these concepts. In particular, they struggled to coordinate the various objects and processes involved (see Leron, Hazzan, & Zazkis, 1995) and they struggled with the level of abstraction (see Hazzan, 1999) of the content. The research described in the chapter by Melhuish and Fagan represents an important new phase of this area of research. Drawing on previous work, Melhuish (2018) developed and validated an instrument for assessing students' conceptual understanding of group theory, producing the first true large-scale investigation of students' understanding of abstract algebra.

More recent research has moved away from a focus on student difficulties and utilized design research methodologies to explore alternative instructional treatments that build on students' thinking to support them in developing abstract algebra concepts through a process of mathematical inquiry (e.g., Cook, 2014; Larsen, 2009). In addition to supporting the creation of new curricular approaches (Larsen, Johnson, & Bartlo, 2013), this research has provided new insights into the complexity of concepts like group and isomorphism (e.g., Larsen, 2009), and has uncovered cognitive challenges that mirror previously reported student difficulties with secondary concepts like associativity and commutativity (e.g., Larsen, 2010). The research described in Cook's Chap. 3 is situated within this body of work and represents a first step in developing an inquiry approach to ring and field theory instruction. His approach builds on students' previous experiences solving equations to develop key abstract algebra concepts, such as zero divisors, units, integral domains, and fields.

While both of the studies described in the two chapters are best understood as fitting within (and advancing) the research on undergraduate students' understanding of abstract algebra concepts, the authors leverage these studies to draw attention to the potential value of an abstract algebra course as part of preservice teachers' mathematical preparation. In the case of Melhuish and Fagan, insights about undergraduate students' struggles with binary operations and functions in abstract algebra suggest important connections to preservice teachers' mathematical knowledge for teaching around the secondary versions of these core concepts. In the case of Cook, his qualitative analysis of students' reinvention of the integral domain concept highlights the relationship between the students' understanding of the zero-product property from secondary mathematics and their development of the abstract algebra notion of a zero divisor. To support our analysis of these two chapters, we now turn to the ongoing efforts of mathematics educators to understand the role of mathematical knowledge in supporting the practice of teaching mathematics.

Mathematical Knowledge in the Practice of Teaching

Likely spurred by Monk's (1994) discouraging finding that taking advanced mathematics courses did not seem to help secondary teachers support their students' learning, there has been much work done in the past couple of decades in order to more fully understand the nature of the mathematical knowledge needed by teachers. Ball et al. (2008) presented a framework that delineated six domains of mathematical knowledge for teaching (MKT). These were divided into two categories, pedagogical content knowledge and subject matter knowledge.

Pedagogical content knowledge (PCK) is the professional subject-specific knowledge for teaching mathematics. Within the MKT framework, PCK contains three subdomains that include such things as knowing which mathematical representations or examples to use in teaching, or being aware of the difficulties student typically encounter when engaging with particular mathematics—part of Shulman's (1986) original descriptions of PCK. *Knowledge of content and teaching* (KCT) includes knowledge required to see the relationship between mathematics and teaching and how instructional decisions are made based on these relationships. For example, strategically sequencing examples or deciding whether to explore or table student contributions, involves drawing on knowledge in the KCT subdomain. The subdomain of *knowledge of content and students* (KCS) combines knowledge of mathematics with knowledge of students' mathematics. Teachers draw on KCS when they anticipate how students are likely to approach a task, or when they interpret their students' solutions. The subdomain of *knowledge of content and curriculum* includes knowledge of available curricular materials and how those approach various mathematical topics. Teachers draw on knowledge of content and curriculum to ascertain how well specific curricular materials are aligned with educational goals and how effective they are likely to be in supporting their students' learning.

The Subject Matter Knowledge domain includes the kinds of knowledge needed for teaching that are more strictly mathematical, in that they do not explicitly involve coordination with pedagogical issues. This includes *common content knowledge* (CCK), which describes mathematics that can be used in a wide variety of settings outside of teaching and is knowledge that is considered common among other fields that use mathematics. In contrast, *specialized content knowledge* (SCK) is knowledge that is needed specifically for the work of teaching mathematics. Teachers draw on CCK when they recognize that a student answer is incorrect or nonstandard while they draw on SCK when they notice patterns in students' errors or determine the viability of a nonstandard approach. Lastly, and of particular importance to this chapter, is *horizon content knowledge* (HCK), which describes an awareness of the mathematical territory and "how mathematical topics are related over the span of mathematics included in the curriculum" (Ball et al., 2008, p. 403). Ball and Bass (2009) point out that teachers draw on HCK when they reconcile pedagogical choices in relation to the larger mathematical landscape. An example of this is when a teacher makes an instructional decision based on knowledge that particular choices may (or may not) align with the mathematics that students' will encounter in later courses.

An important purpose of this framework is to tease apart the different kinds of knowledge that might be important for teachers in supporting their students' learning. The work of finding ways to measure these kinds of teacher knowledge and ascertain how they are (or could be) related to student success is ongoing. For our purposes, we will *assume* that developing knowledge in any of these domains can be useful to preservice teachers and we will draw on the framework in order to characterize the nature of the learning opportunities that are highlighted in the two previous chapters.

In the next section, we briefly summarize the way that each of the previous chapters call explicit attention to the potentially bidirectional relationship between preservice teachers' learning of abstract algebra and their knowledge of secondary mathematics. We then consider this relationship in light of Ball et al.'s (2008) framework in order to characterize in what ways this relationship could provide opportunities to strengthen teachers' MKT. In the subsequent sections, we critically consider the validity of the existence of such opportunities as an argument in support of requiring preservice teachers to study abstract algebra. Finally, we share our thoughts on how to leverage these opportunities along with some recommendations and directions for future research on the topic.

Knowledge of Secondary Mathematics and Learning Abstract Algebra

As Wasserman (this volume) notes in his introduction, there are a number of ways to think about how secondary teachers might benefit from studying abstract algebra. Perhaps the most obvious way to think about this is to focus on the mathematical connections between abstract algebra and school mathematics. Such a focus is not without its dangers, as indicated by the controversial New Math era reforms. Yet the idea behind that reform can still be seen in the CBMS (2012) recommendations for the mathematical preparation of teachers in the form of statements like, "it would be quite useful for prospective teachers to see how $\mathbb{C}$ can be built as a quotient of $\mathbb{R}[x]$" (CBMS, 2012, p. 59). A very reasonable question to ask in response to such a statement is "Why?" Abstract algebra certainly provides a highly sophisticated perspective on a variety of secondary mathematics topics, but it simply does not follow that a teacher's pedagogical practice would (or even could) benefit from studying abstract algebra. Or perhaps, rather, we should say it does not follow simply. In their chapters, both Cook, and Melhuish and Fagan take a much deeper (and focused) look at the connection between abstract algebra and secondary mathematics. In doing so, they are able to make considerably more convincing and specific conjectures about how studying abstract algebra could benefit preservice teachers.

Melhuish and Fagan report on a mixed methods study using the Group Concept Assessment (GCA). The quantitative results from a large-scale administration of the GCA tool demonstrated that abstract algebra students struggle with questions about binary operations and functions, and struggle even more when tasks require the coordination of these concepts with other abstract algebra concepts. The validation interviews conducted as part of the development of the instrument provided richer qualitative data that surfaced a number of critical ways in which students' preexisting understandings of binary operation and function supported and constrained their reasoning about the abstract algebra tasks. Further, these tasks appeared to provide opportunities for students to confront limitations in these understandings and to develop richer knowledge of binary operation and function. For example, students tended to have views of both concepts that limited the kinds of things they were able to perceive as functions or binary operations. Some students expected functions to be injective; others expected binary operations to be presented in E-O-E (element–operation–element) form. These limited conceptions constrained the students' abilities to construct the kernel of a non-injective homomorphism or to determine whether the averaging operation was associative.

Although these observations are relevant to those interested in the teaching and learning of abstract algebra for its own sake (they suggest that students' prior experiences, especially regarding functions and operations, may have an important impact on how they are able to engage with abstract algebra), the authors appropriately focus on the possibility of an influence in the opposite direction. As this volume is interested in the preparation of secondary mathematics teachers, it is worth considering whether grappling with these sorts of abstract algebra tasks might impact students' understanding of the concepts of function and operation (core K-12 concepts) in ways that can support their teaching practice. Melhuish and Fagan frame this in terms of Hohensee's (2014) notion of backward transfer. Ultimately, they claim that there are opportunities, while studying abstract algebra, for preservice teachers to deepen their understandings of secondary concepts, such as functions and binary operations. They then describe a number of different ways that preservice teachers' knowledge could be deepened by studying abstract algebra. For the most part, these are focused on increasing mathematical flexibility and disentangling related concepts.

In his chapter, Cook provides an even richer analysis of a connection between secondary mathematics and abstract algebra. He engages a pair of students in the process of reinventing fundamental concepts of ring and field theory in the context of a laboratory teaching experiment. This context gives him a unique opportunity to watch students grapple with a limited view of the zero-product property (ZPP) as they attempt to reinvent the related abstract algebra concept of a zero divisor, on the way to developing a productive way to classify different kinds of rings. In particular, Cook observed that one of the students came into the reinvention process with a view that the ZPP is a universal truth. As such, this student rejected solutions to an equation that involved two nonzero elements multiplying to produce zero. This perspective was eventually refined as he began to use the label "awkward" to describe products of nonzero elements that equaled zero. This refinement enabled him to build a new understanding, in which (for example) uniqueness of solutions is a property that is enjoyed by systems that don't have zero divisors (awkward ways to get zero). In this way, Cook's study provides an existence proof for the kind of backward transfer that Melhuish and Fagan propose in their chapter.

To What Kinds of MKT Are These Chapters Referring?

The first two chapters of this section make claims about the potential of learning abstract algebra to deepen preservice teachers' knowledge. Our aim in this commentary is to critically examine these claims and then, under the assumption that this potential exists, to consider how to capitalize on this potential in ways that can positively impact preservice teacher education. We start by attempting to characterize the nature of the learning opportunities the authors refer to in terms of the MKT framework.

We begin by observing that while the kinds of knowledge referenced in the two chapters might support preservice teachers in developing pedagogical content knowledge, they cannot be categorized as examples of KCS, KCT, or knowledge of content and curriculum. Each of these categories of MKT involves coordinating mathematics and an aspect of pedagogy (students, teaching, and curriculum). Later, when we discuss recommendations, we will briefly address the possibility of promoting the development of pedagogical content knowledge in the context of learning abstract algebra (drawing inspiration from Cook's Chap. 3). However, until then, we will focus on subject matter knowledge, which more accurately captures the kinds of knowledge the authors explore as it is focused on knowledge of mathematics itself.

Melhuish and Fagan delineate a number of aspects of knowledge related to functions and binary operations that could be developed in the context of studying abstract algebra. Here we will consider two specific aspects as we attempt to identify the kinds of subject matter knowledge that might be developed in an abstract algebra course: (1) symbolic expressions for binary operations; and (2) conceptual understanding of the associative property. Similarly, we will attempt to characterize the primary learning opportunity (involving the zero-product property) that was highlighted in Cook's Chap. 3. However, we first argue that none of these is properly characterized as exemplifying specialized content knowledge (SCK).

Specialized content knowledge refers to knowledge "not typically needed for purposes other than teaching" (Ball et al., 2008, p. 400). This includes things like the ability to analyze a nonstandard student strategy to see if it is likely to work in general, and the ability to recognize patterns in students' errors. Johnson and Larsen (2012) observed that it is likely the case that mathematicians make use of this kind of knowledge in their research activity (e.g., reviewing research papers or collaborating with other mathematicians), which suggests that it is perhaps an overstatement to say that this kind of knowledge is not typically needed outside of teaching. Nevertheless, it is useful to think of this category of knowledge as being related closely to the activity of teaching.

Can one categorize the kinds of learning opportunities described in the previous two chapters as developing SCK? To help tease apart why we do not think so, we consider Melhuish and Fagan's observation that abstract algebra can help preservice teachers develop flexibility in moving between and leveraging alternate function representations. Certainly, when students generate nonstandard approaches, these approaches often come with nonstandard representations. Thus, developing flexibility with representations likely positions a preservice teacher well to engage in the activity of making sense of students' ideas in practice. So, why do we not consider developing such flexibility to be an example of developing SCK? First, as Melhuish and Fagan note, this kind of flexibility is important to mathematics majors in the context of abstract algebra, so it does not qualify as knowledge specific to teaching. Second (and more importantly), having such flexibility is not the same thing as having the ability to leverage it to make sense of student thinking. The argument, that flexibility with representations likely supports teachers in making sense of student thinking, actually suggests that a teacher's ability to develop and apply SCK is likely constrained by the nature of their common content knowledge (and perhaps also by their horizon content knowledge). Given the goals of this book, this is both an important connection and an important distinction to make between SCK and other kinds of subject matter knowledge. The connection suggests the potential usefulness of having teachers study abstract algebra, while the distinction suggests that realizing this potential is a nontrivial proposition. In our summary remarks, we pick up this thread again and consider the possibility of developing SCK in the context of an abstract algebra course. Until that time, we will focus on the potential of studying abstract algebra for supporting the development of preservice teachers' horizon content knowledge and the deepening of their common content knowledge, which we regard as the two categories of MKT most appropriately associated with the learning opportunities highlighted in these two chapters. We begin by considering two issues addressed by Melhuish and Fagan and argue that one exemplifies horizon content knowledge, while the other exemplifies common content knowledge (CCK).

We first consider the observation by Melhuish and Fagan that abstract algebra provides an opportunity to move students beyond limited views of binary operation that would require symbolic expressions to be in the E-O-E form. For a secondary teacher, horizon content knowledge would refer to knowledge of how ideas in secondary mathematics evolve as one moves into undergraduate mathematics and beyond. Certainly, binary operations are not assumed to be given by symbolic formulae or presented in E-O-E form in advanced mathematics. Thus, it is clear that this advance in knowledge about binary operations counts as an instance of developing horizon content knowledge. An examination of the secondary mathematics curriculum suggests that students are confronted infrequently by binary operations that are not expressed in terms of symbolic expressions and, while they may be exposed to binary operations (e.g., averaging) not expressed in E-O-E form, they are unlikely to need to think of them explicitly as binary operations (e.g., to prove averaging is non-associative). In this sense, this deepening of the binary operation concept is probably not best categorized as an instance of common content knowledge, which refers to knowledge that is not specific to teaching, but that would be expected of anyone with a working understanding of secondary mathematics.

However, when Melhuish and Fagan discuss opportunities to develop conceptual understanding of the associative property, they *are* referring to an opportunity to strengthen preservice teachers' common content knowledge (CCK). While anyone with a working knowledge of secondary mathematics would have some understanding of the associative property, research suggests that preservice teachers may still need to strengthen this knowledge by disentangling a number of related concepts (e.g., commutativity, order of operations, bracketing). This raises an important issue related to common content knowledge. While it is classified as knowledge that is not specific to teaching, the strength of a preservice teachers' common content knowledge in terms of connections is likely to be a significant factor in their ability to develop and utilize other kinds of MKT. For example, as Larsen (2010) notes, understanding *how* commutativity and associativity are related to *order* (order of operations versus order of operands) is a nontrivial matter, even for undergraduates. Such an understanding is likely important when teachers are making sense of students' struggles (Kieran, 1979) with bracketing (SCK) or making decisions about what kind of tasks will help their students in understanding associativity (KCT). So while this and other kinds of knowledge highlighted in the previous two chapters are best categorized as exemplifying common content knowledge (CCK), this should not be taken as an indication that they do not represent important potential ways that studying abstract algebra could support future teachers.

Cook focused on the relationship between the zero-product property (an important concept from secondary mathematics) and zero divisors (a foundational idea in ring theory). In his design experiment, Cook was able to identify a deepening of students' understanding of the zero-product property as they developed the concept of zero divisors in the context of exploring uniqueness (or rather nonuniqueness) of solutions to linear equations. We argue that the knowledge gained in this process included both horizon and common content knowledge. On the one hand, the students' understanding of the zero-product property was deepened in the sense that they realized that this is not a basic fact about the additive identity (zero), but rather is dependent on the nature of the operation. Furthermore, the connection between this property and the uniqueness of solutions was strengthened by the awareness that systems that do not have this property contain linear equations with multiple solutions. In this sense, the students were deepening their common content knowledge (CCK). On the other hand, the students were able to build direct connections between secondary algebra (ZPP, equation solving) and abstract algebra (zero-divisors, integral domains) to develop knowledge of the trajectory of these ideas across the mathematics curriculum. In this sense, the students were developing horizon content knowledge.

In summary, we consider the two previous chapters to be compelling arguments that abstract algebra courses provide numerous opportunities to develop preservice teachers' horizon content knowledge and deepen their common content knowledge. We also argue that the nature of a preservice teachers' common and horizon content knowledge likely provides important constraints on and affordances for their development of other kinds of mathematical knowledge for teaching. We now turn to the question of realizing these opportunities provided by abstract algebra courses.

Reasons to Be Skeptical About the Idea of Backward Transfer

Although both chapters make a credible case for the possibility that studying abstract algebra can provide opportunities to deepen preservice teachers' understanding of secondary mathematics, there are good reasons to be skeptical. Lobato (2006) notes that students construct their own connections between situations and that these do not necessarily reflect the mathematical invariants to which experts attend (like algebraic structure). It seems that the prospects for backward transfer are likely to be even worse, since in the context of an abstract algebra course there is little reason to expect preservice teachers to do the work of reconsidering their secondary mathematics knowledge in light of their new knowledge of abstract algebra. Certainly, the research demonstrates that students are very capable of maintaining unconnected and even contradictory ideas about concepts. For example, students may have concept images of the limit of a function that are in conflict with the formal definition of limit without perceiving any contradiction (Tall & Vinner, 1981). This suggests that a preservice teacher could develop a concept image of a secondary concept that contains the more sophisticated aspects developed in an abstract algebra course *without* these being coordinated with the aspects associated with the secondary version of the concept. As a result, while we (as experts) may see a more sophisticated version of a secondary mathematics concept when we observe a preservice teacher reasoning successfully in an abstract algebra context, it does not follow that the way the preservice teacher will reason about this concept in a secondary context will reflect this sophisticated understanding. For example, a preservice teacher who is able to construct an operation on the set $\{1, 2, 4\}$ that defines a group may still only bring to mind operations with explicit formulas when designing and teaching a high school algebra lesson.

Even in the case where preservice teachers do make connections between abstract algebra and secondary mathematics, it does not follow directly that as teachers they will be able (or should even try) to leverage them to support student learning. Some connections may hold more potential for positively impacting practice than others and some means of leveraging such knowledge in practice may be more beneficial than others. For example, it seems unlikely that it would be helpful for a high school teacher to start using terms like zero divisor and integral domain when teaching a lesson on polynomial equations. However, it could be quite helpful for a teacher to make extra time in such a lesson to help students make sense of *why* it is useful to manipulate the equation so that zero appears alone on one side of the equation, while a factored polynomial appears on the other.

In short, one can argue that even with the more focused and nuanced connections between abstract algebra and school mathematics that are elaborated in the two previous chapters, we are still in very much the same place we were when the

New Math reforms were initiated. As experts, we see a connection between abstract algebra and school mathematics. It seems reasonable to us that knowledge of those connections could somehow support teachers in their practice. However, even with the ability to articulate what kind of mathematical knowledge this backward transfer could support, we are still left to figure out *how* to actually promote the occurrence of this kind of learning and *how* to support teachers in productively leveraging these kinds of knowledge in their practice.

A Reason for Optimism about Intentionally Engineering Backward Transfer Opportunities

We start this section by acknowledging that the field of mathematics education is still working to understand connections between MKT, teaching practice, and student learning. It may very well be the case that the field eventually learns that some of the kinds of knowledge described in Ball et al.'s (2008) framework do not actually have a significant impact on teachers' practice or on their students' learning. We will not take on this question in this commentary. Instead, we will operate under the assumption that the domains of MKT are potentially valuable to teachers, and accept the claims of Meluish and Fagan, and Cook that abstract algebra provides an opportunity to promote these kinds of MKT. As such, we will focus for the remainder of the chapter on the question of how to capitalize on these opportunities.

We make two recommendations for teacher educators who find potential value in the idea of supporting the development of MKT in preservice teachers through a process of what Melhuish and Fagan refer to as backward transfer. The first recommendation is to not expect it to happen without intentional intervention. Here we note that preservice teachers sometimes take abstract algebra courses that are not specifically designed for teachers. In this case, there is little reason to hope that the preservice teachers in the course will make these connections for themselves. As Wasserman (this volume) notes in his introduction, research suggests that they are likely to see this course as having little or nothing to do with teaching secondary mathematics. However, it is possible to follow up such a course with an algebra course for teachers that could recreate the situations seen in Melhuish and Fagan's interviews, or Cook's teaching experiment, in order to explicitly design opportunities for preservice teachers to actively engage in actor-oriented (Lobato, 2006) backward transfer. We strongly recommend the inclusion of such courses as part of the mathematical preparation of teachers. Frankly, given the findings of Monk (1994) and others, we see no good argument for requiring preservice teachers to take a regular abstract algebra course if it is not followed up by experiences that are designed to leverage the MKT learning opportunities provided by the study of abstract algebra.

Our second recommendation stems from Cook's experience with his participants. His goal was to develop an approach to teaching ring and field theory featuring a process of guided reinvention. With this approach, the students are explicitly engaged in developing the formal abstract algebra concepts by building on the informal understandings they bring to the process (including their knowledge of secondary mathematics). This approach seems to have the potential to support backward transfer in that it engages the students in a process that requires them to navigate back and forth between their knowledge of secondary mathematics (e.g., ZPP) and abstract algebra (e.g., zero divisors). This kind of process seems well suited to support students in connecting these two domains. For this reason, we argue that engaging students in this type of reinvention process is a promising instructional approach for developing preservice teachers' horizon content knowledge and deepening their common content knowledge.

Additionally, we argue that such an approach also creates opportunities to expand on the potential of studying abstract algebra to develop MKT by also supporting the development of specialized content knowledge (SCK) and knowledge of content and students (KCS). Collectively engaging abstract algebra students—who, in this case, are preservice teachers—in guided reinvention requires an instructor to elicit and then build on students' informal understandings to develop formal concepts. Often these informal understandings represent students' common content knowledge of related secondary concepts. We saw this in Cook's Chap. 3, as Brian's understanding of the secondary concept of the zero-product property provided the starting point for the development of the concept of zero divisors (and eventually integral domains). In a classroom in which students are engaged in the guided reinvention of abstract algebra concepts, students will frequently share their informal understandings and be asked to make sense of and critique other students' strategies and ideas. This provides two kinds of learning opportunities that are not afforded by teacher-centered instructional approaches. First, this context provides them with opportunities to confront common struggles that students (themselves and others) have with core concepts (e.g., they can observe classmates struggle with the notion of a zero divisor or with an operation not given by a formula) and thus develop KCS (a category of knowledge that is focused on understanding typical student approaches and struggles related to a concept). Second, this context provides many opportunities to make sense of other students' ideas, thus providing an excellent opportunity to develop SCK. Essentially, the collective endeavor of guided reinvention has a unique feature in that the students are engaged as participants in what is typically a teaching activity (analyzing student thinking), which affords a rare opportunity to engage in the kind of mathematical thinking that is (typically) uniquely required in teaching. Notice that this offers a second argument in support of providing preservice teachers with experiences, as learners, in the kinds of classrooms (interactive and focused on student thinking) that they are encouraged to create for their own students. Not only do such courses provide important models of teaching for preservice teachers (as noted by Wasserman in his introduction to this volume), but they also engage preservice teachers in the kinds of mathematical activity that can promote the development of SCK and KCS.

Summary and Directions for Future Research

In the previous two chapters, the authors make a credible case that learning abstract algebra can provide opportunities to strengthen preservice teachers' common content knowledge and to develop their horizon content knowledge. However, we cannot assume that these opportunities will be realized simply by requiring preservice teachers to take a traditional abstract algebra course. Instead, we should design experiences for preservice teachers that actively engage them in making connections between their newly acquired knowledge of abstract algebra and secondary mathematics. Furthermore, preservice teachers should be actively engaged in thinking about the consequences of these connections for their practice and how their new and deeper knowledge of secondary mathematics could help them support their students' learning. Finally, we argue that actively engaging preservice teachers in developing the concepts of abstract algebra by building on their understandings of the secondary concepts (perhaps through guided reinvention) is likely to be more effective than traditional approaches in terms of building connections between secondary mathematics and abstract algebra. Additionally, such an approach has the potential to develop aspects of MKT beyond horizon and common content knowledge by actively engaging preservice teachers in making sense of and critiquing the mathematics of others. In summary, Melhuish and Fagan, and Cook provide us with productive ways to think about *why* preservice mathematics teachers should study abstract algebra, but much work will need to be done to learn *how* to consistently realize the potential of abstract algebra courses in preservice teacher education.

We conclude by calling for research and development focused on ascertaining and realizing the true potential of abstract algebra courses to develop preservice teachers' MKT. First, research is needed to determine the extent to which the kinds of learning opportunities described in the two previous chapters are, in fact, realized by preservice teachers who take standard courses in abstract algebra. A study modeled after that of Melhuish and Fagan (but focused more closely on preservice and/or inservice teachers, and with a more intentional emphasis on core concepts of secondary mathematics) could address this issue nicely. We strongly suspect that the results of such research may be discouraging because traditional abstract algebra courses are not typically designed to support these specific kinds of learning. For this reason, we also call for instructional design research focused on developing means of intentionally engineering these kinds of opportunities in algebra courses for teachers. Such research could profitably build on the work described in Cook's Chap. 3.

Finally, it is important that the field continue to take on the very difficult questions regarding how and to what extent different kinds of MKT can have a positive impact on teachers' practice and students' learning. We strongly suspect that it is necessary to explicitly and intentionally support preservice teachers in developing these kinds of knowledge *and* to explicitly and intentionally support them in translating this knowledge to their teaching practice. For this reason, we encourage projects that bring together researchers primarily focused on the mathematical preparation of teachers with researchers primarily focused on the pedagogical training of teachers. Such collaborations hold the most promise for realizing the potential of abstract algebra courses as an important part of preservice teacher education.

References

Asiala, M., Dubinsky, E., Mathews, D. M., Morics, S., & Oktaç, A. (1997). Development of students' understanding of cosets, normality, and quotient groups. *Journal of Mathematical Behavior, 16*(3), 241–309.

Asiala, M., Kleiman, J., Brown, A., & Mathews, D. (1998). The development of students' understanding of permutations and symmetries. *International Journal of Computers for Mathematical Learning, 3*(1), 13–43.

Ball, D. L., Thames, M. H., & Phelps, G. (2008). Content knowledge for teaching: What makes it special? *Journal of Teacher Education, 59*(5), 389–407.

Ball, D. L., & Bass, H. (2009). *With an eye on the mathematical horizon: Knowing mathematics for teaching to learners' mathematical futures*. Paper presented at the 43rd Jahrestagung der Gesellschaft für Didaktik der Mathematik, Oldenburg. Retrieved from www.mathematik.uni-dortmund.de/ieem/BzMU/BzMU2009/BzMU2009-Inhalt-fuer-Homepage.htm

Brown, A., DeVries, D. J., Dubinsky, E., & Thomas, K. (1997). Learning binary operations, groups, and subgroups. *Journal of Mathematical Behavior, 16*(3), 187–239.

Conference Board of Mathematical Sciences (CBMS). (2012). *The mathematical education of teachers II*. Providence, RI: American Mathematical Society and Mathematical Association of America.

Cook, J. P. (2014). The emergence of algebraic structure: Students come to understand units and zero-divisors. *International Journal of Mathematical Education in Science and Technology, 45*(3), 349–359.

Dubinsky, E., Dautermann, J., Leron, U., & Zazkis, R. (1994). On learning fundamental concepts of group theory. *Educational Studies in Mathematics, 27*(3), 267–305.

Dubinsky, E., & Leron, U. (1994). *Learning abstract algebra with ISETL*. New York, NY: Springer.

Hazzan, O. (1999). Reducing abstraction level when learning abstract algebra concepts. *Educational Studies in Mathematics, 40*(1), 71–90.

Hohensee, C. (2014). Backward transfer: An investigation of the influence of quadratic functions instruction on students' prior ways of reasoning about linear functions. *Mathematical Thinking and Learning, 16*(2), 135–174.

Johnson, E. M., & Larsen, S. P. (2012). Teacher listening: The role of knowledge of content and students. *Journal of Mathematical Behavior, 31*(1), 117–129.

Kieran, C. (1979). Children's operation thinking within the context of bracketing and the order of operations. In D. Tall (Ed.), *Proceedings of the third international conference for the psychology of mathematics education* (pp. 128–133). Coventry: Warwick University, Mathematics Education Research Centre.

Larsen, S. (2009). Reinventing the concepts of group and isomorphism: The case of Jessica and Sandra. *Journal of Mathematical Behavior, 28*(2–3), 119–137.

Larsen, S. (2010). Struggling to disentangle the associative and commutative properties. *For the Learning of Mathematics, 30*(1), 37–42.

Larsen, S., Johnson, E., & Bartlo, J. (2013). Designing and scaling up an innovation in abstract algebra. *Journal of Mathematical Behavior, 32*(4), 693–711.

Leron, U., Hazzan, O., & Zazkis, R. (1995). Learning group isomorphism: A crossroads of many concepts. *Educational Studies in Mathematics, 29*(2), 153–174.

Lobato, J. (2006). Alternative perspectives on the transfer of learning: History, issues, and challenges for future research. *Journal of the Learning Sciences, 15*(4), 431–449.
Melhuish, K. (2018). Three conceptual replication studies in group theory. *Journal for Research in Mathematics Education, 49*(1), 9–38.
Monk, D. H. (1994). Subject area preparation of secondary mathematics and science teachers and student achievement. *Economics of Education Review, 13*(2), 125–145.
Shulman, L. S. (1986). Those who understand: Knowledge growth in teaching. *Educational Researcher, 15*(2), 4–14.
Tall, D., & Vinner, S. (1981). Concept image and concept definition in mathematics with particular reference to limits and continuity. *Educational Studies in Mathematics, 12*(2), 151–169.

Part II
Exploring Mathematical Connections Between Abstract Algebra and Secondary Mathematics

Chapter 5
Excavating School Mathematics

William McCallum

Introduction

The school curriculum can be viewed as an archeological record of the history of mathematics and of previous efforts at reforming school instruction, eroded in places by the winds of policy and covered in others by the sands of neglect. Some topics in the school curriculum are like encrusted relics, difficult to make sense of because they are no longer connected to a larger structure that once gave them meaning. Other areas may be more coherent, but no longer serve the originally intended purpose.

Of course, the purposes of the archeologist and the purposes of the curriculum designer are quite different; the one has a goal of understanding the past, whereas the other has a goal of building for the future. However, in order to make decisions about where and whether to implement changes in the curriculum, some archeological reconstruction and analysis is needed, both tasks in which a higher standpoint, in the sense of Klein, Menghini, and Schubring (2016), is useful.

In this chapter we give some examples of how the higher standpoint of abstract algebra might be useful in excavating the secondary school algebra curriculum and help in building a coherent architecture of school mathematics.

W. McCallum (✉)
Department of Mathematics, The University of Arizona, Tucson, AZ, USA
e-mail: wmc@math.arizona.edu

N. H. Wasserman (ed.), *Connecting Abstract Algebra to Secondary Mathematics, for Secondary Mathematics Teachers*, Research in Mathematics Education,
https://doi.org/10.1007/978-3-319-99214-3_5

Three Moments in the History of Quadratic Equations

As an example of a secondary school algebra topic that could benefit from some excavation, let us consider quadratic equations. One of the tools an archeologist brings to an excavation is historical knowledge, and we will start by examining three historical documents that provide hints about the way the topic is currently situated in the matrix of school mathematics. Another tool is an eye for form and function, which in this case is represented by the higher standpoint of abstract algebra.

The first historical document illustrates the current status of quadratic equations as a topic whose very existence in the school curriculum is questioned. The second shows that it was once an area at the forefront of mathematical research—the abstract algebra of its time. The third points at a possible intermediate status between the first two, when knowledge of quadratic equations became a mark of distinction—no longer an area of advanced research, and not yet embedded in the routine of school mathematics.

A Debate in the House of Commons

We start with a record of a debate about quadratic equations in the UK House of Commons in 2003 (UK Parliament 2003).

The debate was ignited by Terry Bladen, president of the National Association of Schoolmasters/Union of Women Teachers, a teachers union in the United Kingdom, who called for eliminating quadratic equations from the school curriculum (BBC News 2003a):

> For a significant number of pupils things like quadratic equations have absolutely no relevance to them, they can't see how they would ever use them and in fact they probably never would use them.
>
> So for these pupils what I think should be there is something like a numeracy curriculum that's designed to equip them with the number concepts that they'll need in their adult life.

In the debate, Tony McWalter, member of parliament for the Labour Party, defended quadratic equations (UK Parliament 2003):

> A quadratic equation is not like a bleak room, devoid of furniture, in which one is asked to squat. It is a door to a room full of the unparalleled riches of human intellectual achievement. If you do not go through that door ... much that passes for human wisdom will be forever denied you.[1]

[1]Although it is not relevant to our current purpose, it is hard to resist quoting the exchange that followed: "Hear, hear," cried Eleanor Laing, member of parliament for the Conservative Party, to which McWalter replied "Oh dear. I would like to have support from elsewhere as well."

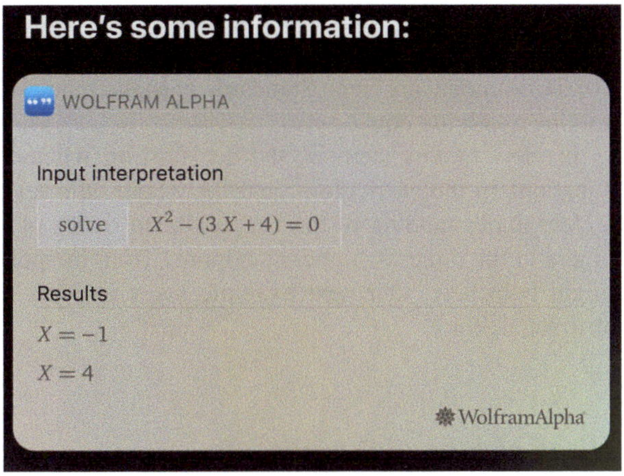

Fig. 5.1 What is the solution to x squared plus 3 x plus 5 equals zero

Education Secretary Alan Johnson replied:

> In preparing for this debate, the DFES [Department for Education and Skills] conducted a straw poll involving a 16-year-old who had just sat maths GCSE [General Certificate of Education], a head of maths and an experienced chemical engineer. The 16-year-old thought that quadratic equations were logical and fairly straightforward because "you substitute stuff into a formula." He did say, however, that his opinion might have been influenced by having a good teacher. The head of maths said that quadratic equations formed an important step in students' ability to solve equations, taking them from simple—one unknown—and simultaneous—two unknowns—and paving the way for more advanced work in mechanics and complex number theory. The engineer said that he did not use quadratic equations now, but had in the past in detailed design applications. Where he works, the chemists use them to explain multiple reactions.

In Terry Bladen's view, school mathematics is divorced from reality: "how often do the majority of people need or use mathematical concepts once they have left school?" (BBC News 2003b). Even those who defend quadratic equations—such as McWalter, Laing, and Johnson—cannot always find the words to do so; the topic has become an eroded shell, leading to endless dry exercises ("you substitute stuff into a formula"), and the *reductio ad absurdum* of omnipresent technology (see Fig. 5.1 for Siri's response). The headmaster's defense refers the purpose of quadratic equations to a higher purpose, solving equations in general, without specific justification. The engineer does not use quadratic equations but is pretty sure somebody else does.

Calls for change such as Terry Braden's run up against custom, and there is some evidence that customary support remains strong. In a 2009 survey (ACT 2009), Algebra II teachers ranked quadratic equations as the most important topic they teach. In contrast, College Algebra instructors did not include it in their top ten. This lower ranking could reflect the often remedial nature of College Algebra; in another survey (Conley, Drummond, de Gonzalez, Rooseboom, & Stout 2011) of postsecondary instructors, which included Alan Johnson's chemists and engineers among other higher education faculty, 79% thought solving quadratic equations was either critical or important for success in the courses they taught. This importance is reflected in standards aimed at college and career readiness, including the 1997 California standards, the 2000 Massachusetts standards, the 2004 American Diploma Project standards, the 2007 Minnesota standards, and the 2010 Common Core State Standards in Mathematics (CCSSM), all of which require solving quadratic equations.

In view of this support, the decision on whether or not to include quadratic equations in the curriculum, and for whom, is a serious one. It should depend on a clear understanding of the "unparalleled riches of intellectual achievement" that inhere in the topic. A higher standpoint, from the point of view of abstract algebra, might help here. Our next example takes us back to a seminal early treatise on abstract algebra.

Hisab al-jabr w'al-muqabala

Our second historical document is a passage from Al-Khwarismi's *Hisab al-jabr w'al-muqabala* published in 825, discussing the solution to the equation that we would write in modern notation as $x^2 + 10x = 39$ (Struik 2014).

> ...what is the square which combined with ten of its roots will give a sum total of 39? The manner of solving this type of equation is to take one-half of the roots just mentioned. ...Therefore take 5, which multiplied by itself gives 25, an amount which you add to 39 giving 64. Having taken then the square root of this which is 8, subtract from it half the roots, 5 leaving 3.

Here the "square" is x^2, the "root" is x, and "ten of its roots" is $10x$. The method described is completing the square:

$$x^2 + 10x = 39$$

$$x^2 + 2 \cdot 5x = 39 \qquad \text{"take one-half of the roots"}$$

$$x^2 + 2 \cdot 5x + 25 = 39 + 25 = 64 \qquad \text{"which multiplied by itself gives 25, an amount which you add to 39 giving 64"}$$

$$(x + 5)^2 = 64$$

$$x + 5 = 8 \qquad \text{"Having taken then the square root ... which is 8"}$$

$$x = 3. \qquad \text{"subtract from it half the roots, 5 leaving 3"}$$

Why did we call this a treatise on abstract algebra? Although the method is given for specific numbers, the numbers are chosen to be indicative of a general algorithm for finding whole number solutions to quadratic equations. And although the description is procedural, the accompanying diagram as shown in Fig. 5.2 suggests a mathematical justification for the algorithm. It seems clear that solving this particular equation is not the point. Al-Khwarizmi is not "substituting stuff into a formula" but rather is explaining a general method which leads to the formula. [It should be noted that he was not the first to do so; an Indian manuscript 200 years earlier gives an explicit formula for solving quadratic equations (Katz 2007, p. 431).]

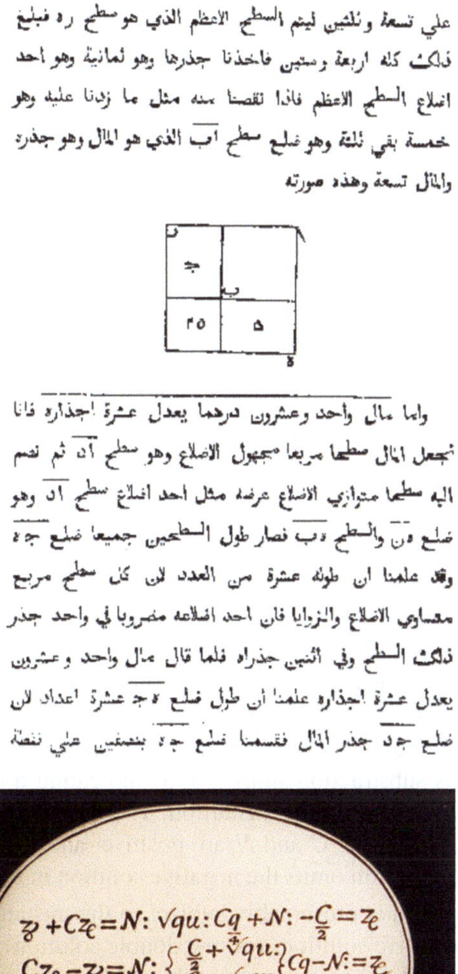
علي تسعة وثلثين ليتم السطح الاعظم الذي هو سطح ره فبلغ
ذلك كله اربعة وستين فاخذنا جذرها وهو ثمانية وهو احد
اضلاع السطح الاعظم فاذا نقصنا منه مثل ما زدنا عليه وهو
خمسة بقي ثلثة وهو ضلع سطح اب الذي هو المال وهو جذره
والمال تسعة وهذه صورته

واما مال واحد وعشرون درهما يعدل عشرة اجذاره فانا
نجعل المال سطحا مربعا مجهول الاضلاع وهو سطح اد ثم نضم
اليه سطحا متوازي الاضلاع عرضه مثل احد اضلاع سطح اد وهو
ضلع ون والسطح هب فصار طول السطحين جميعا ضلع جه
وقد علمنا ان طوله عشرة من العدد لان كل سطح مربع
متساوي الاضلاع والزوايا فان احد اضلاعه مضروبا في واحد جذر
ذلك السطح وفي اثنين جذراه فلما قال مال واحد وعشرون
يعدل عشرة اجذاره علمنا ان طول ضلع هج عشرة اعداد لان
ضلع جد جذر المال فنقسم ضلع جه بنصفين على نقطة

Fig. 5.2 Page from *Hisab al-jabr w'al-muqabala* (Wikimedia Commons 2006)

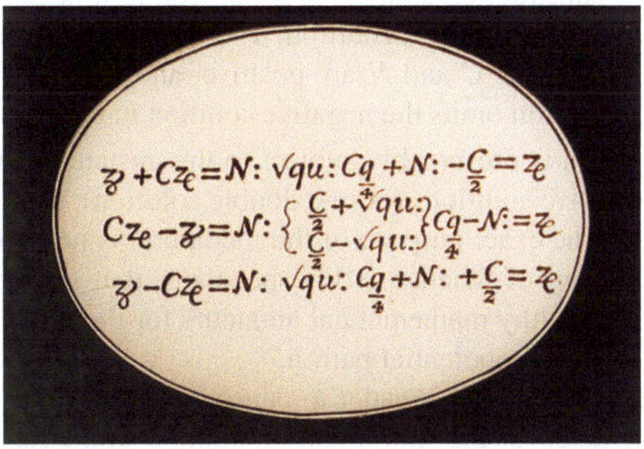

Fig. 5.3 Seventeenth century English medallion, ©Museum of the History of Science, University of Oxford, inventory no. 14977

A Mathematical Medallion

Moving forward to seventeenth century England, we discover the beautiful object in Fig. 5.3 (Museum of the History of Science 2017). There is just enough resemblance between this and the quadratic formula for us to be able to decipher the notation.

The notation used on the medallion is a hybrid. The symbol for the unknown (modern x) is a form of r (for root, radix or res), while the z for square comes from zensus or census, literally meaning wealth or excess. These symbols had been in use from at least the early

16th century, and had first been published in English in Robert Recordes Whetstone of Witte (London, 1557). They began to be displaced from the 1590s firstly by Franois Viétes notation of Aq, Ac, etc (for squares and cubes) and then latterly (after 1637) by the modern index notation. While the medallion does not use index notation, two of its solutions do use Cq for C^2.

The modern root symbol $\surd$ appears, qualified by a q to show that it is a square root. In the sequence $\surd$qu: the u stands for universal and indicates that the square root operates over all the terms between the first and second colons (Museum of the History of Science 2017).

In modern notation (how hard won it is!), the medallion reads:

$$x^2 + Cx = N, \qquad \sqrt{\frac{C^2}{4} + N} - \frac{C}{2} = x$$

$$Cx - x^2 = N, \qquad \frac{C}{2} \pm \sqrt{\frac{C^2}{4} - N} = x$$

$$x^2 - Cx = N, \qquad \sqrt{\frac{C^2}{4} + N} + \frac{C}{2} = x$$

Why are there three formulas here rather than one? There seems to be an aversion to negative numbers. If the coefficients C and N must always be positive, then we no longer have a single form for a quadratic equation, $x^2 + Cx + N = 0$, but rather four forms, obtained by arranging the terms on each side of the equal sign so that no negative coefficients are necessary (we must view the subtraction signs as substraction signs, not as indicating a negative coefficient). Three of the forms are given on the medallion. The missing form, $x^2 + Cx + N = 0$, has no positive solutions if C and N are positive, and is therefore not of interest. Furthermore, the medallion omits the negative solution in the first and third cases. In the second case, which is presumably subject to the unstated condition that $\frac{C^2}{4} \geq N$, there are two positive solutions (or one double solution), so both are given.

The exact purpose of the medallion is not known, but Johnston (Museum of the History of Science 2017) speculates that such objects were "commissioned either by wealthy mathematical amateurs for their own use or perhaps by a practitioner as a gift for a potential patron."

We invite the reader to pause for moment to think of what joy it must have been to discover the simplification under a single formula afforded by allowing negative coefficients. School children must have breathed a sigh of relief when that happened.

These three documents show a striking evolution in the status of the topic of quadratic equations. In Al-Khwarizmi's time, solving quadratic equations was an area of advanced mathematics. In seventeenth century England it was popular science, something you could be proud enough of knowing to have inscribed on a medallion. In twentieth century England it has become a topic of school mathematics, a topic possibly headed for extinction.

Quadratic Equations from a Higher Standpoint

We use Klein's term "higher standpoint" (Klein et al. 2016) to refer to a way of looking at topics in school algebra which holds a knowledge of abstract algebra at the top of the mind and seeks to find connections with that knowledge. In this section we look at quadratic equations as they currently sit embedded in the matrix of (US) school mathematics, and provide some ways in which a higher standpoint can remove their encrustations and reveal hidden "unparalleled riches." It is not our purpose here to adjudicate whether the topic should be in the curriculum; that is a matter for debate between many stakeholders, including those who are neither mathematicians nor educators. Indeed, as we have seen, such a debate took place in the UK House of Commons in 2003. Rather it is our purpose to show how a higher standpoint can inform the debate, to allow the participants to decide with more information whether the topic is indeed truly obsolete, or whether, with a new perspective, it can be refurbished into something worthwhile.

We start by examining how the topic is taught in order to see if the higher standpoint of abstract algebra has any insights to offer. One point immediately stands out: the primary role of factoring as a method for solving quadratic equations. It is often the first method in the textbook [see New York State Education Department (2017) for one recent and widely used example] and, in all but one of the set of standards listed earlier, it is the first method mentioned, before completing the square or the quadratic formula, and before graphing, if that method is mentioned at all. The exception is the Common Core State Standards in Mathematics (CCSSM), which puts factoring last:

> A-REI.B.4b. Solve quadratic equations by inspection (e.g., for $x^2 = 49$), taking square roots, completing the square, the quadratic formula, and factoring, as appropriate to the initial form of the equation. (National Governors Association Center 2010)

A Google search on "solving quadratic equations by factoring" yields 715,000 hits; the same search with "completing the square" or "the quadratic formula" yields 114,000 and 241,000, respectively. All this is strange if one's goal is simply to solve quadratic equations; factoring is not a good general method, whereas completing the square, or its codification in the quadratic formula, always works.

The situation can be seen more clearly from a higher standpoint. On the one hand, in abstract algebra one studies the arithmetic of polynomial rings over various fields, including the question of whether a given polynomial can be factored, and how that depends on the field of coefficients. This leads to a study of algebraic number fields, Galois theory, and solvability of equations in radicals. The activity of factoring in school belongs to this domain of study, although the domain itself is not generally visible from the standpoint of school mathematics [an important exception being the CME curriculum (Pearson 2013)]. The quadratic polynomials that students are given to factor generally live in the ring of polynomials over the rational numbers, or over the integers, and the activity of factoring amounts to developing some "number sense" with these rings.

On the other hand, in applied mathematics and the sciences one studies quadratic functions as models for relationships between physical quantities. The chances of such a function having zeros easily obtainable by factoring as it is taught in schools are small (one might say zero, since the probability of a real number being rational is zero), unless the function is artificially constructed to be amenable to this method. In any authentic context from science or engineering, it makes much more sense to find the zeros by graphical or numerical estimation, or to express them using the quadratic formula, in particular when the coefficients are expressed in terms of other variables.

The tension between these two perspectives is not necessarily visible to the student, but manifests itself, on the one hand, in a lack of meaningful purpose for all those factoring worksheets, and, on the other hand, artificial modeling problems designed to be solvable by factoring.

One way to give purpose to factoring is think of it not as a procedure for solving equations—it is neither efficient nor general as a procedure—but rather as a demonstration of equivalence between two forms, the standard form and the factored form. Seeing this equivalence does have a purpose: if a quadratic function on the real numbers is expressed in factored form you can read its zeros, and if you know its zeros you can express it in factored form up to multiplication by a constant.[2]

This view of factoring leads to some interesting insights not often explored in school mathematics. The method of factoring depends on Viètes formulas, which give the relationship between the zeros and coefficients of a quadratic expression:

$$\text{If } x^2 + bx + c = (x - r)(x - s), \text{ then}$$

$$r + s = -b \quad \text{and} \quad rs = c.$$

This applies whatever the number system in which the roots lie: rational, real, or complex. In particular, we can take the (possibly irrational) zeros given by the quadratic formula (with $a = 1$):

$$r = \frac{-b + \sqrt{b^2 - 4c}}{2} \qquad s = \frac{-b - \sqrt{b^2 - 4c}}{2}.$$

It is a lovely exercise in looking for algebraic structure to try to see mentally why these two numbers add to $-b$ and multiply to c. The quadratic formula is exquisitely designed to make that happen.

Another important connection is the connection between the quadratic formula and completing the square. The proof of the formula often proceeds by following the steps in completing the square, with coefficients represented by letters. But, even with numerical coefficients, completing the square can become a tedious thicket of

[2]Note that this is not necessarily true for polynomial rings over other base rings, such as are studied in a college abstract algebra course.

procedures for many students, and so the proof with letters remains a mystery, or perhaps a section of the curriculum that the instructor decides to skip.

However, there is a beautiful structural insight behind the method of completing the square: every quadratic equation can be expressed in the form

$$(x+p)^2 = q, \tag{5.1}$$

which we already know how to solve by taking square roots[3]:

$$x = -p + \sqrt{q} \quad \text{or} \quad x = -p - \sqrt{q}. \tag{5.2}$$

A derivation of the quadratic formula can draw on this insight. First one could see the basic structure of (5.1) and (5.2), perhaps using specific examples, and then, as a cleaning up operation, once and for all go through the details of figuring out exactly what p and q are in terms of a, b, and c:

$$p = \frac{b}{2a}, \quad q = -\frac{c}{a} + \left(\frac{b}{2a}\right)^2. \tag{5.3}$$

These equations, together with (5.2), are already the quadratic formula, in an expanded form. It is nice to notice that you can write it in a more compact form, but the usual way of doing this actually obscures the connection with completing the square.

The details of deriving (5.3) are where students can get bogged down; but from the structural point of view they are indeed merely details. If students remember nothing but the structure, that is better than an inaccurate memory of befogged confusion. Only two of the standards mentioned above, California and the Common Core, explicitly require students to understand the connection between completing the square and the quadratic formula.

In summary, completing the square, seen as a procedure, is a complicated trick. But it can also be seen as a structural maneuver that reveals hidden information, a surprising transformation that suddenly makes easy what seemed intractable. Mathematicians talk about tricks with respect and admiration because of this flash of enlightenment, but in discussions of school mathematics the word "trick" can be a term of derogation, often contrasted with "big ideas." Some ideas in mathematics are not big, but small and consequential, and completing the square is one of them.

In this section, we have seen that the higher standpoint of abstract algebra leads to a profound shift in the view of the topic of quadratic equations from a set of procedures for finding answers, often without purpose, to a set of structural insights about equivalence of expressions; about factored and standard form; about the relationship between these forms and solutions to an equation; and about completing

[3]The higher standpoint also has something to say about the existence of square roots, and why there are only two of them, a story we leave aside for now.

the square as a statement about the existence of a canonical form for an equation, rather than as an algebraic procedure. In this latter view the procedures are still important, but they are the fine grain of the structural insights, rather than blind ends in themselves.

The Role of the Higher Standpoint in the Common Core State Standards in Mathematics (CCSSM)

The author of this chapter was one of the lead writers of CCSSM. The writing of the CCSSM involved many different groups of stakeholders, drew on many different sources of evidence, and synthesized many different perspectives. [See Zimba (2014) and McCallum (2015) for a description of the process and an overview of the standards.] In this section we consider how one of those perspectives, which we have been calling the higher standpoint in this chapter, is reflected in the standards.

As a first example, the higher standpoint exemplified in the previous discussion of quadratic equations is visible in the structural view of factoring and completing the square expressed in the standard A-SSE.B.3ab:

> Choose and produce an equivalent form of an expression to reveal and explain properties of the quantity represented by the expression.
>
> a. Factor a quadratic expression to reveal the zeros of the function it defines.
> b. Complete the square in a quadratic expression to reveal the maximum or minimum value of the function it defines.

Note the word "choose," and the emphasis on the purpose of the choice, to "reveal and explain." Students studying a curriculum based on this standard will be factoring, expanding, and completing the square, as they have always done. But a curriculum that went no further than telling them to do so would not satisfy the standard. It would be required, in addition, to help them understand why they might want to perform those procedures, and what they could get out of them.

Next we consider some broad features of the standards that reflect the higher standpoint.

Properties of Operations

The properties of operations played a unifying role in the standards, starting in the earliest grades. Formalized in higher mathematics as the field axioms, the properties of operations are fundamental to students' work with arithmetic, starting in grade 1, as exemplified by the following standard from CCSSM:

> 1.OA.B. Understand and apply properties of operations and the relationship between addition and subtraction.

3. Apply properties of operations as strategies to add and subtract. *Examples: If* $8+3 = 11$ *is known, then* $3 + 8 = 11$ *is also known. (Commutative property of addition.) To add* $2 + 6 + 4$, *the second two numbers can be added to make a ten, so* $2 + 6 + 4 = 2 + 10 = 12$. *(Associative property of addition.)*
4. Understand subtraction as an unknown-addend problem. *For example, subtract* $10-8$ *by finding the number that makes 10 when added to 8.*

A footnote on the first standard says "Students need not use formal terms for these properties." This is an important point: the purpose of emphasizing the properties of operations is not to impose undue formality on the early grades, but to "make arithmetic a rehearsal for algebra" (Zimba 2011).

For example, consider the following task that might be given to a student in grade 1:

> Lucy has 3 apples. Julie has 5 apples. How many more apples does Julie have than Lucy?

A student who understands that there are two ways of writing an equation for this problem—$3+? = 5$ or $? = 5 - 3$—does not, when they get to algebra in middle school, need to "subtract 2 from both sides" to see that $x + 2 = 5$ means $x = 5 - 2$.

Or consider a student who adds 46 and 37 by thinking

$$40 + 6 + 30 + 7 = 40 + 30 + 6 + 7 = 70 + 13,$$

using the "any grouping any order principle," an amalgam of the commutative and associative properties of addition. Such a student may naturally see

$$4x + 6 + 3x + 7 = 4x + 3x + 6 + 7 = 7x + 13,$$

without needing to conceive of this as a new and separate operation called "collecting like terms." Just as M. Jourdain in Molière's *Le Bourgeois Gentilhomme* had been speaking prose all his life, such students have been collecting like terms all their life. This is not to say they shouldn't hear or use the term; just that it is possible to have a view of algebra that makes it an extension of what they have already been doing rather than an extra piece of mental baggage.

In contrast with the emphasis on properties of operations, CCSSM relegates "order of operations" to its proper place, in a footnote, as a mere convention for reading expressions rather than a fundamental fact about the way arithmetic operations work. This also has important consequences for the role of arithmetic in preparing students for algebra. Students who have been drilled in order of operations as a command about how calculations must be performed, rather than as a convention on how a numerical expression is to be read, can find themselves in a difficult position when they get to algebra and find numbers represented by letters. For example, the following problem was given to 8th grade algebra students as preparation for a workshop "Making Connections" at the University of Arizona (2006, June 7–9).

You are simplifying

$$7 - 2(3 - 8x).$$

Which of the expressions is a correct next step?

	Correct	Notcorrect
$5(3 - 8x)$		
$7 - 2(-5x)$		
$7 - 6 - 16x$		
$7 - 6 + 16x$		

For each expression, explain why you made the choice you did.

One student, under the unfortunate influence of PEMDAS, responded: "$7 - 2(-5x)$ is correct. Always do the parentheses first!" The problem is that in algebra you cannot always do what is in the parentheses first. The word "order" in "order of operations" is best understood as referring to order in the sense of hierarchy. The expression $a(b + c)$ has a hierarchical structure: is parsed as a product, the second factor of which is a sum. You can either perform the sum first, or use the distributive property to rewrite it as a sum of two products, $ab + ac$. If b and c are not explicit numbers, then the second is the only option. Students who interpret PEMDAS as ruling out that option are like deer in headlights and might leap in the wrong direction.

In summary, the design purpose in basing the arithmetic standards on the properties of operations was to make the study of arithmetic a preparation for a view of algebra in which letters are seen as standing for numbers and algebraic expressions are seen as meaningful calculations with numbers, not as mysterious hieroglyphs. Giving meaning and purpose to the procedures of algebra organizes them into a coherent whole, which may be easier to retain than a jumble of isolated and seemingly unrelated moves and mnemonics.

The Standards for Mathematical Practice

A successful practitioner of mathematics uses reasoning, precision, and an ability to see structure, among other skills. These practices are described in the CCSSM Standards for Mathematical Practice. Two of these standards are particularly relevant to the connection between school algebra and abstract algebra:

MP7. Look for and make use of structure.
MP8. Look for and express regularity in repeated reasoning.

Viewing integers and polynomials (or rational numbers and rational functions) as instances of the same structure is an example of MP7; distilling all the ways in which we calculate with elements of those structures into a list of properties of operations is an example of MP8.

These two standards may also be viewed as tools for excavating the curriculum. As we saw in the discussion of quadratic equations in the previous section, the habit of looking for and making use of structure brings us to see factoring or completing the square as purposeful activities designed to reveal hidden properties of a quadratic expression. In seeing the topics this way, we also lift them from their dusty worksheets and see them in their original state and purpose. By the same token, looking for and expressing regularity in repeated reasoning brings us to see the derivation of the quadratic formula as an extension of the method of completing the square. In so doing we restore the connection between two topics that seemed to be lying separately in the curriculum.

The Structure of the Standards

Another way in which the higher standpoint is manifested in the standards is in the structure of the standards.

> In the Common Core State Standards, individual statements of what students are expected to understand and be able to do are embedded within domain headings and cluster headings designed to convey the structure of the subject. "The Standards" refers to all elements of the design—the wording of domain headings, cluster headings, and individual statements; the text of the grade level introductions and high school category descriptions; the placement of the standards for mathematical practice at each grade level (Daro, McCallum, & Zimba 2012).

Critics have often complained about the US curriculum being a mile wide and an inch deep (Schmidt, Houang, & Cogan 2002). One contributing factor is the proliferation of forms and procedures to be memorized separately. The curriculum is a mile wide in part because it has been scattered a mile wide. Just as an archeologist can reveal the lost streets and walls that explain the scattered potsherds, a higher standpoint can reveal the unity of seemingly different forms and procedures. This achieves a sort of focus, a compacting of what must be learned into a smaller number of underlying principles.

In the high school standards pertaining to algebra, many domain level headings display a reasoning and structural perspective, for example

- Seeing Structure in Expressions
- Creating Equations
- Reasoning with Equations and Inequalities
- Interpreting Functions
- Building Functions

These headings describe broad cross-cutting habits of mind that apply across many different specific topics in algebra. The same goes for some of the cluster headings. For example, consider the following cluster in the high school algebra standards:

A-SSE.A. Interpret the structure of expressions

1. Interpret expressions that represent a quantity in terms of its context.
   a. Interpret parts of an expression, such as terms, factors, and coefficients.
   b. Interpret complicated expressions by viewing one or more of their parts as a single entity. *For example, interpret* $P(1+r)^n$ *as the product of* P *and a factor not depending on* P.
2. Use the structure of an expression to identify ways to rewrite it. *For example, see* $x^4 - y^4$ *as* $(x^2)^2 - (y^2)^2$, *thus recognizing it as a difference of squares that can be factored as* $(x^2 - y^2)(x^2 + y^2)$.

The heading is designed to indicate the unity of the individual standards, which represent topics that might be scattered across the entire high school curriculum. The underlying principle is "don't just do something, stand there." That is, before plunging into an algebraic manipulation, ask yourself what your purpose is and choose appropriately. This habit of mind can help students develop algebraic foresight, the ability to anticipate where a calculation might go and what it might be good for.

As a final example of the unity afforded by a structural point of view, consider the different forms of the equation of a line that students learn: slope-intercept form, point-slope form, point-point form. A structural perspective sees all of these as instances of the same general principle: you can find the coordinates of a point on the line by starting at some fixed point and using the slope. The starting point might vary, the slope might be calculated in different ways, but the forms are structurally the same.

Concluding Remarks

An archeologist brings to the dig a set of mental structures with which to interpret the results of their excavation—tools, ornaments, fragmentary walls, dikes, ditches, half-revealed roads—all born of a knowledge of history and culture. A mathematician brings to the examination of school mathematics a set of structural sensibilities born of the accumulated architecture of the subject itself.

Who is served by these structural sensibilities? What are the implications for secondary teacher preparation? A teacher who can take the higher standpoint, who can see the structure of the subject and the connections between different topics, may have a greater ability to convey to their students the reasons for any lesson, the connections with later lessons, and the structural similarities between different forms and procedures. They may be able to make more informed decisions about whether and how to change curriculum. In the end, they might agree with Terry Bladen about quadratic equations. But the decision would be made after careful excavation and a balanced consideration of the relative merits of renovation versus removal.

References

ACT. (2009). *Focusing on the essentials for college and career readiness*. Iowa City, IA: Author. Retrieved from www.act.org

BBC News. (2003a, July 2). Debate over maths teaching. Retrieved from news.bbc.co.uk

BBC News. (2003b, April 21). Maths 'should be optional.' Retrieved from news.bbc.co.uk

Conley, D. T., Drummond, K. V., de Gonzalez, A., Rooseboom, J., & Stout, O. (2011). *Reaching the goal: The applicability and importance of the common core state standards to college and career readiness*. Eugene, OR: Educational Policy Improvement Center.

Daro, P., McCallum, W., & Zimba, J. (2012). The structure is the standards. Retrieved from mathematicalmusings.org

Katz, V. (2007). *The mathematics of Egypt, Mesopotamia, China, India, and Islam: A sourcebook*. Princeton, NJ: Princeton University Press.

Klein, F., Menghini, M., & Schubring, G. (2016). *Elementary mathematics from a higher standpoint*. Berlin/Heidelberg: Springer.

McCallum, W. (2015). The common core state standards in mathematics. In S. J. Cho (Ed.), *Selected regular lectures from the 12th international congress on mathematical education* (pp. 547–560). Berlin/Heidelberg: Springer.

Museum of the History of Science, Oxford University. (2017, November 27). An English mathematical aide-memoire of the 17th century. Retrieved from http://www.mhs.ox.ac.uk

National Governors Association Center for Best Practices and Council of Chief State School Officers. (2010). Common Core State Standards for Mathematics. Retrieved from corestandards.org

New York State Education Department. (2017). EngageNY Algebra I. Module 4. Retrieved from https://www.engageny.org

Pearson. (2013). *CME mathematics common core*. New York, NY: Author.

Schmidt, W., Houang, R. & Cogan, L. (2002). A coherent curriculum: the case of mathematics. *American Educator, 26*, 10–26.

Struik, D. J. (2014). *A source book in mathematics, 1200–1800*. Princeton, NJ: Princeton University Press.

UK Parliament. (2003, June 26). Debate on quadratic equations. Retrieved from publications.parliament.uk

Wikimedia Commons. (2006). File:The Algebra of Mohammed ben Musa (Arabic).png. Source: Rosen, F. (1831). *The Algebra of Mohammed ben Musa*. Retrieved from https://commons.wikimedia.org

Zimba. J. (2011, August 19). *Operations and algebraic thinking* [Video File]. Retrieved from www.youtube.com

Zimba, J. (2014). The development and design of the common core state standards for mathematics. *New England Journal of Public Policy, 26*(1), Article 10. Retrieved from https://scholarworks.umb.edu/nejpp/vol26/iss1/10

Chapter 6
Foreshadowing Ideas in Abstract Algebra: Reflections from the Secondary School Mathematics Curriculum Improvement Study

J. Philip Smith

Introduction

The 1960s and early 1970s was a challenging period for those who lean toward gradual curricular change and thoughtful, incremental, curricular innovation. Part of the American reaction to the Soviet Union's launch of the earth's first orbital satellite was to question the quality of the country's science and mathematics preparation, followed shortly thereafter by the creation of a number of curriculum writing projects aimed primarily at "updating" middle and secondary school mathematics programs.

One such project, The Secondary School Mathematics Curriculum Improvement Study (SSMCIS), aimed to produce material for "mathematically capable," college-intending students in grades 7 through 12. The project was not designed to become a national curriculum—or even a broadly adopted curriculum—but was conceived as an exploration of what might be achievable by those interested in bringing the curriculum of the upper grades into the modern era and incorporating the mathematical structures, rigor and applications perceived as missing in more widely available textbooks.

What does such a curricular endeavor written over 45 years ago, with materials long out of print, have to offer teachers and other educators today? Most importantly, the SSMCIS material demonstrates that it is possible to incorporate sophisticated mathematical content into high school and junior high school curricula. It was possible, for instance, for a seventh grader to begin studying SSMCIS algebra and by the end of his secondary school studies to have demonstrated, say, the

J. P. Smith (✉)
Department of Mathematics, Science and Technology, Teachers College, Columbia University, New York, NY, USA
e-mail: jps164@tc.columbia.edu

N. H. Wasserman (ed.), *Connecting Abstract Algebra to Secondary Mathematics, for Secondary Mathematics Teachers*, Research in Mathematics Education,
https://doi.org/10.1007/978-3-319-99214-3_6

impossibility of trisecting an angle with the classic Euclidean tools. This is heady stuff! The SSMCIS writers, too, offer intriguing examples and replicable pedagogical strategies for teaching, as they weave together spiraling approaches to topics considered to be interconnected by the SSMCIS, although frequently not regarded as such elsewhere.

Funded initially by the US Office of Education and later by the National Science Foundation and Teachers College, Columbia University, the SSMCIS personnel included about 30 mathematicians, mathematics education specialists, and advanced graduate students who met each summer at Teachers College to hash out the upcoming year's writing goals. Prominent among the mathematicians were Edgar Lorch (Columbia University), Lennart Råde (Chalmers Institute of Technology, Sweden), Marshall Stone (University of Chicago), and Albert Tucker (Princeton University). The mathematics educators included Howard Fehr, Jeremy Kilpatrick, Myron Rosskopf, Bruce Vogeli (all from Teachers College), Howard Levi (City University of New York), James Fey (University of Maryland), and Hans-Georg Steiner (University of Karlsruhe, Germany). Following the yearly summer conference, designated writers (frequently graduate students at Teachers College) would disperse to their respective locations and produce drafts which would be reviewed, revised, and published in time for the next year's conference. Teachers College Press produced the early materials, which were little more than copies of typewritten pages with rudimentary graphics. (Later, some commercially available material was produced, but this chapter is based on the original SSMCIS publications.)

The SSMCIS was led by Howard F. Fehr, a professor of mathematics education at Teachers College. Fehr had been the principal author of the Organization for European Economic Co-operation's Fehr, 1961 report, *New Thinking in School Mathematics*, which criticized "traditional" mathematics teaching as not providing adequate preparation for an increasingly technical, computer-oriented, society. That report, along with *Goals for School Mathematics*, the 1963 report of The Cambridge Conference on School Mathematics, impelled Fehr to argue for a more unified approach to the mathematics curriculum. In fact, the curricular materials produced by SSMCIS were entitled *Unified Modern Mathematics*. No longer were there to be separate courses of study focusing on algebra, geometry, trigonometry, or solid geometry.

The curricular approach employed a "spiral" design: concepts and skills introduced at one level reappear and are used in subsequent study. A number of fundamental concepts and structures served as a basis for the program's "unifying ideas": sets, operations, mappings, relations, groups, rings, fields, and vector spaces. The structures and mappings of abstract algebra played a large role in generating and organizing the "unified" curriculum.

The SSMCIS courses of study were labeled Course 1, Course 2, and so on, up through Course 6, rather than the more traditional Grade 7, Grade 8, ..., Grade 12, in order to avoid tying users to a particular educational level without regard for student interests or abilities. However, it is reasonable to interpret Course 1 as Grade 7 materials, etc. In this chapter, I will refer to these materials as Courses 1, 2, etc.

(and not as Grades 7, 8, etc.). Also, the last materials produced by the SSMCIS were a series of five booklets, conceived of as a second part to Course 6, and which intended to give instructors flexibility in choosing some additional topics of study during the latter part of Grade 12: Booklet A—*Introduction to Statistical Inference*; Booklet B—*Determinants, Matrices and Eigenvalues*; Booklet C—*Algebraic Structures, Extensions, and Homomorphisms*; Booklet D—*Introduction to Differential Equations*; and Booklet E—*Geometric Mappings and Transformations.* A "*Teachers Commentary*" was also produced for each of the texts and for the booklets. These included lists of objectives, answers to exercises, suggested instructional time for each unit, sample tests, and pedagogical advice and remarks. (See SSMCIS 1967a, 1967b, 1970, 1971, 1972a, 1972b, 1972c, 1973.)

Given the many writers and the ever-changing curriculum writing groups, it is not surprising that the SSMCIS materials, while carefully written and edited, differ internally somewhat in tone, in pedagogical approach, and in the level of mathematical sophistication that was, apparently, expected of the reader. The writers were quite successful, however, at avoiding a "silo" approach to their subject matter. Mappings that "preserve structure" appear in every year's material, even though the structures themselves may differ; the language of group theory abounds; geometry is accompanied by talk of isometries; the Course 2 chapter "Fields" has as its goal "to extend the study of abstract operational systems to fields and, in the process, to deepen student insights into the algebraic structure of number systems," etc. The Course 2 chapter headings themselves illustrate the spiral nature of the texts, as well as the modern (for the 1960s and 70s) subject matter and the advanced level of the material: "Mathematical Logic and Proof;" "Groups;" "Affine Geometry;" "Fields;" "The Real Numbers;" "Coordinate Geometry;" "Real Functions;" "Descriptive Statistics;" "Transformations in the Plane: Isometries;" "Length, Area, Volume."

In thinking about the role of abstract algebra in the SSMCIS materials, we should keep in mind that the subject itself was only a recent addition to the undergraduate curriculum in the mid 1960s. I. N. Herstein's classic work was first published in Herstein, 1964, and the great twentieth century text on the subject, van der Waerden's *Modern Algebra* (based on lectures by Emil Artin and Emmy Noether in the 1920s), had only been available in English since van der Waerden, 1949. Student-friendly texts like John Fraleigh's (Fraleigh, 1967) *A First Course in Abstract Algebra* did not exist when the SSMCIS began its collaborations. So much of what we might regard today as fairly routine, in a curricular sense, was adventurous and groundbreaking for a secondary school audience in the late 1960s.

The exposition that follows in this chapter is what we might call a "mathematical analysis" of the SSMCIS material. The project's designers took concepts that had been a traditional part of the 7–12 curriculum, joined with those certain modern mathematical notions, and then "re-imagined" how the material might be presented in a unified way, yet free of traditional sequencing and content requirements. We shall explore in the pages that follow the result of this "re-imagining" insofar as it produced connections to abstract algebra. That is, we explore how ideas from abstract algebra might be incorporated into the secondary school curriculum in ways that could support, unify, and elaborate on standard school mathematics topics.

We will begin our exploration with a very brief look at the Grade 7 and 8 materials, followed by an examination of algebra-related content for grades 9 through 12. We conclude with an exploration of relevant sections in the senior year booklets.

Courses 1 Through 6

The algebraic notions in SSMCIS course material for grades 7 through 12 begin, logically enough, with an exploration of basic definitions and language, accompanied by numerous examples. The examples lead to the concepts of number systems, binary operations, relations, groups, fields, and mappings (particularly isomorphisms), and eventually to a study of the algebra of matrices. Through matrices, distance-preserving and other mappings of the Euclidean plane are explored, since SSMCIS offers no single-year focused treatment of geometry. Even basic ideas, such as matrix multiplication, flow fairly naturally from a discussion of real-world examples. Isomorphic mappings and a particular field of 2×2 matrices are used to introduce the complex number system. Vector spaces are presented as another example of mathematical structure, and the notion of vector space isomorphisms is compared to linear mappings between vector spaces, as part of the emphasis on structure-preserving mappings. The notion of an "extension" of an algebraic structure is introduced in Grade 12, leading to a discussion of integrals and limits involving infinity.

Courses 1 and 2: Basic Definitions, Groups, Fields, and Isomorphisms

SSMCIS Course 1 provides little of immediate relevance to our concerns, but its focus on the language of number systems, sets, operations, relations, and mappings sets the stage for subsequent development. Course 2 continues introducing the "cast of characters" that will populate the later courses.

Abstract algebra makes its formal entrance in the Course 1 and 2 chapters on groups and on fields. Many examples lead to a definition of a mathematical group. Some basic theorems are proved, permutation groups are introduced, and the basic idea of an isomorphism and isomorphic groups are explored. The examples include some finite groups and finite fields. The *Teachers Commentary* for the Course 1 chapter on fields remarks, "Though the development of the theory of fields and ordered fields is rigorous—in the sense of formally stated axioms and carefully proven theorems—students should not be expected to *memorize* proofs to any particular theorems. There will be opportunities to discuss concepts of logic and proof illustrating ideas from Chap. 1, but keep in mind...that almost all topics in this chapter will be developed further in future courses." In a typical effort to broaden student insight into a fundamental concept, the *Commentary* suggests that teachers explore the distributive property by considering a case that is not distributive. Namely, they explore addition "distributing over" multiplication by trying out a case: $1 + (1 \cdot 0) = (1 + 1) \cdot (1 + 0) \Rightarrow \ldots\ldots \Rightarrow 1 = 2$.

Course 3: Matrices, Transformations of the Euclidean Plane, Rings

A simple listing of the nine chapters of the SSMCIS grade 9 text illustrates both the "cosmopolitan" nature of the course of study and the strong algebraic focus: "Introduction to Matrices;" "Linear Equations and Matrices;" "The Algebra of Matrices;" "Graphs and Functions;" "Combinatorics;" "Probability;" "Polynomial and Rational Functions;" "Circular Functions;" and "Informal Space Geometry."

A glance through the opening pages of the material illustrates the frequent focus on applications (which, today, we would probably file under "mathematical modeling"). Examples include problems about keeping track of equipment parts in a manufacturing facility, examining a pay-off matrix in game theory, describing bus routes connecting four cities, sending coded messages, tracking population movements via transition matrices, etc. These applications are all discussed while introducing the notions of matrix addition, scalar multiplication and matrix multiplication. The "unified" aspect of the curriculum is illustrated by a discussion entitled "Matrices and Transformations." Such mappings of the Euclidean plane as reflections, half-turns, rotations, dilations, shears, and so on are explored through matrix multiplication. No proofs appear in this section, whose purpose is motivational and introductory, but phrases like "It would seem that multiplication on matrices is associative" do appear. A typical problem in this section is:

> The population of a city at the end of 1968 is 3,000,000, and that of its suburb is also 3,000,000. Assume that 70% of the city people in any year remain in the city and 30% of them move to the suburb, while 80% of the suburban population remain in the suburbs and 20% of them move to the city. Using matrices, calculate the population in both places at the end of 1971 (p. 35).

The chapter offers an interesting departure from the more traditional way of introducing matrix multiplication, as well as illustrating the texts' general care in presenting new ideas in intuitively appealing ways. Rather than just presenting the definition or taking, say, two or three linear equations and re-casting them in matrix form (as matrix times column vector = column vector), the text focuses on a man who builds two models of houses in three different towns. The Model A home requires six doors and eight windows, whereas the Model B home has five doors and seven windows. In one town, he is going to build six Model A homes and three Model B homes. In another town, the respective figures are 2 and 7, and in the third town, 4 and 3. How many doors and windows does the builder need in each town?

The answers are calculated one at a time as problems in arithmetic, but the answers are ultimately listed in matrix form. The text advises, "These calculations involve multiplication and addition on scalars. But, we regard the entire calculation as our matrix operation, called *multiplication on matrices*." The operation is depicted this way:

$$\begin{pmatrix} 6 & 3 \\ 2 & 7 \\ 4 & 3 \end{pmatrix} \cdot \begin{pmatrix} 6 & 8 \\ 5 & 7 \end{pmatrix} = \begin{pmatrix} 6 \cdot 6 + 3 \cdot 5 & 6 \cdot 8 + 3 \cdot 7 \\ 2 \cdot 6 + 7 \cdot 5 & 2 \cdot 8 + 7 \cdot 7 \\ 4 \cdot 6 + 3 \cdot 5 & 4 \cdot 8 + 3 \cdot 7 \end{pmatrix} = \begin{pmatrix} 51 & 69 \\ 47 & 65 \\ 39 & 53 \end{pmatrix}.$$

The example is followed by several partially multiplied matrices, where students are invited to search for patterns before any general rules for matrix multiplication are stated.

The second chapter's focus on solving linear equations using matrices offers a fairly standard approach to the so-called elementary row operations of Gauss–Jordan elimination as a way to uncover solution sets. The chapter concludes with explanations of how to use elementary row operations to find the inverse of a given matrix, as well as a thoughtful example of attempting the same method for a non-invertible matrix.

Full-blown, formal, abstract algebra constructs make their first appearance in the third chapter of the Course 3 text. The general $n \times m$ matrix is defined, as is equality of matrices. The addition of matrices is given a general definition, as is the zero matrix, the additive inverse of a matrix, and multiplication by a scalar. The various properties of scalar multiplication are proved. Matrix multiplication is defined in general and the fact that it is associative is proved in the 2×2 case. The multiplicative identity is defined and inverses are given a full treatment. The uniqueness of identity and inverses is proved. In the chapter's final section, the notion of a ring is introduced. Because prior to Course 3 the curriculum had defined a mathematical group, it is possible here to make such statements as, "We find that we have a richer structure than a group since we have two operations." The general definition of a ring is produced (as is the definition of a ring with unity) and the proof that the set of 2×2 matrices form a ring is given. (That the entries come from the field of real numbers is assumed.)

The chapter concludes with its most unusual section: "A Field of 2×2 Matrices." In the preceding courses, students have been introduced to the notion of a field, and here it is observed that the ring of 2×2 matrices over the field of real numbers is not a field "because multiplication is not commutative." Further, "the subset of invertible matrices is not a field for the same reason, and also because this subset does not contain the identity element for addition." So the students are told, "If we take the set of 2×2 invertible matrices and add to them the identity matrix for addition, we will have a set which may have a subset in which multiplication is commutative." The ensuing discussion leads to showing that the set of matrices of the form $\begin{bmatrix} x & -y \\ y & x \end{bmatrix}$, where x and y are real numbers, is, in fact, a field (foreshadowing the following course's development of the complex number system).

This entire section of Course 3 offers a near-seamless integration of basic arithmetic notions with fundamental structures of linear and abstract algebra.

Course 4: Vectors Spaces, Subspaces, Field of Complex Numbers

Although the major focus of Course 4 is on quadratic, circular, exponential, and logarithmic functions, as well as the complex number system, the latter part of the course offers an introduction to the algebra of vectors, with a discussion of vector spaces and subspaces.

Using terminology already familiar to the students, the vector space notion is developed from a study of n-tuples and, later, branches to a consideration of vector spaces of matrices and polynomials.

Employing an unusual approach, at least for secondary school students, Course 4 introduces the complex number system by considering the field S of 2×2 matrices of the form $\begin{bmatrix} x & -y \\ y & x \end{bmatrix}$ (x and y real numbers). Students are invited to consider the matrix $\begin{bmatrix} 1 & 0 \\ 0 & 1 \end{bmatrix}$ as I and the matrix $\begin{bmatrix} 0 & -1 \\ 1 & 0 \end{bmatrix}$ as J. If we let $A = \{a \bullet I \mid a \in \mathbb{R}\}$, the text proves that the mapping $f : A \rightarrow \mathbb{R}$ defined by $f(a \bullet I) = a$ is an isomorphism and that the matrix I and the real number 1 correspond. Next to be considered is $J \in S - A$. In this case, $J^2 = \begin{bmatrix} -1 & 0 \\ 0 & -1 \end{bmatrix} = -1 \cdot I$. This means that J^2 is a solution of $X^2 = -1 \cdot I$, and it is argued that J cannot correspond to any real number. Thus, if we are to extend the original isomorphism from A to $\mathbb{R}$ to an isomorphism that maps all of S to some set "super $\mathbb{R}$," we must "adjoin to $\mathbb{R}$ another kind of number whose square is -1."

The discussion continues by inventing the number i, whose square is -1, and then considering the correspondence $a \cdot I + b \cdot J \leftrightarrow a + bi$. The algebraic properties of the complex number system are now derived by examining the behavior of the corresponding matrices in S.

Course 5: Linear Mappings, Structure-Preserving Mappings

The abstract algebra present in SSMCIS grade 11 generally serves a subsidiary role, providing notation and language, rather than major new concepts or insights. In fact, the major focus of Course 5 is analysis. We do, however, find a chapter entitled "Linear Mappings and Linear Transformations," with a note that among the many fields of study within mathematics, the current chapter would be classified under linear algebra; "However, there is much interrelation among these branches of mathematics. The notion of a mapping, which preserves a structure on a set of objects, is one major concept which acts as a unifying thread woven through the fabric of mathematics. Linear mapping, as we shall see, is one such mapping. This chapter highlights both the theoretical and applied nature of linear mapping."

The introduction is followed by a four-equation, five-variable model of trade among four South American countries. The ensuing discussion leads naturally to the mathematical definition of a linear mapping from $\mathbb{R}^n$ to $\mathbb{R}^m$.

Soon, the text returns to structure-preserving mappings. I will summarize the discussion with a bit of detail because of the light it sheds on the SSMCIS algebra-related curriculum up to this point. Three structures are ventured as examples familiar to students: groups, fields, and vector spaces. Among examples of the respective structures are the group of isometries of the plane, the field of complex numbers under the usual addition and multiplication, and the vector space of continuous functions over $\mathbb{R}$ under addition.

The major concern of the section is with linear mappings. The students are reminded that in Course 4 they established that the field of complex numbers was isomorphic to the set of 2×2 matrices of the form $\begin{bmatrix} x & -y \\ y & x \end{bmatrix}$ $(x, y \in \mathbb{R})$ and the general definition of an isomorphism between fields is given. [The authors demonstrate an old, classic way of illustrating the value of isomorphic maps. Question to student: "Which arithmetic problem would you prefer to solve: (a) $2.40 \cdot 2.75$ or (b) $.3802 + .4393$?" It is noted that $\log_{10}2.40 = .3802$ and that $\log_{10}2.75 = .4393$, and that the use of logarithms to solve a problem like $2.40 \cdot 2.75$ is actually an application of an isomorphism from the group of positive real numbers under multiplication to the group of real numbers under addition.]

In the case of two vector spaces U and V over the same field, a one-to-one mapping F from U onto V is defined as an isomorphism if it preserves both addition and scalar multiplication. Following the definition of vector space isomorphism, the almost identical definition of a linear mapping from vector space U to V is given, with the remark that, "if a linear mapping between vector spaces is one-to-one and onto, then it is an isomorphism."

The range and kernel of a linear mapping is defined and the fact that both are subspaces is proved. The discussion concludes with a proof of the theorem: Let $T : U \to V$ be a linear mapping from vector space U to vector space V. (1) The range of T is V if and only if T is an onto mapping; (2) the kernel of T is $\{0\}$ if and only if T is a one-to-one mapping; (3) the range of T is V and the kernel of T is $\{0\}$ if and only if T is invertible.

The remainder of the chapter is devoted to a lengthy presentation of linear programming methods and applications, making use of the linear algebra just discussed.

Course 6: Absorbing Elements, Identity Elements, Extended Real Number System

After an initial discussion entitled "Infinity," the first part of the grade 12 material focuses on circular, exponential and logarithmic functions and their properties and derivatives. A section on conics and polar equations, a section on probability (involving infinite outcomes set and applications), and a final section on problem solving round out the text.

A more detailed look at the opening chapter, "Infinity," illustrates the use of abstract algebra concepts in a surprising way: to introduce a discussion which will lead to the definition and exploration of neighborhoods of ∞ and the consideration of expressions such as $\lim_{x\to 0}\frac{1}{x^2}$ and $\int_{-\infty}^{5}\frac{1}{x^2}dx$. The authors wish eventually to consider the set $\mathbb{R}^*$, the set of real numbers augmented by the elements $+\infty$ and $-\infty$, to work with "neighborhoods of infinity" in a rigorous way. But the opening discussion has little to do with analysis.

The introductory paragraphs tell us that "in the operational system $(\mathbb{R}, \cdot)$, the number 0 is an absorbing element. This is another way of describing a familiar property of 0: for every $x \in \mathbb{R}$, $x \cdot 0 = 0 \cdot x = 0$." The definition of "absorbing element" follows. Next appears a familiar group table (the elements 2, 4, 6, 8 under multiplication modulo 10), which is then extended by the addition of a new element, b. The table entries under "b" are all bs. The new five-element set is called an "extension" of the original one, and a lengthy discussion of "absorbing-element extensions" follows. Students are asked to observe that in each example, the original structures were groups, but that the extended structures were not. Question for the reader: "Is it always the case that appending an absorbing element to a group destroys the group structure?"

Next up for discussion is an "identity-element extension." The prime example here is to consider the set of whole numbers as an extension of the natural numbers by appending 0. By now, students are probably wondering what these lengthy discussions have to do with infinity, the presumed topic of the chapter. Fear not, write the authors, and consider this example: The system $(\mathbb{N}, +)$ of natural numbers under addition has no absorbing element. Suppose $\mathbb{N}$ is extended to $\mathbb{N}^*$ "by introducing an absorbing element—call it ∞—so that $\infty + x = x + \infty = \infty$ for all $x \in \mathbb{N}^*$."

What follows, appearing at first to be a non sequitur, is a discussion of idempotent elements, culminating with the observation that, "It is an easy matter to prove that identity elements and absorbing elements are always idempotent, simply by using the definition of such elements."

We cite four exercises from the opening section of "Infinity" to help capture the flavor of the exposition.

1. Is $(\mathbb{Z}_5, \cdot)$ an extension of $(\mathbb{Z}_4, \cdot)$? Why or why not?
2. Using the operation of composition of mappings, is the system of translations in a plane an extension of the system of translation on a line in the plane?

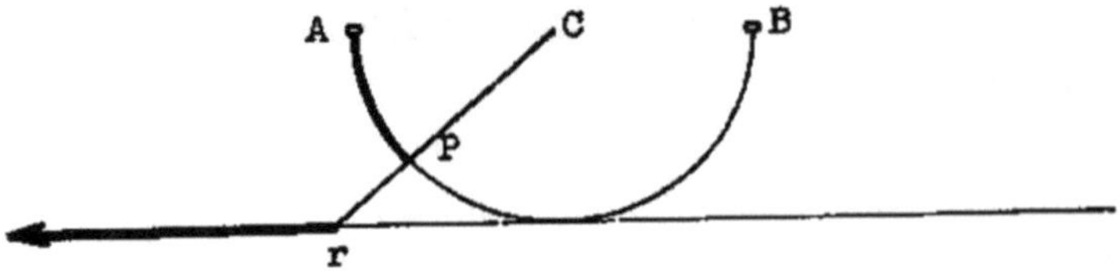

Fig. 6.1 One-to-one mapping between neighborhood of A on semicircle and neighborhood of infinity on the real number line

3. Show that any two identity-element extensions of an operational system must be isomorphic.
4. Let $\mathbb{N}$ be the set $\{1, 2, 3, 4, \ldots\}$ and consider the system $(\mathbb{N}, lcm)$ where "lcm" assigns to each pair the least common multiple of the numbers in the pair. For example, $4\ lcm\ 6 = 12$. Is there an identity element in $(\mathbb{N}, lcm)$? Is there an absorbing element?

The chapter continues with discussions of infinity as an absorbing element, neighborhoods of infinity (Fig. 6.1), infinity and improper integrals, asymptotes, and limits of sequences. Ensuing chapters return to more standard calculus and pre-calculus topics with little material that could be considered strictly algebraic.

Grade 12 Booklets

Of the five SSMCIS booklets, written to give teachers some flexibility in choosing a curricular focus for the second half of the 12th grade, only two (Booklets E and C) contain substantial algebraic content. Both are natural extensions of material covered in Courses 1 through 6 and provide interesting, albeit challenging, explorations for those seeking deeper insights into the particular topic.

Geometric Mappings and Transformations (Booklet E) uses notions from cartography to motivate the study of various projections from three-dimensional to two-dimensional surfaces. Here it is geometric, rather than algebraic, structures that are (or are not) preserved by various mappings. The discussion of several families of transformations leads naturally to a consideration of algebraic properties that the families might have under the operation of composition of mappings. The familiar group structure emerges and leads to natural ways of classifying the mappings.

Algebraic Structures, Extensions, and Homomorphisms (Booklet C) begins by developing the machinery to show that a line segment of length $\sqrt[3]{2}$ cannot be constructed with the classic Greek tools of straightedge and compass. The booklet concludes by discussing homomorphisms, quotient rings, and quotient fields. Unlike the case in Courses 1–6 and the other senior year booklets, the student here is expected to play a heavy role in progressing through the material. The text, for instance, may state definitions, offer some examples and proofs, but not infrequently the student must tackle a series of exercises to complete a proof or to reach a culminating point in the discussion.

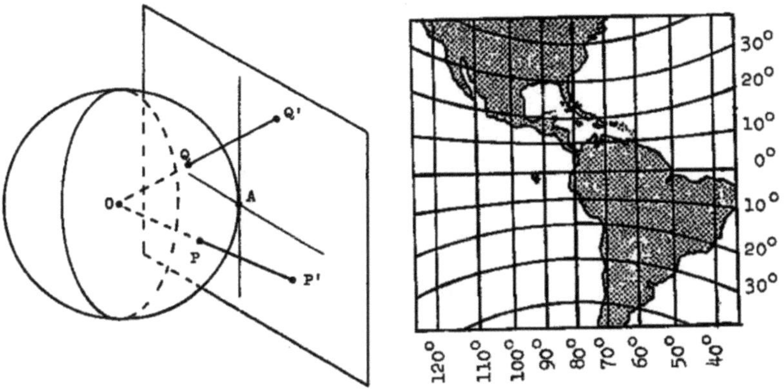

Fig. 6.2 Gnomonic projection and gnomonic map of western hemisphere

Booklet E: Geometric Mappings and Transformations

Geometric Mappings: Mappings from Sphere to Plane

The recurrent, unifying SSMCIS theme of mappings that preserve structure is evident in the opening pages of Booklet E. The discussion of mappings from sphere to plane begins by asking how terrestrial maps are constructed: "How does one go about transferring the globe's information to a flat surface?" Latitudes, longitudes, the prime meridian, and great circle routes are all explored on the way to contemplating a projection from a sphere to a plane tangent to that sphere: a gnomonic projection. A gnomonic map of the western hemisphere is shown (Fig. 6.2).

Tangents, secants, and cosines abound as images of given points on the sphere are calculated on the tangent plane.

The gnomonic projection is easily shown to preserve shortest distances (in the sense that the shortest distance between two points on a globe—a great circle arc—will correspond to a line segment connecting the images of those points on the map). Great! But an imaginary navigator has weighed in with another problem: gnomonic projections do not preserve angles, so if he uses a gnomonic projection map, he needs to constantly readjust his compass setting, a major annoyance. So now we consider another type of projection: the stereographic projection, which contemplates the sphere bisected by the plane to which it is being mapped. This "conformal mapping" does preserve angles, but presents another problem: areas are distorted. Two regions on the projection that appear to have the same area may, in fact, be of different areas on the sphere.

The solution? Consider a cylindrical projection, which maps a sphere inside and tangent to a right circular cylinder onto that cylinder. Open up the cylinder and you have an equal-area map! But alas, the standard cylindrical projection does not preserve shapes. Regions near the poles of the sphere are altered almost beyond recognition.

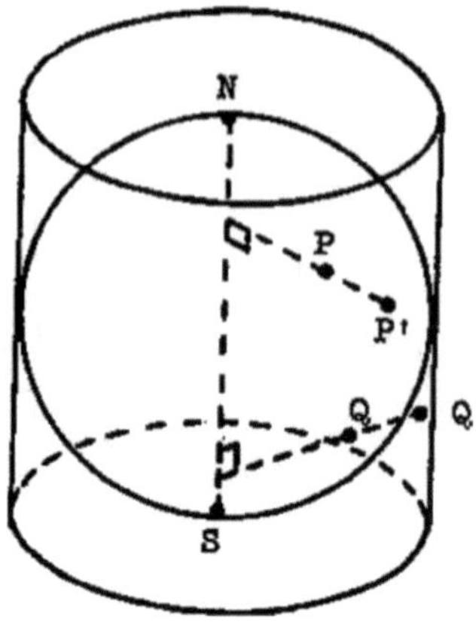

Fig. 6.3 Archimedean cylindrical projection

Hence, the entire discussion that opens Booklet E is devoted to mappings and properties that are or are not preserved by those mappings, a topic first broached in the abstract algebra discussions in earlier grades.

Three typical problems:

1. What is the latitude and longitude of the point on the earth antipodal to Chicago?
2. A sea captain wishes to sail his vessel from Sitka, Alaska, to Brisbane, Australia. He has only a gnomonic map whose point of tangency is the North Pole. Why will the captain's map not suffice?
3. Archimedes knew how to find the surface area of a cylinder. He is reputed to have known how to make use of a property of the "Archimedean" cylindrical projection in discovering the formula for the surface area of a sphere of radius r. How do you think he did this?

To answer 3, for instance, we can consider, as Archimedes did, a sphere inside and tangent to a right circular cylinder (Fig. 6.3). The projection from sphere to cylinder proceeds as follows: From point P on the sphere, drop a perpendicular to the diameter of the sphere that is parallel to the cylinder's sides. The perpendicular is extended through P to meet the cylinder at P'. Thus the Archimedean projection maps P to P'. Archimedes projected a hemisphere onto the cylinder and then "unrolled" the cylinder to get a rectangular "map" of half the sphere's surface. Let us call it the "western hemisphere." If the sphere has radius r, then the map's dimensions are $2r \cdot \left(\frac{1}{2}C\right)$, where C is the circumference of a circle of radius r. So the area of the "map" is $2r \cdot (\pi r) = 2\pi r^2$. Since Archimedean projections are "equal area" projections, the area of the hemisphere is $2\pi r^2$ and so the surface area of the entire sphere is $4\pi r^2$.

Geometric Mappings: Mappings by Projection from Plane to Plane

In considering projections from one plane to a second (or image) plane, the text makes use of both parallel and central projections. For *parallel* projections, the projecting rays are parallel to each other; the central projection rays emanate from a single point. In projecting from one plane to another, under either parallel or central projections, we can consider the two planes to be parallel or intersecting. Of interest to us is, once again, the focus on properties preserved by such mappings. In fact, the authors comment on Felix Klein's insight that a particular geometry can be thought of as a study of those properties preserved under a particular type of mapping.

"The study of properties preserved under parallel projections on parallel planes...might be called a 'congruence geometry.'" On the other hand, a *central* projection from a plane to a parallel plane preserves similarity, and the associated geometry might be called a "similarity geometry." These latter mappings preserve straight lines and the property of being parallel, but a circle might map to an ellipse or a square to a parallelogram. "The study of properties preserved under either a central or parallel projection of a plane to a parallel plane has traditionally been called Euclidean geometry." It is noted that parallel projections from a plane to an intersecting plane still preserve lines and the property of being parallel, but no longer preserve sizes or angles. The study of properties preserved under all three types of projections is labeled "affine geometry." Finally, central projections from a plane to an intersecting plane are considered, leading to a discussion of conic sections and a definition of "projective geometry."

Groups of Transformations of the Inversive Plane

Booklet E's second chapter is entitled, "Inversion and Circle-Preserving Transformations." Before we consider the group structures appearing in that chapter, a few words about inversions are in order.

Inversions, like isometries and similarities, are one-to-one mappings of the Euclidean plane onto itself. Isometries preserve length; similarities preserve shape; inversions preserve neither length nor shape, but still maps angles onto congruent angles and circles onto circles. In a plane, consider a circle C of radius r and center O. Let $\ell(OP)$ represent the length of line segment OP. Under what is called an "inversion in C," the image of any point P other than O is defined to be that point P' such that P' lies on the ray OP and $\ell(OP) \cdot \ell(OP') = r^2$. The point O is called the "center" of the inversion and P and P' are called "inverse points."

The text notes, "Clearly an inversion interchanges the inside and outside of the circle C: if $\ell(OP) < r$ then $\ell(OP') > r$. The circle itself is fixed pointwise: if $\ell(OP) = r$ then $\ell(OP') = r$. Because of the definition's symmetry, it is clear that if $P \to P'$ then $P' \to P$." Under the definition of inversion, the circle's center O has no image. However, as the point P moves closer and closer to O, the image P' recedes further and further away from O, without bound. This behavior, together with the earlier discussions of infinity, suggests augmenting the standard plane by adding a "point at infinity" which would be the image of O under the inversion. The new point is considered to belong to each line in the plane, and lines are regarded as circles that pass through the point at infinity! The ordinary plane with the point at infinity appended is called the "inversive plane."

Consider a stereographic projection on an inversive plane. The sphere is bisected by the inversive plane. Let us call the great circle where the plane and sphere meet "the equator." We project from, say, the "north pole" by using the line through that pole and any other point S on the sphere. The point of intersection of that line and the plane is the image of S. The pole itself is considered to map to the inversive plane's "point at infinity." Points close to the "north pole" map to points far out on the plane. Points below the "equator" map to points in the plane within the circle of intersection of the sphere and plane. How satisfying! We now have a one-to-one correspondence, without exception, between points on the sphere and points on the inversive plane.

The SSMCIS devotes two pages to the development of a coordinate rule for the images of points mapped by inversion in a circle of radius k centered at the origin. The question soon arises: What happens to lines under an inversion mapping? What happens to circles? The question is easily answered: the image of a line through the origin is that same line; the image of a line not through the origin is a circle passing through the origin; the image of a circle through the origin is a line not passing through the origin; and the image of a circle not through the origin is a circle not through the origin. Great symmetry here! Even better, some circles are self-inverses, in the sense that every point on the circle maps to a (probably different) point on the same circle. Such matters are investigated analytically in Booklet E.

The discussion concludes with a re-statement of the definition of a group and of a commutative group. The text notes, "In earlier work we studied several groups of similarities: all translations, all isometries, all rotations having a common center, etc. These are also groups of transformations of the inversive plane." The first example, with justification provided, is that the set of all inversions with center O together with all dilations with center O and positive scale factor is a group under composition.

The chapter ends by considering whether or not the set of all inversions, similarities, and their compositions form a group—a gigantic group encompassing many of the previously considered groups of mapping of the plane. The response is "yes"—and such a group is called "the group of circle-preserving transformations."

Illustrative problems:

1. The direct transformation T leaves fixed the points $(0, 0)$, $(1, 0)$, and ∞. Prove that all such maps form a group.
2. A member of the group of circle-preserving transformations consists of an inversion in $(x + y)^2 + y^2 = 1$ followed by the translation $(x, y) \rightarrow (x + 1, y)$ followed by inversion in the circle $x^2 + y^2 = 1/4$. Under this mapping, what is the image of ∞?

3. Show by means of a diagram the group-subgroup relationships among the following groups: isometries, rotations with center P, all translations, circle-preserving transformations, all direct similarities, the identity mapping, all similarities.

Booklet C: Algebraic Structures, Extensions, and Homomorphisms

The high point of SSMCIS's venture into abstract algebra appears in Booklet C, which could well form a short course in itself. Earlier core definitions are repeated and earlier discussions summarized so that the booklet is largely understandable by those unfamiliar with previous course material. Here is the booklet's introduction:

> Mathematics today is the study of structure on sets of objects. There is no natural way to separate the problems and methods of the subject into disjoint independent compartments. Nevertheless, it is common and often helpful to identify algebraic, geometric, or probabilistic aspects of problems and theories. In this booklet, you will explore several mathematical techniques and concepts that are primarily algebraic—they involve the study of operational systems. The methods will be applied to problems ranging from solutions of equations to geometric construction and trigonometry. Chapter 1 deals with problems that originated in the problems of Euclid (c. 300 B.C.) and his contemporaries; the solutions you will obtain were produced by algebraists 2000 years later. Chapter 2 deals with problems in group theory and homomorphisms that are of nearly current development.

The Classic Greek Ruler and Compass Constructions

The first task of Chap. 1 is to discuss some standard geometrical constructions with straight-edge and compass: a line perpendicular to a given line from a given point not on that line, a triangle congruent to a given triangle, an angle congruent to a given angle. Students are offered 17 construction problems interspersed throughout the text's discussions. Eventually, the text turns to stating four construction problems that the ancient Greeks were unable to solve: (1) divide a given angle into thirds (i.e., trisect an angle); (2) construct a square that has the same area as a given circle (i.e., square a circle); (3) construct a regular seven-sided polygon whose sides are congruent to a given segment; (4) given a line segment AB, construct a second line segment CD such that a cube of side CD has twice the volume of a cube of side AB (i.e., double a cube).

The text notes that the four constructions are now known to be impossible, and it leads the reader through a series of explanations and exercises to prove that the last-stated construction problem is, in fact, impossible. Unlike the standard SSMCIS texts, where problems appear at the end of each section, problems in Booklet C are interspersed throughout the text, sometimes singly, sometimes in a series, so that the student is actually developing much of the exposition, with the text often just commenting on or summarizing what the previous string of problems has established.

Because the classic (and impossible) construction problems were bound up with the idea of constructing the real line via straightedge and compass, the text begins by considering the problem of which real numbers can actually be constructed—or located—on a line by use of the two standard tools. (While developing constructible numbers, the text also considers the Greek question of procedures for constructing the sum, difference, product and quotient of any two numbers.) Having identified points to be labeled as 0 and 1, it is an easy matter to "construct" the integers. The chapter's first problem was to construct a perpendicular bisector of a given segment, so it follows easily that 0.5, 0.25, 0.75, … are constructible and, in fact, that any number of form $m/2^n$ can be constructed where m and n are integers. The standard construction of dividing a given line segment into n congruent parts leads to the realization that any rational number can be constructed.

By considering the diagonal of a square, Problem 23 enables students to see that $\sqrt{2}$ is constructible, and Problem 24 asks them to construct $\sqrt{5}$, $\sqrt{10}$, $\sqrt{13}$, $\sqrt{17}$, $\sqrt{4}$ (ha, ha), $\sqrt{6}$ and $\sqrt{5}$. Then: "Your solution to Problem 24 might have restored hope that all real numbers could be constructed." Problem 25 asks for a demonstration that $\sqrt{n}$ can be constructed for any positive integer n. ("Hint: Use an inductive approach.") The section concludes by asking students to construct the ungainly looking $\sqrt{\sqrt{3\sqrt{7}+\sqrt{6}}}$.

The next section of Chap. 1, "Algebraicizing the Problem," moves toward giving students an understanding of at least a bit of the abstract algebra that, in the first half of the nineteenth century, actually solved the classic Greek problems. Wasting no time, the text defines the set of numbers that can be obtained through the usual arithmetic operations on the set of rationals, $\mathbb{Q}$, together with $\sqrt{3}$, as $\mathbb{Q}\left(\sqrt{3}\right) = \left\{a+b\sqrt{3}\right\}$ where a and b are rational. We are shown that $\mathbb{Q}\left(\sqrt{3}\right)$ is an ordered subfield of $\mathbb{R}$ and are told that $\mathbb{Q}\left(\sqrt{3}\right)$ is called "an extension of $\mathbb{Q}$ by adjoining $\sqrt{3}$" or "a quadratic extension of $\mathbb{Q}$."

In a nice demonstration of the course's "unified" approach, the text pauses to remark that the present discussion should remind students of some earlier topics of exploration: "Problem 32: If k is a positive element in field F, show that $F\left(\sqrt{k}\right)$ is a vector space over F. Show further that if $\sqrt{k}$ is not an element of F, then $\left\{1, \sqrt{k}\right\}$ constitutes a basis for $F\left(\sqrt{k}\right)$."

In a thorough discussion of constructible numbers and field extensions, the text leads to a proof that a real number is constructible if and only if the number lies in a chain of quadratic extensions emanating from $\mathbb{Q}$, the field of rationals. Finally, in a section entitled, "How to Fail in Doubling the Cube," a combination of text and exercises shows that: (1) given a cube of side length 1, a cube with twice the volume would have a side of length $\sqrt[3]{2}$; (2) $\sqrt[3]{2}$ is not rational; (3) if $\sqrt[3]{2}$ were constructible then $x^3 - 2$ would have at least two real roots; (4) however, $f(x) = x^3 - 2$ has been shown to be monotone increasing and, therefore, can have at most one real root. Conclusion: Given a segment of length 1, we cannot construct a segment of length $\sqrt[3]{2}$ using classic Euclidean tools.

The chapter closes by giving hints, suggestions, and references for students wishing to understand why the remaining classic constructions are also impossible.

Groups, Rings, and Homomorphisms

Although basic notions and definitions of group theory appeared in Course 2, this section of Booklet C opens by restating the definition of binary operation, while asking students to complete the definition of a group by filling in various blanks in an accompanying problem. Six examples of a group appear, illustrating the broad nature of the preceding course material: (1) real numbers under addition; (2) the units 1, 3, 5, 7 under multiplication modulo 8; (3) the symmetries of a square under composition; (4) the non-constant affine functions (having form $f(x) = ax + b$, $a \neq 0$) under composition; (5) the set of dilations centered at some fixed point under composition; and (6) the set of 2×2 lower triangular matrices under multiplication. The text encourages students to "make a list of all groups you know of or can make up. You might want to organize a class contest. If you stop with fewer than 50 on your list, you haven't been trying!"

A short review of fundamental group theorems, and a few hints for spotting non-groups, follows the question of "Why search for groups?" The suggested answer is that by verifying the existence of a group structure, one can immediately conclude many things about the system of interest, by making use of many known group theorems. Why reinvent the wheel?

How does one introduce the concept of a group isomorphism? Certainly not by launching the mathematical definition, which is mysterious on its face. By saying something like "isomorphic groups are groups that, except for the names of elements and operations, are identical"? Possibly. The SSMCIS writers of Booklet C chose what one might call a "pragmatic" approach. Students are reminded of how they probably verified that $\{1, 3, 5, 7\}$ formed a group under multiplication modulo 8 and then are confronted with a table for the group of symmetries of a rectangle and with a standard Klein four-group table with elements a, b, c, d. Would you, the text asks, need to go through the entire verification process for these new tables? Certainly not—a quick glance reveals that the tables are the same except for names, so if one of them represents a group, all of them represent groups. The text superimposes the tables, one on another, to illustrate the matching.

Next, four small group tables are provided (the complex numbers $\{1, i, -1, -i\}$, $\{0, 1, 2, 3\}$ under addition modulo 4, the rotations of a square, and $\{1, 2, 3, 4\}$ under multiplication modulo 5) and explored via text and exercises, leading to the question, "What is an acceptably precise mathematical formulation of the phrase 'same structure'?" The standard definition is given, followed by a list of five different group pairs with an invitation to check that $f(a \times b) = f(a) \circ f(b)$ for all appropriate elements a, b. Once again, the group pairs illustrate the breadth of previous course material: (1) $(\mathbb{Z}_2,+)$ and $(\mathbb{Z}_3 - \{0\}, \cdot)$; (2) $(\mathbb{Z}_6,+)$ and the rotations of a hexagon; (3) the integers under addition and the even integers under addition; (4) $\left(\mathbb{Q}\left(\sqrt{2}\right) - \{0\}, \cdot\right)$ and the set of matrices $\begin{bmatrix} a & 2b \\ b & a \end{bmatrix}$ where a, b are rational and not both 0; and (5) the symmetries of an equilateral triangle and the permutations of three elements.

The following problems from the isomorphism section illustrate the level of understanding that the text strives for:

1. Show that if one of two isomorphic groups is commutative, so is the other.
2. Show that the inverse of an isomorphism is an isomorphism.
3. Show that "is isomorphic to" is an equivalence relation on the set of all groups.
4. Show that there are only two non-isomorphic groups of four elements. (Guided hints given for this problem.)
5. Formulate a satisfactory definition of isomorphic vector spaces and isomorphic rings. Give examples of each.

Working toward an example of a "valuable isomorphism," the text provides illustrations and discussions of three types of slide rules. First, a slide rule for addition: essentially two identical rulers placed side-by-side. Place the origin of one opposite the 2 of the other. Now locate the 3 on the former rule: voila! It is opposite the 5 on the second rule. Thus: $2 + 3 = 5$. A similar computation involving the use of an infrequently seen circular slide rule is used to compute $3 + 4$ in $(\mathbb{Z}_6,+)$. (You can create such a slide rule for yourself by employing two concentric disks marked with 0, 1, 2, 3, 4, 5.) The final illustration is of the classic slide rule being used to compute $2 \cdot 3$ using the fact that $\log 2 + \log 3 = \log (2 \cdot 3) = \log 6$. A subsequent problem asks students to investigate the isomorphic mapping $x \rightarrow \log x$ to explain the classic slide rule function. A type of circular slide rule (Fig. 6.4) is used to examine the isomorphism between the groups $(\mathbb{Z}_6, +)$ and $(\mathbb{Z}_6 - \{0\}, \cdot)$. Students are asked to use the slide rule to compute table entries showing the one-to-one correspondence between computations in the respective groups.

The following four pages (two of which are taken up with complete Cayley tables for the group of symmetries of the square, $(\mathbb{Z}_{12},+)$, $(\mathbb{Z}_{11} - \{0\}, \cdot)$, and the 12-element group of symmetries of the hexagon lead the student from the definition of a subgroup to the proof of Lagrange's theorem. This is a fast pace, indeed!

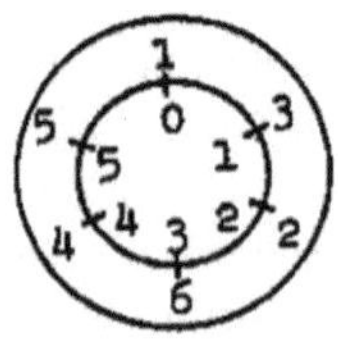

Fig. 6.4 Isomorphism illustration: circular slide rule

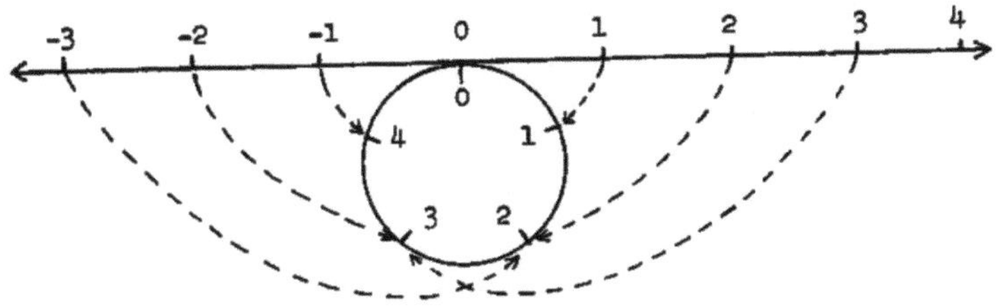

Fig. 6.5 Isomorphism illustration: circular slide rule

The text is minimal, but is accompanied by ten problems and two definitions (that of a "coset" and that of the "order" of a group). Another 17 problems and three definitions ("order" of an element, "cyclic" group, "generator" of a group) later, the text turns to what it calls "applications" of the material.

The notion of an isomorphism, if not the mathematical definition of one, seems easy for students to grasp intuitively, and many abstract algebra texts follow a discussion of isomorphism with a near-immediate definition of homomorphism, urging students to recognize the inherent similarities between the two types of maps. Despite the generally sophisticated nature of its presentations and the rapid pace of textual development, however, the SSMCIS booklet devotes several pages of intuitively clear many-to-one mappings before actually giving a definition of "homomorphism."

The opening mapping is accompanied by an illustration of a circle tangent to the real number line at the origin. The circle is to be interpreted as a "five digit clock" of the type used when calculating modulo 5. Equally spaced, the numerals 0, 1, 2, 3, and 4 are placed on the "clock," with 0 adjacent to the origin. Dashed arcs depict 1,2,3, ... mapping to 1,2,3, ... on the circle while -1, -2, -3, ... map to 4, 3, 2, ... (see Fig. 6.5). Although the diagram is a static one, the accompanying text says, "It is helpful to imagine a five clock rolling along the number line assigning images" [as in the diagram]. A series of problems allows students to see that the five sets of pre-images of the map act as a group themselves.

The introductory example is followed by a similar diagram depicting the real number line "wrapping" around a circle as a way of transferring the domain of the sine and cosine functions from the circle to the line. Here, for instance, one can see that if a point A on the circle corresponds to $\pi/2$, then it will also correspond to $-3\pi/2$, $5\pi/2$, etc. Students are ultimately given four instances of a group mapping onto another group, where the particular mapping acts like an isomorphism (in that operations are preserved), but is not one-to-one. This leads naturally to the definition of homomorphism. The definition is followed by exploring the notion of the kernel K of a homomorphism from a group G to a group H, closely followed by a problem which defines the "quotient group" G/K and then asks for a proof that the mapping $x \to xK$ is a homomorphism of G to G/K.

The penultimate section focuses on quotient rings and vector spaces. The notion of an "ideal" is introduced. Associated problem: "Show that if f is a homomorphism of ring S to ring T, the kernel of f is an ideal in S."

The final section of Booklet C returns to two classic algebra problems: (1) the equation $x^2 = 2$ has no solution in the field of rational numbers; and (2) the equation $x^2 = -1$ has no solution in the field of real numbers. With an eye toward developing "a general method for attacking these and similar problems," the text proceeds to consider the first problem, leaving the interested student to explore the second via the booklet's final problem: "Problem 82. In $R[x]$ the polynomial $x^2 + 1$ has no real zeros. Follow the procedure [developed in the preceding four pages] to produce a field in which [a generalized equation] has two solutions." The solution involves defining an operation on certain equivalence classes of polynomials, which become a quotient ring under appropriate definitions.

Conclusion

The creations of the SSMCIS can be considered as a type of existence proof. The project set out to create materials "free of any restrictions of traditional sequence or content," and, by and large, achieved its goals. For its time, the content was quite radical. The field trials of Courses 1–6 also showed that it was possible to use such materials with students. The few studies that examined the performance of SSMCIS-taught students concluded that they performed as well as or better than their peers (e.g., Jones, Rowen, & Taylor, 1977; Kansky, Hiatt, & Odom, 1974; Williams & Jones, 1973). For instance, Williams and Jones (1973) found superior performance for the SSMCIS students on the PSAT, even after matching for background and performance on the PSAT verbal section.

We have focused in this chapter on the algebra within SSMCIS, where the significant use of groups, fields, and vectors spaces in precollege curriculum was groundbreaking. But it is also worth noting that even with somewhat more traditional content, like the beginnings of calculus, the SSMCIS approach was not traditional. (Continuity was taught prior to the concept of a mathematical limit, for example.) Although SSMCIS was not intended to be adopted broadly, its curricular approach was instrumental in leading New York State to adopt a three-year unified approach to teaching mathematics in secondary schools in the late 1970s and the 1980s. Probability and statistics, appearing throughout all six SSMCIS courses, was emphasized for the first time in the New York curriculum.

In reading the materials themselves, some 45 years after their creation, one is struck by a sense of both exhilaration and adventure in the exposition. After all, if one wishes to be "free of any restrictions" in most areas of human endeavor, one might be forgiven for feeling a bit elated. Chapter headings like "Isomorphism or Haven't I Met You Somewhere Before" or "How to Fail at Doubling the Cube" illustrate the sense of play the written materials can reflect. Problems, too, can be both non-routine and fun:

1. What type of projection forms the basis of the element of horror in Edgar Allen Poe's short story *The Sphinx*? (Booklet E: *Geometric Mappings and Transformations*).
2. A farmer has determined that in one of his fields, cow manure appears to be distributed according to a two-dimensional Poisson process with intensity $\lambda = 0.01$ per square foot. The area of the bottom of one of the farmer's boots is 36 square inches. If the farmer takes 100 steps to cross the field and does it without watching the ground, what is the probability that he does not step in any manure? (Booklet E: *Geometric Mappings and Transformations*).

It is clear that a certain amount of audacity and creativity went into exploring the "realm of the possible" through the SSMCIS publications. The era in which it functioned may well have been too ambitious and too naive in its curricular goals, but the SSMCIS, at least, still offers lessons for those creative souls among us today.

References

Educational Services, Inc. (1963). *Goals for school mathematics: The report of the conference on school mathematics*. Watertown, MA: Author (ERIC Document No. ED 015 140).

Fehr, H. F. (1961). *New thinking in school mathematics*. Paris: Organisation for European Economic Co-operation, Office for Scientific and Technical Personnel.

Fraleigh, J. B. (1967). *A first course in abstract algebra*. London: Addison-Wesley.

Herstein, I. N. (1964). *Topics in algebra*. New York, NY: Blaisdell Publishing.

Jones, C. O., Rowen, M. R., & Taylor, H. E. (1977). An overview of the mathematics achievement tests offered in the admissions testing program of the college entrance examination board. *The Mathematics Teacher, 70*(3), 197–208.

Kansky, R. J., Hiatt, A. A., & Odom, M. M. (1974). New programs: Accelerated mathematics instruction. *The Mathematics Teacher, 67*(8), 729.

Secondary School Mathematics Curriculum Improvement Study (SSMCIS). (1967a). *Unified modern mathematics: Course I* (Revised ed.). New York, NY: Teachers College, Columbia University.

Secondary School Mathematics Curriculum Improvement Study (SSMCIS). (1967b). *Unified modern mathematics: Course II* (Revised ed.). New York, NY: Teachers College, Columbia University.

Secondary School Mathematics Curriculum Improvement Study (SSMCIS). (1970). *Unified modern mathematics: Course III* (Revised ed.). New York, NY: Teachers College, Columbia University.

Secondary School Mathematics Curriculum Improvement Study (SSMCIS). (1971). *Unified modern mathematics: Course IV* (Revised ed.). New York, NY: Teachers College, Columbia University.

Secondary School Mathematics Curriculum Improvement Study (SSMCIS). (1972a). *Unified modern mathematics: Course V* (Revised ed.). New York, NY: Teachers College, Columbia University.

Secondary School Mathematics Curriculum Improvement Study (SSMCIS). (1972b). *Unified modern mathematics: Course VI* (Revised ed.). New York, NY: Teachers College, Columbia University.

Secondary School Mathematics Curriculum Improvement Study (SSMCIS). (1972c). *Unified modern mathematics: Booklet C "algebraic structures, extensions, and homomorphisms"*. New York, NY: Teachers College, Columbia University.

Secondary School Mathematics Curriculum Improvement Study (SSMCIS). (1973). *Unified modern mathematics: Booklet E "geometric mappings and transformations"*. New York, NY: Teachers College, Columbia University.

van der Waerden, B. L. (1949). *Modern algebra*. New York, NY: Frederick Ungar Publishing.

Williams, S. I., & Jones, C. O. (1973). *Performance of SSMCIS-trained students on college board mathematics examinations. Test development report*. Princeton, NJ: Educational Testing Service (ERIC Document No. ED 081 847).

Chapter 7
Understanding School Mathematics in Terms of Linear Measure and Discrete Real Additive Groups

Hyman Bass

Introduction

I report here on a capstone mathematics course for secondary mathematics teachers,[1] developed experimentally over the past several years. In this course I have attempted, among other things, to incorporate ideas of abstract algebra that are situated in the fundamental mathematical structures of school mathematics, and to use them to illuminate those structures, and reveal often-unseen connections. For example, rather than develop general group theory, I focus on the structure of the additive and multiplicative groups of the basic rings of school mathematics, and use the group theoretic perspective to illuminate and connect many school topics. I contend that this exposure to "abstract algebra in context" contributes significantly to mathematical knowledge for teaching (Ball, Thames, & Phelps, 2008), perhaps an aspect of "horizon knowledge." But this might be less clear, or less accessible, with other important abstract algebra topics (Sylow Theorems, Galois Theory, etc.).

In addition to making curricular connections across mathematical topics, I have designed some novel problem-solving formats to prompt making connections. These connections have, primarily, been of one of two forms: (1) connections across mathematical domains, and (2) structural connections between apparently unrelated problems. Examples of these also are presented below.

[1]Though addressed to secondary teachers, much of the course content deepens understanding of mathematics in the early grades as well. The course is also suitable for regular mathematics majors.

H. Bass (✉)
Department of Mathematics and School of Education, University of Michigan, Ann Arbor, MI, USA
e-mail: hybass@umich.edu

N. H. Wasserman (ed.), *Connecting Abstract Algebra to Secondary Mathematics, for Secondary Mathematics Teachers*, Research in Mathematics Education,
https://doi.org/10.1007/978-3-319-99214-3_7

I begin below with a brief overview of abstract algebra and its origins. Following that, I outline the structure of my course. The content of the course represents my own mathematical perspective on school mathematics, aimed at conceptual coherence. It centrally features the real number line, with its geometric and arithmetic structures. Linear measurement, expressed through division with remainder, leads directly to place value and modular congruence. Abstract algebra enters through the study of discrete additive groups of real numbers, from which multiplicative arithmetic and commensurability (irrationality) naturally emerge. Brief treatments of polynomial algebra and combinatorics then culminate in discrete calculus, the natural generalization of the "pattern generalization" activities in school mathematics. Finally, I present and discuss some problem-solving designs, which I regard as a way to cultivate important mathematical practices in the course.

What Is Abstract Algebra, and Where Did It Come from?

"Algebra" signifies many things. First, it provides a compact and efficient (but sometimes opaque) symbolic notational system to compose mathematical expressions and relations involving abstract quantities. An early appearance of this is the (x, y) Cartesian coordinate system, which supported the algebraic formulation of many geometric ideas. The origins of algebra as a substantive mathematical domain came from efforts to solve polynomial equations in one variable, for example finding higher degree analogues of the quadratic formula (degree 2, "known" already by the ancient Babylonians and Egyptians). Italian mathematicians in the sixteenth century succeeded with cubic and quartic equations (degrees 3 and 4), but the quintic (degree 5) resisted. The deep proof (by Abel) that the general quintic equation could *not* be solved by radicals ushered in some of the most fundamental ideas of modern mathematics, group theory in particular, via algebraically permissible permutations of the roots of a polynomial. This strand culminates today in Galois Theory, the capstone of some abstract algebra courses.

"Modern" or "abstract" algebra, with its emphasis on general structures (groups, rings, and fields), axiomatically defined, took hold in the twentieth century, starting with the lectures of Emmy Noether and Emil Artin, and captured in van der Waerden's classic text, *Moderne Algebra* (van der Waerden, 1930, 1931). These structures arise from both number theory (studying generalizations of the arithmetic of integers and the rational numbers) and from efforts to understand solutions of polynomial equations (with real or complex coefficients)—eventually, several equations of arbitrary degree in several variables. One linear (i.e., degree 1) equation in one variable reduces to elementary arithmetic with the four operations (typically learned in grades 1–6); the solution of $ax + b = 0$ is $x = -b/a$. One equation in one variable of *higher degree* is a big jump in sophistication and culminates with Galois Theory (undergraduate math majors). Many *linear* equations in several variables are the core of linear algebra, for which there is a fairly complete and accessible algebraic theory (undergraduate STEM students). This is fundamental

Table 7.1 The "geography" of algebraic equations

Domain	Degree	Variables	Learners
Elementary arithmetic	1	1	Grades 1–6
Linear algebra	1	>1	STEM majors
Field and Galois theory	>1	1	Math majors
Algebraic geometry	>1	>1	Research mathematicians

since, whereas the world is mostly nonlinear; many problems are first approached by linear approximation, calculus being a preeminent example of this. The vast general case (many equations, many variables, high degree) is algebraic geometry, one of the most advanced areas of contemporary mathematics (research mathematicians) (see Table 7.1).

The power of axiomatic methods became clear in the twentieth century: mathematicians were able to solve many longstanding problems[2] that had earlier seemed intractable. As such, abstract algebra gained a prominent place in the professional training of mathematicians. This standing gradually trickled down into the curriculum of general mathematics majors, thereby affecting secondary mathematics teachers, who are required to be mathematics majors. The "New Math" reforms of the 1950s and 1960s even witnessed (eventually aborted) efforts to have abstract approaches infiltrate the early school grades.

Abstract Algebra for Secondary Teaching

The power of mathematical abstraction is its generality, thus having the potential to conceptually unify many apparently distinct mathematical contexts. But abstraction also exacts a cost for the novice learner since, by its nature, abstraction purges the concepts of the concrete individual contexts that it generalizes, and these are a main source of sense making.

Thus, one may reasonably ask, "What is the abstract algebra easily accessible to secondary teachers that deeply relates to their work as teachers?" This question is hard to manage since teachers' mathematics content courses are generally taught in mathematics departments, and it makes sense for these departments to require their majors to learn abstract algebra. At the same time, the work of teaching requires a depth and flexibility of mathematical understanding of the fundamental structures of the school curriculum (place value, the number line, algebraic equations, and

[2]For example, Hilbert's proof, using Noetherian conditions, that the ring of invariant polynomials of a reductive group action is finitely generated, thus eclipsing years of laborious, inconclusive computation.

functions) that is often not fully achieved in typical academic mathematics courses, and is even lacking among many mathematics majors.[3]

While the main ideas and constructs of abstract algebra are arguably relevant to school mathematics, their typical academic presentation contains much that is remote from the needs of K-12 teachers, in two respects. First, its generality and basic examples quickly exceed the context of school mathematics. This is a pragmatic consideration since teachers' exposure to mathematics is limited in time. Second, the rich algebraic structures already present in school mathematics (for example the algebraic (and geometric) structure of additive and multiplicative groups of real numbers and of modular rings) are often not treated, in abstract algebra courses, in sufficient detail to illuminate and add depth and coherence to related school mathematics topics.

Consider the case of group theory. In an abstract algebra course, among the first examples after cyclic groups are the symmetric groups. While they are of major mathematical importance, they are not familiar objects to school students, and computations in them are difficult and notationally complex. Non-commutative groups first arise most naturally as groups of geometric transformations (symmetries), a rich and beautiful development. While geometric transformations appear in high school curricula, they are not generally treated from a group theoretic perspective.

The Conceptual Foundations of My Course

I take the real number line, with its combined arithmetic and geometric structures, to be a central object of in-depth study throughout. In this regard, I am influenced by the ideas of Davydov (1990), who emphasized the centrality of quantities and their measurement in the teaching and learning of mathematics, starting in the earliest grades. In particular, I feature *division with remainder* as the far-reaching *conceptual foundation of linear measurement*, not as a computational task in whole number arithmetic. It leads directly to both the detailed development of place value expansion (in any base) of a positive real number, and to modular congruence.

In the treatment of the four arithmetic operations, school mathematics emphasizes algorithmic computation in the place value, or fraction, notational systems. While concrete computation is vitally important, what is relatively neglected in school mathematics is the *conceptual geometric meaning of the operations on the number line*. For example, addition and subtraction can be interpreted in terms of translations, and multiplication in terms of 0-centered dilation and reflection. These understandings are at once intuitively accessible, *pre-computational*, and they are the foundation for proving some striking and consequential theorems about

[3]I say this from having taught the course several times to successful mathematics majors, who consistently note that they are learning things about place value, and about the number line, that they had never known before.

real additive groups, starting with division with remainder itself. Interestingly, this treatment leads naturally to the notion of commensurability, providing a deepened understanding of the distinction between rational and irrational numbers, as well as group theoretic definitions of gcd and lcm, thus making their basic properties all easily derivable, prior to prime factorization theory.

Group theory enters through the study of the additive and multiplicative groups of the basic rings of school mathematics—integers, rationals, reals, and, eventually, complex numbers, and modular rings. Many of the early theorems in arithmetic (Fermat's Little Theorem, Cauchy's Theorem, multiplicativity of the phi-function, quadratic residues, and even gcd and lcm) are essentially group theoretic in nature. Division with remainder is used to give a complete description of discrete additive groups of real numbers. The multiplicative group $\mathbb{R}^*$ (with $\mathbb{R}^* = \mathbb{R}\{0\}$) of nonzero real numbers exhibits torsion, $\{\pm 1\}$, and a direct product structure, $\mathbb{R}^* = \{\pm 1\} \times (0, \infty)$, with corresponding *homomorphisms*, *sign*, and *absolute value*. Moreover we have an order preserving group *isomorphism*, $\exp : \mathbb{R} \to (0, \infty)$, from the additive to the positive multiplicative group, with inverse log. The latter permits us to transport additive group theorems to the multiplicative group, where they would be much more difficult to discover and prove directly. So these real additive and multiplicative groups provide a direct and organic connection of significant school mathematics to some substantial, yet "familiar," abstract algebra—even prior to the axiomatic definition of a group. And this arises even before number theoretic ideas, which are captured in the study of additive and multiplicative groups of modular rings. The eventual study of additive and multiplicative groups of complex numbers leads to two-dimensional vector algebra and to two-dimensional geometric transformations.

The course also includes some fundamental "classical algebra," for example a brief treatment of the basic properties of polynomials (root theorems, Binomial Theorem, interpolation, etc.), coupled naturally with an introduction to combinatorics. The proof of the Binomial Theorem is derived from a more basic algebraic identity ("A product of sums is a sum of products" or "distributivity on steroids") that yields also a proof of the inclusion–exclusion formula, even in geometric measure settings. Combining the root theorems with group theory, we show that any finite multiplicative group of a field (e.g., the multiplicative group of a finite field) is cyclic.[4]

Many of these ideas converge in the chapter on Discrete Calculus. Many secondary curricula now introduce functions largely as an activity of "finding a pattern" in some finite sequence of numbers (or figures), on the basis of determining the next term, and, eventually, the general term. The data presented are equivalent to a table pairing the term number in the sequence with the corresponding value, or

[4] This is derived from the following theorem about a finite group G of order n. For each whole number d, let $G_d = \{x \in G | x^d = 1\}$. If: $(*)$ [For each divisor d of n, $\#G_d \leq d$], then G is cyclic. If G is in a field, then $(*)$ follows from the root theorem for $X^d - 1$.The theorem is proved by close comparison of G with $\mathbb{Z}/\mathbb{Z}n$.

equivalently, a finite list of points in the plane. An expression for the general term then simply defines a function whose graph passes through the given points. In other words, this is essentially an interpolation problem. The "pattern" comes into play in suggesting a simplest function solution, linear in the case of constant difference, exponential in the case of a constant ratio. If the differences are not constant, one can look at differences of differences, etc., thus reaching quadratic functions, for example.

The method of looking at successive differences is just the "differential" aspect of Discrete Calculus, a topic that is built on the basic algebraic foundations described above. It provides an elegant and accessible formal analogue of Calculus itself, and it leads quickly to a number of interesting results and connections to school mathematics—e.g., the Binomial Theorem, the formulas for sums of powers of consecutive natural numbers, etc. Above all, Discrete Calculus provides, in my view, a much more comprehensive and powerful framework for initial understanding of functions. It is perhaps a good alternative to high school calculus, one that would strengthen the often-underdeveloped algebraic skills of high school graduates, even those having taken AP Calculus.

A consistent aim of my course, as I hope can be discerned in the above overview, is conceptual coherence, building on *connections* between mathematical ideas, something I feel is often lost in the fragmentation and shallow treatments of many curricular materials.

Detailed Description of the Course

The outline of a *book project* (Chaps. 1–12) based on the course is presented in Table 7.2 below. The book's contents are more comprehensive than those of the (one term) course, and would be more suitable for a full-year course. Though oriented toward secondary teachers, the course also could be of interest to more general mathematics majors (and would not substantially duplicate their typical advanced undergraduate mathematics coursework). At the same time much of the course content is highly relevant to the elementary math curriculum, and so it could be of value to mathematically well-motivated and prepared elementary teachers. For the present, necessarily brief, discussion of the course, I will simply highlight some of the novel ideas and approaches that each of the topics (i.e., chapters) of the book offers.

Division with Remainder (DwR) and Place Value

Dividing a by b ($b \neq 0$), $a = qb + r$, yields an *integer* quotient q, and a remainder r *smaller than* b. I view the problem of dividing a by b not as a piece of integer arithmetic, but rather as the *foundation of linear measurement*, where a and b are

Table 7.2 Outline of a book developed from the capstone course

Chapter	Topics
1	**Division with Remainder and Place Value** Division with Remainder (*DwR*): Given $a, b \in \mathbb{R}, b \neq 0, \exists\, !\, q \in \mathbb{Z}, r \in \mathbb{R}, 0 \leq r < \lvert b \rvert$ such that $a = qb + r$ $q = q_b(a), r = r_b(a)$. The case $b = 1$: $[a] = q_1(a), a = r_1(a)$ Place value: $b \in \mathbb{Z}, b \geq 2, a \in \mathbb{R}$: $a = \sum_h d_h(a)b^h$, with $d_h(a) = r_b([ab^{-h}])$ Order of magnitude
2	**Modular Congruence** Given $a, b, m \in \mathbb{R} : a =_m b$ means $a - b \in \mathbb{Z}m \iff r_m(a) = r_m(b)$ Divisibility tests; "Fast track to the remainder" Base-b representation of a fraction, $N/D : d_{-h}(N/D) = q_D(r_D(Nb^{h-1})b)$; eventually periodic; Wait time t; period p; Write $D = D_0D_1$, so that D_0 divides b^e for some e, $gcd(D_1, b) = 1$. Then $t =$ least e such that $b^e =_{D_0} 0$, and $p =$ least e such that $b^e =_{D_1} 1$
3	**Rules of Arithmetic: Commutative Rings** Quick axiomatic definitions of (semi-)groups, rings, commutativity. Unit groups, zero divisors, etc. Examples, including modular rings and polynomials. Some geometric series polynomial identities. It is shown that for positive relatively prime integers a, b, the largest integer <u>not</u> in the semi-group $\mathbb{N}a + \mathbb{N}b$ is $ab - a - b$.
4	**Geometry of the Number Line** Metric spaces (X, d). Ball and spheres. Isolated points. Discrete sets. Closure; density. Detailed study of (often arithmetic) examples on the number line
5	**(Discrete) Additive Groups of Real Numbers** Let A be an additive group of real numbers. Theorems: I. If 0 is isolated in A then A is uniformly discrete; II. A is either discrete or dense in $\mathbb{R}$; III. If A is discrete then $A = \mathbb{Z}a$ for a unique $a \geq 0$
6	**Commensurability: gcd and lcm** Theorem IV. $\mathbb{Z}a + \mathbb{Z}b$ is discrete iff a and b are commensurable. In this case, $d = \gcd(a, b)$ and $m = \operatorname{lcm}(a, b)$ are defined to be the nonnegative generators of $\mathbb{Z}a + \mathbb{Z}b$ and $\mathbb{Z}a \cap \mathbb{Z}b$, respectively. Chinese Remainder Theorem
7	**Primes and Factorization** Definition; infinitely many. Prime (Power) Factorization: $a = \pm \Pi p^{e(p, a)}$. Mersenne, perfect, and Fermat numbers. Appendices: Multiplicative groups of modular rings; applications to cryptography
8	**Combinatorics** Brief introduction. "n-choose-d," ${}_nC_d$; Pascal's Triangle
9	**Polynomials** Degree, leading term, roots. $f(a) = 0$ iff $(x - a)$ divides f. f has $\leq \deg(f)$ roots Interpolation. Cor. $f(\mathbb{Q}) \subseteq \mathbb{Q}$ iff all coefficients of f are in $\mathbb{Q}$. Binomial Theorem. A finite multiplicative group in a field is cyclic. Appendix: Inclusion–Exclusion Formula
10	**Discrete Calculus** Sequences as (the ring $\mathcal{F}$ of) functions $f : \mathbb{N} \to \mathbb{R}$. The discrete derivative Δ and integral S, as linear operators $\Delta, S{:}\mathcal{F} \to \mathcal{F}$. Fundamental Theorem: $\Delta Sf = f$ and $S\Delta f = f - f(0)$. Binomial polynomials $B_d(x) = {}_xC_d$. Polynomials $f = \sum a_dB_d$. $f(\mathbb{Z}) \subseteq \mathbb{Z}$ iff all $a_d \in \mathbb{Z}$. Formulas for sums of consecutive dth powers

(continued)

Table 7.2 (continued)

Chapter	Topics
11	**Complex Numbers** Beyond the course
12	**Plane Transformation Geometry Using Complex Numbers** Beyond the course
	Appendix: Some Problem-Solving Activity Designs
	1. Finding a structure common to a diverse set of problems 2. Showing that a diverse set of problems are all modeled by the same equation

real numbers, with b the unit of measure. Note that q (but not r in general) is still an integer. Thus, Division with Remainder (*DwR*) is a theorem about linear measure, and whose meaning is easily pictured on the number line, partitioned by the integer multiples of b, between two of which a lies. The uniqueness of *DwR* says that $q = q_b(a)$ and $r = r_b(a)$ are multivariate functions of (a, b). Here, dividing by 1 is interesting! $q_1(a) = [a]$ is the "integer part of a," and $r_1(a) = \langle a \rangle$ is the "fractional part of a." If a is a whole number, written in base-10, then $r_{10}(a)$ is the unit (or ones) digit of a. This observation leads to the following explicit formula for the coefficient d_h of b^h in the base-b expansion (with b an integer greater than 1) of a real number $a \geq 0$: $d_h = r_b([ab^{-h}])$. We also define the order of magnitude of an integer N as the number of significant digits of N, and show that this is sub-multiplicative. I pose questions showing that bases $b \neq 10$ commonly and naturally occur, for example: (1) What is a base-1000 representation of 48,279,506,371,817? Answer: You're staring at one! In fact, we recite the number in base-1000! (2) Calculating a^{89} from the definition requires 88 multiplications. How many multiplications are required using iterated squaring? This depends on the base-2 expansion of 89, and needs only nine multiplications. (3) The best time ever in the NY marathon is: 2:05:06 = 2 h, 5 min, 6 s. Maria ran her first NY marathon in 3:05:02. How much longer than the record was her time? This is a problem of 3-digit subtraction in base 60.

Modular Congruence

"$a =_m b$" is my notation for what is usually written in the more cumbersome form, "$a \equiv b \pmod{m}$."[5] Importantly, I allow a, b, m to be *real numbers*, and take the congruence to mean that $a - b$ is an *integer multiple* of m; equivalently, $r_m(a) = r_m(b)$. (This is the equivalence relation corresponding to the additive group $\mathbb{R}/\mathbb{Z}m$.) It is shown that modular congruence preserves sums, but it preserves products in general only when a, b, m, etc. are integers. This chapter also includes a systematic discussion of so called "divisibility tests," such as "casting out nines"

[5]After 6 years of use, this notation has not encountered any mathematical difficulties or conflicts.

(N is divisible by 9 iff the sum of its (base-10) digits is so). Instead, for a given divisor D, we show more generally how one can construct a function $S(N)$ with two properties: (1) $S(N)$ is, in general, much smaller than N; and (2) $S(N) =_D N$. It follows that iterating S gives "a fast track to $r_D(N)$."

Base-b Expansion of a Proper Reduced Fraction $N/D = \sum_{h>0} d_{-h} b^{-h}$

It is shown that $d_{-h} = q_D(r_D(Nb^{h-1})b)$. Note that this involves quotients and remainders only of division by the denominator D (not N or b). This formula essentially encodes base-b long division of N by D. From this, one shows that for some "wait time t," and "period p," we have $d_{-h} = d_{-k}$ whenever $t < h < k$ and $h =_p k$. Write $D = D_0 D_1$ so that D_0 divides some power of b, and $\gcd(D_1, b) = 1$. Then t is the least e such that $b^e =_{D_0} 0$, and p is the least e such that $b^e =_{D_1} 1$.

Rules of Arithmetic: Commutative Rings

Commutative rings are introduced at this point, presented as formalizing the "Rules of Arithmetic," accompanied by numerous examples. Starting with one binary operation, I define semi-groups and groups, particularly the group of units of a semi-group. Among the interesting (additive) examples are: the set of fractions with square-free denominator; and the set of periods of a function of a real variable. We also derive the "Frobenius number," $ab - a - b$, of the additive semi-group generated by relatively prime integers $a, b > 0$. Next come rings, with the distributive law linking multiplication and addition. In this general context we prove familiar identities, like $0 \bullet a = 0$, $-(-a) = a$, $(-a) \bullet (-b) = a \bullet b$, etc. Then we construct the modular rings $\mathbb{Z}/\mathbb{Z}m$, and polynomial rings $A[x]$.

Geometry of the Number Line

The geometry of the number line is treated in the context of general *metric spaces*, a topic initially independent of the preceding material (except for the examples used), but then merged with the algebra in the next chapter. Though metric spaces may seem excessively general, I find this context both mathematically and pedagogically advantageous. Mathematically, it is natural and useful to understand the line as a (one-dimensional) case of Euclidean geometry. Pedagogically, geometric ideas (like the "neighborhood" of a point) are intuitively easier to process in dimensions greater than 1. Also, the geometry of $\mathbb{R}$ is defined by its order structure, but this feature, unique to dimension 1, does not convey a general geometric intuition. This is a territory notorious for definitions with multiple quantifiers, which are often difficult for students to manage. To mitigate this, I have found it helpful to use the concept of a point being *isolated* in a metric space (having "room of its own," or having "its private space"). For example, a point $x \notin Y$ is in the *closure* of Y (in a metric space X) if x is not isolated in $Y \cup \{x\}$. The main concepts we need for our applications are *discreteness* of a metric space (all points are isolated), and *density* of one metric space Y in another space X (X is the closure of Y). Here is a sample problem: If a sequence of fractions converges to an irrational number, show that their denominators go to infinity.

(Discrete) Additive Groups of Real Numbers

The (discrete) additive groups of real numbers are a central, but, at this point, deceptively easy part of the course. We prove the following three important theorems about any additive group A of real numbers, as relatively simple applications of *DwR* (division with remainder). (I) If 0 is isolated in A, then A is uniformly discrete. (II) A is either discrete or dense in $\mathbb{R}$. (III) If A is discrete, then $A = \mathbb{Z}a$ for a unique $a \geq 0$.

Commensurability: gcd and lcm

Real numbers a and b are commensurable if $b = 0$ or if a/b is rational. Also using *DwR*, we prove the following theorem: (IV) (Commensurability Theorem) $\mathbb{Z}a + \mathbb{Z}b$ is discrete iff a and b are commensurable. In this case, $d = \gcd(a, b)$ and $m = \operatorname{lcm}(a, b)$ are defined to be the nonnegative generators of $\mathbb{Z}a + \mathbb{Z}b$ and $\mathbb{Z}a \cap \mathbb{Z}b$, respectively. From this, one can directly develop extensive properties of gcd and lcm, all prior to prime factorization (see Appendix 2).

Primes and Prime Factorization

Here I emphasize the uniqueness of the factorization, which gives rise to the functions $e_p(a)$, which is the exponent of p in the *prime power factorization* of an integer $a \neq 0$. These functions extend to rational numbers $a \neq 0$, and define homomorphisms from the multiplicative group of $\mathbb{Q}$ to the additive group of $\mathbb{Z}$. This easily gives the traditional characterizations of gcd and lcm, the divisor function, and the Euler phi function, as well as the criteria for a dth root of a rational number to be rational. There is also discussion of Mersenne, perfect, and Fermat numbers.

Combinatorics

The formula ${}_nC_d = n(n-1)\ldots(n-d+1)/d!$, for the number of d-sets in an n-set, is derived by first showing that the numerator counts the number of *ordered* d-sets, while the denominator counts the number of orderings of a d-set. I emphasize that it is worth noting the non-obvious fact that this fraction is an integer. In the next class I ask the class to show that the product of d consecutive integers is always divisible by $d!$, to find out if they see the connection between this question and the formula. We also verify several identities that lead to Pascal's Triangle.

Polynomials

After the basics on the degree and leading term of polynomials, the "root theorems" are proved; in particular, f has at most deg(f) roots if the coefficient ring is an integral domain. This, combined with some group theory and combinatorics, is used to show that *a finite multiplicative group in a field is cyclic*. We then demonstrate *interpolation*: Given a_j, b_i $(1 \leq i \leq n)$, with the a_j twice distinct, there is a unique f of degree less than n such that $f(a_i) = b_i$, $(1 \leq i \leq n)$. It follows that if, for an infinite field $Q, f(Q) \subseteq Q$, then the coefficients of f belong to Q. The analogue for $\mathbb{Z}$ in place of a field is false, but this situation is precisely analyzed later using Discrete Calculus. We also prove the Binomial Theorem using combinatorial arguments. Similar arguments give a proof of the Inclusion–Exclusion Formula.

Discrete Calculus

The section on "Discrete Calculus" vastly generalizes the part of the high school curriculum focused on "finding patterns" in a partially given sequence. I view the set of (infinite) sequences as the ring $\mathcal{F}$ of functions $f : \mathbb{N} \rightarrow \mathbb{R}$. The discrete derivative Δ (where $\Delta f(x) = f(x+1) - f(x)$) and the discrete integral S (where $S(x) = f(0) + f(1) + \ \ldots \ + f(x-1)$) are linear operators (i.e., Δ, $S : \mathcal{F} \rightarrow \mathcal{F}$). The Fundamental Theorem of Discrete Calculus is then a pair of simple algebraic identities: $\Delta Sf = f$, and $S\Delta f = f - f(0)$. Then I introduce the Binomial Polynomials $B_d(x)$= "x-choose-d." These are perfectly suited for discrete calculus, because: $\Delta B_d = B_{d-1}$ (Pascal's Relation), and $SB_d = B_{d+1}$. Any polynomial f can be uniquely written as $f = \sum a_d B_d$, and it is shown that $f(\mathbb{Z}) \subseteq \mathbb{Z}$ iff all $a_d \in \mathbb{Z}$. By expressing x^n as a linear combination of $B_d (0 \leq d \leq n)$, and integrating, we obtain formulas for the sums of consecutive nth powers.

Problem-Solving Activities that Emphasize Connections

Complementary to the topical connections developed in the construction of the course curriculum, I have tried to design instructional problem-solving activities that explicitly involve seeing and using (sometimes subtle) mathematical connections. In this section, I discuss two different types of activities: (1) connections across mathematical domains, and (2) structural connections between apparently unrelated problems.

Connections Across Mathematical Domains

A simple example of connecting across mathematical domains would be the problem discussed above, about a base-1000 representation of 48,217,589,625,903. Another example asks for a description of all real functions $f(x)$ of a real variable such that $|f(x) - f(y)| = |x - y|$ for all x and y, to see if students recognize that this is asking for all isometries of the line.

For another example, one that spans several topics of the book, after deriving the combinatorial formula, $n(n - 1)(n - 2)\ldots(n - d + 1)/d!$, for ${}_nC_d$, in the next class I ask students to show that any product of d consecutive integers is divisible by d! The exercise is intended to prompt them to see this connection between combinatorics and arithmetic. Relatedly, the ${}_xC_d$ (note, x in place of n) function gives examples of polynomials (with non-integer coefficients) that have integer values at integers. In our later discussion of Discrete Calculus, we show that integer linear combinations of the ${}_xC_d$ function are, in fact, the only such polynomials. This contrasts with our earlier conclusion, from the interpolation formula, that a polynomial taking rational values at rational numbers must have rational coefficients. All of these examples provide an opportunity for students to identify and reflect on mathematical connections that span different areas of mathematics—discrete mathematics, arithmetic, algebra, etc.

Structural Connections Between Apparently Unrelated Problems

In a more structured design activity, I focus deliberately on the goal of "seeing and using mathematical structure" (one of the eight mathematical practices in the Common Core State Standards of the United States (CCSSM, 2010)). Here, I use some designs and examples also presented in Bass (2017). The general idea is to give the class sets of diverse problems for which the task is to seek, or to discover, unexpected connections between them. This work is done collaboratively in small groups, over a period of two or more classes.

Table 7.3 An elementary sorting task (from Bass, 2017)

1. What are all three-digit numbers that you can make using each of the digits 1, 2, 3, and using each digit only once?
2. Angel, Barbara, and Clara run a race. Assuming there is no tie, what are all possible outcomes of the race (first, second, third)?
3. You are watching Angel, Barbara, and Clara playing on a merry-go-round. As the merry-go-round spins, what are all the different ways that you see the three of them from left to right?
4. You want to choose from among Angel, Barbara, and Clara, a two-person rowing team, one for the bow, the other for the stern. What are all ways to do this?
5. In a 3×3 grid square, color three of the (unit) squares blue, in such a way that there is at most (or at least) one blue square in each row and in each column. What are all ways of doing this?
6. Find all of the symmetries of an equilateral triangle

In one format for these problems, *sorting*, I provide students with a list of diverse problems and ask them to group together problems which are "essentially" the same, and also indicate other significant but perhaps weaker connections, using a "connection network" representation. In another format, *common structure*, I provide a set of problems that, although quite diverse, all share a common mathematical structure. The students are asked not only to solve the problems, but also to identify and articulate the common structure, and to demonstrate how it is involved in each of the problems. The problems in these common structure sets often seem, on the surface, to be unrelated. This is a challenging and engaging activity that is best enacted over several class sessions. Students have reacted quite positively to these activities, noting specifically the elements of "mathematical surprise" involved.

Tables 7.3 and 7.4 present two examples. (Other examples can be found in Bass (2017).)

It is easy to see that problems 1 and 3 are "isomorphic." And problem 5 is likewise, though the connection is more subtle. Problems 2 and 6, though giving the same kind of answer, are fundamentally different from the others, and from each other. One way to see this is to replace the "3" in these problems by an $n > 3$. Problem 4 is an outlier, essentially unrelated to the other problems.

The next example is an expansion of a set of six problems (Ar1-4, R1, and G1) published by Usiskin (1968). The thirteen problems are grouped into four different subject areas. I formed my class into small groups, each one assigned to solve, and relate, the problems in a single one of the subject areas, and to prepare a class presentation of its work the following week. In the course of these collective presentations, the class discovers, to their great surprise, that in fact *all* of the problems are modeled by a simple variant of the same Diophantine equation, $1/n + 1/m = 1/2$ (Ar1). I have used this activity in professional development settings, with teachers and mathematics educators, with the same effect. As a follow up to this activity, it can be shown how the combinatorial classification of Platonic solids reduces, using Euler's formula, to finding all solutions (V, E, F

Table 7.4 The expanded Usiskin set

ARITHMETIC
Ar1 Find all ways to express 1/2 as the sum of two unit fractions (i.e., fractions of the form $1/n$, n a positive integer)
Ar2 Find all rectangles with integer side lengths whose area and perimeter are numerically equal
Ar3 The product of two integers is positive and twice their sum. What can these integers be?
Ar4 For which integers $n > 1$ does $(n - 2)$ divide $2n$?
RATES
R1 Which pairs of positive integers have harmonic mean equal to 4? [The harmonic mean h of n numbers $a_1, a_2, \ldots, a_n$ is defined so that $1/h$ is the average of the reciprocals, $\frac{1}{a_1}, \frac{1}{a_2} \cdots \frac{1}{a_n}$]
R2 Nina can paint a house in n days, and Maria can paint it in m days (n and m positive integers). Working together they can paint the house in 2 days. What are the possible values of n and m?
R3 A turtle travels up a hill at n miles per hour, and returns down the hill at m miles per hour ($n \le m$, n and m positive integers). The average speed for the round trip is 4 miles per hour. What are the possible values of (n, m)?
GEOMETRY
G1 Given a point P in the plane, find all n such that a small circular disk centered at P can by covered by nonoverlapping congruent tiles shaped like regular n-gons that have P as a common vertex.
G2 Two vertical poles, N and M, have heights n meters and m meters, respectively, with n and m being integers. A wire is stretched from the top of pole N to the base of pole M, and another wire is stretched from the top of pole M to the base of pole N. These wires cross at a point 2 m above the ground. What are the possible values of (n, m)?
G3 The base b and corresponding height h of a triangle are integers. A square is inscribed in the triangle with one side on the given base. Suppose that the side length of the square is 2. What are the possible values of (b, h)?
ALGEBRA
Al1 For which numbers s does $p(x) = x^2 - sx + 2s$ have positive integer roots?
Al2 Let u be a positive real number. Find all solutions (n, m, v) with n and m positive integers, and $v > 0$, of the equations: $(uv)^2 = u^n = v^m$
A13 Let (r, b) be positive integers. In a bin containing $r \bullet b$ balls, r of them are red and b of them are blue. For which (r, b) is there a 50–50 chance that a randomly chosen ball will be either red or blue?

being integers greater than or equal to 3) of the (related) Diophantine equation: $1/V + 1/F = 1/2 + 1/E$. Analysis of these problems can be found in Appendix 1.

Conclusion

The course I have described aims to achieve coherence by emphasizing mathematical *connections*, in two ways: *curricular*, by making explicit some often undeveloped connections among different topics; and *cognitive*, through the design of novel cross-domain and common structure problem-solving activities. Abstract algebra fundamentally enables both aspects of this mathematical coherence, via the group theoretic study of the additive and multiplicative groups of the basic rings of school mathematics: integers, rationals, reals, complex numbers, and modular rings. The connections are most dramatic in the last chapters, which exhibit a confluence of ideas from combinatorics, number theory, algebra, and calculus.

The course provides a number of useful enlargements of the topics of typical school curricula that would likely be new for both school students and their teachers. But these could enrich and deepen their understanding of the basic structures of school mathematics, while still being mathematically accessible, and also challenging. I regard a pedagogy that features collaboration and group work, and emphasizes mathematical exposition, explanation, and justification, as an important component of this course. In addition to a course structure, various units could also be adapted and used in professional development settings. In sum, the course provides one example of how ideas from abstract algebra—and other areas of mathematics—can be developed from and connected to the mathematics of the school curriculum.

Appendix 1: Analysis of the Extended Usiskin Problem Set

Problems **Ar1**, **Ar2**, **Ar3**, **Ar4**, in order, lead directly to the following Diophantine equations ("Diophantine" because one seeks (positive) *integer* solutions):

$$1/n + 1/m = 1/2 \tag{7.1}$$

$$2\,(n + m) = nm \tag{7.2}$$

$$2n = m\,(n - 2) \tag{7.3}$$

$$\text{For which } n > 1 \text{ does } n - 2 \text{ divide } 2n \tag{7.4}$$

Version (7.4) is essentially a verbal expression of the Eq. (7.3). Moreover, it is not difficult to see how Eqs. (7.1–7.3) are algebraically equivalent. For example, multiply (7.1) by $2nm$ to get (7.2); and subtract $2m$ from (7.2) to get (7.3). Hence, solving any one of them provides solutions to the others.

My students generally preferred to use (7.3) to express m in terms of n:

$$m = 2n/(n - 2). \tag{7.5}$$

They then did numerical experiments to find those n for which $2n/(n-2)$ is an integer. (Some students even graphed m in (7.5) as a function of $n > 0$, and highlighted the integer points on the graph.) The solutions they found were:

$$(n, m) = (4, 4)\,, (3, 6)\,, \text{or}\ (6, 3)\,. \tag{7.6}$$

None of the students tried to work directly with (7.1), which is my preferred approach. Using the symmetric roles of m and n, we can assume that $n \leq m$. Then $n \geq 3$; otherwise $1/n \geq 1/2$. Also $n \leq 4$; otherwise $1/n + 1/m < 1/2$. Thus either $n = 3$ (and so $m = 6$) or $n = 4$ (and so $m = 4$).

Problem **R1** corresponds to the equation,

$$1/4 = (1/2)\,(1/n + 1/m)\,, \tag{7.7}$$

which is (7.1) multiplied by 1/2.

For Problem **R3**: If one travels distance d at speed v in time t, then: $d = vt$ and $t = d/v$. Now suppose that one travels distance d at speed v_1 in time t_1, and then returns at speed v_2 in time t_2. What is the average speed for the whole trip? It is

$$\begin{aligned} v_{ave} &= \text{(total distance) / (total time)} \\ &= 2d/\,(t_1 + t_2) \\ &= \frac{2d}{\frac{d}{v_1} + \frac{d}{v_1}} = \frac{2}{\frac{1}{v_1} + \frac{1}{v_2}} \end{aligned}$$

Thus,

$$\frac{1}{v_{\text{ave}}} = \frac{1}{2}\left(\frac{1}{v_1} + \frac{1}{v_2}\right).$$

In other words, v_{ave} is the harmonic mean of v_1 and v_2. In problem **R2**, d would be the work of painting the house, and n and m are the rates at which Nan and her Mom do that job. The rate of doing it together (analogous to average speed) is the harmonic mean of the two rates.

The **geometry problems** are less obviously related, but they too lead to the same Diophantine equations. In Problem **G1**, let $\alpha(n)$ denote the (equal) interior angle(s) of a regular n-gon: then it is known that $\alpha(n) = \frac{n-2}{n} \cdot 180^\circ$. For some number, say m, of these regular n-gons to fit together to cover the area around a point P, we would need: $m \cdot \frac{n-2}{n} \cdot 180^\circ = 360^\circ$, i.e.,

$$m\,(n-2) = 2n, \text{ as in } \textbf{Ar3}. \tag{7.8}$$

For **G2** (also framed as the “crossing ladders problem”), consider the diagram:

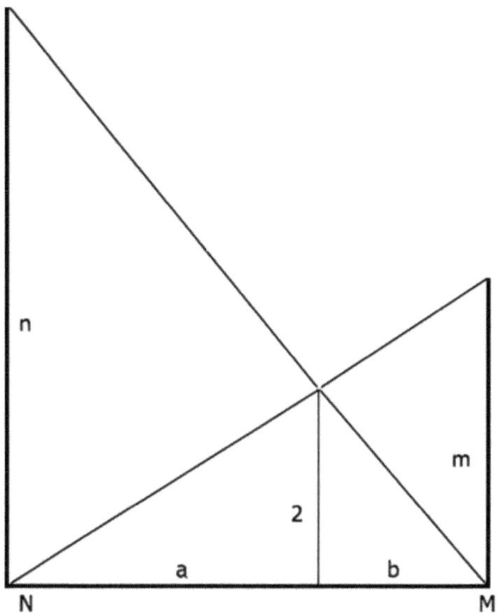

Using similar triangles we have:

$$(a + b) / n = a/2, \text{ and}$$

$$(a + b) / m = b/2$$

Adding these equations, and then dividing by $a + b$, gives

$$1/n + 1/m = 1/2, \text{ as in } \mathbf{Ar1}. \tag{7.9}$$

For **G3**, consider the diagram:

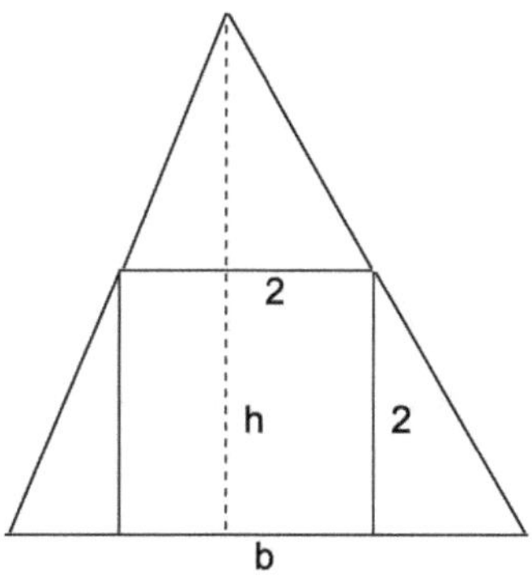

The big triangle and the one above the square are similar (corresponding sides are parallel), and so $h/b = (h - 2)/2$, whence, multiplying this by $2b$, the equation

$$2h = b\,(h - 2)\,,\ \text{as in } \mathbf{Ar3}. \tag{7.10}$$

In **Al1**, if we formally factor p:
$p(x) = x^2 - sx + 2s = (x - n)(x - m)\ldots(n, m$ being integers)
we find that

$$n + m = s,\ \text{and}$$

$$nm = 2s$$

whence n and m are positive, since s is, and so we have the equation

$$nm = 2\,(n + m)\,,\ \text{as in } \mathbf{Ar2}. \tag{7.11}$$

Then s $(=n + m) = 8$ $(=4 + 4)$ or 9 $(=3 + 6)$.
In **Al2**, the mathematics is mainly happening in the exponents: $(uv)^2 = u^n = v^m$
We first get, from $(uv)^2 = v^m$, that $u^2 = v^{m-2}$, so

$$v = u^{2/(m-2)}.$$

Then, substituting for v in $(uv)^2 = u^n$ gives: $(u \bullet u^{2/(m-2)}) = u^n$
Equating exponents then gives:

$$n = 2\,[1 + 2/\,(m - 2)] = 2m/\,(m–2)\,,$$

whence, again, equation

$$2m = n\,(m - 2)\,. \tag{7.12}$$

In **Al3**, the conditions on (r, b) are that, $r + b = rb/2$. Dividing this by rb gives,

$$1/b + 1/r = 1/2,\ \text{as in } \mathbf{Ar1}. \tag{7.13}$$

Appendix 2: Group Theoretic Derivation of Properties of gcd and lcm

Note that all of what follows precedes, and does not depend on, prime factorization.
(**DM0**): Let a and b be real numbers. Then we have proved that:

$\mathbb{Z}a + \mathbb{Z}b$ is discrete $\iff$ a and b are commensurable

In what follows we shall assume that a and b are commensurable, unless the contrary is indicated. In this case we define

$d =$ gcd $(a, b) \geq 0$ and $m =$ lcm $(a, b) \geq 0$

by

$\mathbb{Z}a + \mathbb{Z}b = \mathbb{Z}d$ and $\mathbb{Z}a \cap \mathbb{Z}b = \mathbb{Z}m$

Note that a and b are integers if and only if d is an integer. When $d = 1$, we say that a and b are "relatively prime." We say that a fraction a/b is "reduced" if gcd$(a, b) = 1$.

To ease writing we shall here abbreviate:

$(a, b) =$ gcd (a, b) and $[a, b] =$ lcm (a, b)

We now record some basic properties. We omit proofs if they follow easily from the definitions. We have bolded the items that are especially important and/or useful.

(**DM1**): There exist integers r, s such that $d(=(a, b)) = ra + sb.$

(**DM2**): (d) $d \mid a$ and $d \mid b$. If $d' \mid a$ and $d' \mid b$, then $d' \mid d$

"d is the greatest common divisor of a and b"

(m) $a \mid m$ and $b \mid m$. If $a \mid m'$ and $b \mid m'$, then $m \mid m'$

"m is the least common multiple of a and b"

(**DM3**): Suppose that $a \mid A$ and $b \mid B$. Then $(a, b) \mid (A, B)$ and $[a, b] \mid [A, B]$.

Put $d = (a, b)$ and $D = (A, B)$. Then $d \mid a$ and $a \mid A$, so $d \mid A$. Similarly $d \mid B$, and so $d \mid D$, by (DM3(a)). In similar fashion one shows that $[a, b] \mid [A, B]$.

(**DM4**): $(a, b) = (b, a) = (|a|, |b|)$, and

$[a, b] = [b, a] = [|a|, |b|]$ "**Absolute Symmetry**"

(**DM5**): $(a, 0) = |a|$ and $[a, 0] = 0$.

(**DM6**): For any real number c, $(\boldsymbol{ac}, \boldsymbol{bc}) = (\boldsymbol{a}, \boldsymbol{b}) \cdot |\boldsymbol{c}|$, and

$[\boldsymbol{ac}, \boldsymbol{bc}] = [\boldsymbol{a}, \boldsymbol{b}] \cdot |\boldsymbol{c}|$ "**Multiplicative scaling**"

This follows from the easily verified relations:

$$\mathbb{Z}ac + \mathbb{Z}bc = (\mathbb{Z}a + \mathbb{Z}b) \cdot \text{c, and}$$

$$\mathbb{Z}ac \cap \mathbb{Z}bc = (\mathbb{Z}a \cap \mathbb{Z}b) \cdot c$$

Multiplicative scaling is a very useful property. For example, when a and b are commensurable, we know that there is a nonzero number c such that ca and cb are integers. Then, for example, $(a, b) = |c^{-1}|(ca, cb)$, so this reduces the calculation of (a, b) to the case of integers. Similarly for $[a, b]$.

(**DM7**): Let $d = (a, b)$ and $a = a'd$ and $b = b'd$. Then $(a', b') = 1$.

In fact, $d = (a, b) = (a'd, b'd) = (a', b') \cdot d$, by (DM6), and so $(a', b') = 1$.

(**DM8**): If $(a, b) = 1$ then $[a, b] = |a \cdot b|$

Proof. Write $1 = ra + sb$, $r, s \in \mathbb{Z}$. Let m' be a common multiple of a and b: $m' = ua = vb$, with $u, v \in \mathbb{Z}$. Then $m' = m'ra + m'sb = vbra + uasb = (vr + us)ab$, so $ab \mid m'$. Since ab is visibly a common multiple of a and b, it follows that $|ab| = [a, b]$.

(DM9): $|\boldsymbol{a} \cdot \boldsymbol{b}\ | = (\boldsymbol{a}, \boldsymbol{b}) \cdot [\boldsymbol{a}, \boldsymbol{b}]$.

Proof. $m = [a^{'}d, b^{'}d] = [a^{'}, b^{'}] \cdot d$ by (DM7)
$=| a' \cdot b' | \cdot d\, by$ (DM8 and 9)
so

$$d \cdot m =| a' \cdot b' \cdot d^2 |=| a'd \cdot b'd |=| a \cdot b\ | .$$

(DM10): If $(a, b) = 1 = (a, c)$, then $(a, bc) = 1$.

Proof. Note that a, b, $c \in \mathbb{Z}$. Write $1 = ra + sb = ua + vc$, with r, s, u, $v \in \mathbb{Z}$. Then

$$1 = (ra + sb)\,(ua + vc) = (rua + rvc + sbu)\,a + (sv)bc.$$

(DM11): Given $a_1a_2,\ \ldots, a_n$ and $b_1b_2,\ \ldots, b_m$ with
$(a_ib_j) = 1$ for $1 \le i \le n$ and $1 \le j \le m$,
it follows that $(a_1a_2 \ldots a_n, b_1b_2 \ldots b_m) = 1$.

This follows from (DM10) by induction on *max*(n, m), as follows. In the case $n = m = 1$ there is nothing to prove. Suppose that $m \ge 2$. Then, by induction, we have that
$(a_1a_2 \ldots a_n, b_1b_2 \ldots b_{m-1}) = 1 = (a_1a_2 \ldots a_n, b_m)$
and so the result follows from (DM11).

(DM12): Suppose that $(a, b) = 1 = (c, d)$. Then $(ad, bc) = (a, c) \cdot (b, d)$.

Proof. Let $=\mathbb{Z}ad + \mathbb{Z}bc$. We want to show that:

$$A = (\mathbb{Z}a + \mathbb{Z}c) \cdot (\mathbb{Z}b + \mathbb{Z}d) = \mathbb{Z}ab + \mathbb{Z}ad + \mathbb{Z}bc + \mathbb{Z}bd$$

Clearly the right side contains the left side. For the reverse inclusion, we must show that ab, $bd \in A$. Write $1 = ra + sb$ and $1 = uc + vd$ with r, s, u, $v \in \mathbb{Z}$. Then

$$ab = abuc + abvd = (au)(bc) + (bv)(ad) \in A.$$

Similarly, $bd \in A$.

(DM13): Suppose that a/b and c/d are <u>reduced</u> fractions.
Then $(\boldsymbol{a}/\boldsymbol{b}, \boldsymbol{c}/\boldsymbol{d}) = (\boldsymbol{a}, \boldsymbol{c})\,/\,[\boldsymbol{b}, \boldsymbol{d}]$
= gcd (**numerators**) /lcm (**denominators**)

Proof. $|bd\ |\ (a/b, c/d) = (ad, bc)$ by (DM7)
$= (a, c) \cdot (b, d)$ by (DM13)

so

$$(a/b, c/d) = (a, c) \cdot (b, d)\,/\,|\ b \cdot d\ |$$

$$= (a, c) \cdot (b, d)/(b, d) \cdot [b, d] \text{ by (DM9)}$$

$$= (a, c)\,/\,[b, d]$$

For the next items, we shall use the following notation: If a and b are real numbers we shall write: $\mathbb{Z}(a, b) = \mathbb{Z}a + \mathbb{Z}b$. If a and b are commensurable, then $\mathbb{Z}(a, b)$ is a discrete additive group, generated by (a, b). If a and b are incommensurable then we have shown that $\mathbb{Z}(a, b)$ is dense in $\mathbb{R}$.

(DM14): Let a and b be real numbers, and let t be an integer. Then

$$\mathbb{Z}\,(a, b + ta) = \mathbb{Z}\,(a, b)$$

In case a and b are commensurable, it follows that:

$(\boldsymbol{a}, \boldsymbol{b} + \boldsymbol{ta}) = (\boldsymbol{a}, \boldsymbol{b})$ **"Additive translation"**

Proof. Let $A = \mathbb{Z}(a, b)$, and $B = \mathbb{Z}(a, b + ta)$. We want to show that $A = B$. Since a, $(b + ta) \in A$, it follows that $B \subseteq A$. Writing $b = (b + ta) - ta$, we see that $a, b \in B$, and so $A \subseteq B$. Hence, $A = B$, as claimed.

Example. The Fibonacci sequence F_n is defined recursively by: $F_0 = 0$, $F_1 = 1$, and, for $n \geq 2$, $F_n = F_{n-1} + F_{n-2}$ (i.e., 0, 1, 1, 2, 3, 5, 8, 13, 21, 34, 55, 89, 144, ...)

It follows from (DM14) that, for all $n \geq 1$, $\gcd(F_n, F_{n+1}) = 1$.

(DM15): Given real numbers $a \geq b > 0$, put $c = r_b(a) < b$. Then $\mathbb{Z}(b, c) = \mathbb{Z}(a, b)$. Hence, if a and b are commensurable, then so also are b and c, and $(b, c) = (a, b)$.

Proof. By *DwR*, $a = qb + r$ with q an integer and $0 \leq r = r_b(a) < b$. Thus, $0 \leq r = a - qb = c < b$, and it follows then from Additive Translation (*DM*14) that: $\mathbb{Z}b + \mathbb{Z}c = (a - qb) = \mathbb{Z}a + \mathbb{Z}b$.

(DM16): **The Euclidean Algorithm (EA)**. Let a and b be commensurable real numbers, not both equal to 0. The Euclidean Algorithm (**EA**) is an algorithm to produce $(a, b) =$ gcd (a, b). Without loss of generality (DM5), we can assume that $a \geq b \geq 0$, and we shall then write $a_0 = a$ and $a_1 = b$. Then there is an integer $n = n(a_0, a_1) \geq 0$, and a sequence, $a_0 \geq a_1 > a_2 > \ldots > a_n > a_{n+1} = 0$ such that $(a_j, a_{j+1}) = (a_0, a_1)$ for all $j \leq n$. In particular, $(a_0, a_1) = (a_n, a_{n+1}) = a_n$.

Proof. If $a_1 = 0$, we set $n = 0$, and all is clear. So suppose that $a_1 > 0$. Suppose that $j \geq 1$, and that we have constructed $a_0 > a_1 > a_2 > \ldots > a_j > 0$.

Then, with the notation (of DM16), we set $a_{j+1} = r_{aj}(a_{j-1})$ and we have $(a_j, a_{j+1}) = (a_{j-1}, a_j)$.

We continue in this fashion if $a_{j+1} > 0$. If $a_{j+1} = 0$, then we set $n = j$, and stop. All of the properties above follow from (DM16). The process must stop in a finite number of steps since $\mathbb{Z}(a, b)$ is uniformly discrete, so it cannot contain an infinite decreasing sequence of positive numbers.

(DM17): Suppose that $a > b > 0$ are incommensurable real numbers. Then we can still apply the Euclidean Algorithm process, but it won't stop in finitely many steps. Explicitly, set $a_0 = a$ and $a_1 = b$. Then we can produce an infinite sequence of positive numbers $a_0 > a_1 > a_2 > \dots > a_n > a_{n+1} > \dots$ such that $\mathbb{Z}(a_j, a_{j+1}) = \mathbb{Z}(a_0, a_1)$ for all $j \geq 0$.

This follows inductively from (DM16). We can't have $a_{j+1} = 0$ since $\mathbb{Z}(a_0, a_1)$ is not discrete, in fact it is dense in $\mathbb{R}$. It can further be shown that $a_n \to 0$ as $n \to \infty$.

(DM18): Multiple gcds and lcms. Define commensurability for a sequence $(a_1, a_2, \dots, a_n)$ of real numbers to mean that $\mathbb{Z}a_1 + \mathbb{Z}a_2 + \dots + \mathbb{Z}a_n$ is discrete.

Then, as above, we can define $\gcd(a_1, a_2, \dots, a_n)$ and $\operatorname{lcm}(a_1, a_2, \dots, a_n)$ to be the nonnegative generators of $\mathbb{Z}a_1 + \mathbb{Z}a_2 + \dots + \mathbb{Z}a_n$ and of $\mathbb{Z}a_1 \cap \mathbb{Z}a_2 \cap \dots \cap \mathbb{Z}a_n$, respectively. These are clearly symmetric functions of their n variables. Moreover, we have recursive descriptions,

$\gcd(\gcd(a_1, a_2, \dots, a_{n-1}), a_n) = \gcd(a_1, a_2, \dots, a_n)$
$\operatorname{lcm}(\operatorname{lcm}(a_1, a_2, \dots, a_{n-1}), a_n) = \operatorname{lcm}(a_1, a_2, \dots, a_n)$.

To simplify writing we shall put

$$\delta(a_1, a_2, \dots, a_n) = \gcd(a_1, a_2, \dots, a_n), \text{ and}$$

$$\mu(a_1, a_2, \dots, a_n) = \operatorname{lcm}(a_1, a_2, \dots, a_n)$$

Note that the a_j are all integers if and only if $\delta(a_1, a_2, \dots, a_n)$ is an integer. Moreover, $\delta(a_1, a_2, \dots, a_n)$ and $\mu(a_1, a_2, \dots, a_n)$ satisfy the analogue of multiplicative scaling (DM7).

(DM19): Writing $d = \delta(a_1, a_2, \dots, a_n)$ and $a_j = a_j' \cdot d$ for each j, we have

$$\delta\left(a_1', a_2', \dots, a_n'\right) = 1.$$

(DM20): Assume that all $a_j \neq 0$. If $\delta(a_1, a_2, \dots, a_n) = 1$, then $\mu\left(\frac{1}{a_1}, \frac{1}{a_2}, \dots, \frac{1}{a_n}\right) = 1$. (i.e., if $\mathbb{Z}a_1 + \mathbb{Z}a_2 + \dots + \mathbb{Z}a_n = \mathbb{Z}$, then $\mathbb{Z}{a_1}^{-1} \cap \mathbb{Z}{a_2}^{-1} \cap \dots \cap \mathbb{Z}{a_n}^{-1} = \mathbb{Z}$.)

Proof. Put $G = \mathbb{Z}{a_1}^{-1} \cap \mathbb{Z}{a_2}^{-1} \cap \dots \cap \mathbb{Z}{a_n}^{-1}$. Clearly G contains $\mathbb{Z}$. It remains to show that any a in G is an integer. For each j we can write $a = s_j/a_j$ with s_j an integer. By hypothesis we can write $1 = r_1a_1 + r_2a_2 + \dots + r_na_n$, with all r_j integers. Then

$$a = r_1a_1a + r_2a_2a + \cdots + r_na_na$$

$$= \frac{r_1a_1s_1}{a_1} + \frac{r_2a_2s_2}{a_2} + \cdots + \frac{r_na_ns_n}{a_n}$$

$$= r_1s_1 + r_2s_2 + \cdots + r_ns_n \in \mathbb{Z}.$$

(DM21): Put $A = a_1 \cdot a_2 \cdot \dots \cdot a_n$. Then $A = \delta(a_1, a_2, \dots, a_n) \cdot \mu\left(\frac{A}{a_1}, \frac{A}{a_2}, \dots, \frac{A}{a_n}\right)$.

This is a nice, but nonobvious, generalization of the case $n = 2$, which is just (DM10): $a \cdot b = \text{gcd}\,(a, b) \cdot \text{lcm}\,(a, b)$.

Proof. Writing $a_j = a_j' d$, with the notation of (DM20), we can apply (DM21) to obtain:

$\delta\left(a_1', a_2', \ldots, a_n'\right) = 1 = \mu\left(\frac{1}{a_1'}, \frac{1}{a_2'}, \ldots, \frac{1}{a_n'}\right).$

Multiplying the right side by $\frac{A}{d}$, and using multiplicative scaling, we get $\mu\left(\frac{A}{a_1}, \frac{A}{a_2}, \ldots, \frac{A}{a_n}\right) = \frac{A}{d}$, whence the result.

References

Ball, D. L., Thames, M. H., & Phelps, G. (2008). Content knowledge for teaching: What makes it special? *Journal of Teacher Education, 59*(5), 389–407.

Bass, H. (2017). Designing opportunities to learn mathematics theory-building practices. *Educational Studies in Mathematics, 95*(3), 229–244.

Common Core State Standards Initiative (CCSS-M). (2010). *Common core state standards for mathematics.* Retrieved from http://www.corestandards.org/math

Davydov, V.V. (1990). *Types of generalization in instruction: Logical and psychological problems in the structuring of school curricula* (*Soviet Studies in Mathematics Education* (Vol. 2)) (J. Teller, Trans.). Reston, VA: National Council of Teachers of Mathematics.

Usiskin, Z. (1968). Six nontrivial equivalent problems. *The Mathematics Teacher, 61*(4), 388–390.

van der Waerden, B. L. (1930). *Moderne algebra (I).* New York, NY: Springer-Verlag.

van der Waerden, B. L. (1931). *Moderne algebra (II).* New York, NY: Springer-Verlag.

Chapter 8
Abstract Algebra and Secondary School Mathematics Connections as Discussed by Mathematicians and Mathematics Educators

Ashley L. Suominen

Introduction

Educational committees and professional organizations agree in recommending that secondary teacher preparation programs require coursework in the study of abstract algebraic structures (Conference Board of the Mathematical Sciences [CBMS], 2001, 2012; Leitzel, 1991; National Council of Teachers of Mathematics [NCTM], 2000). The importance of a course in abstract algebra lies in the opportunities it allows prospective secondary mathematics teachers to develop accurate concept images and concept definitions (i.e., Tall & Vinner, 1981) that can help them explain and unite concepts found in school mathematics. For instance, Gallian (1990) explained the importance of abstract algebra in terms of exposing undergraduate students to the "terminology and methodology of algebra" (p. xi). Similary, Findell (2001) asserted:

> When the population of students in an abstract algebra course includes future teachers (which may be almost always), these big ideas, such as inverse and identity, are particularly important because they can help teachers connect advanced mathematics with high school mathematics in ways that can strengthen and deepen their understandings of the mathematics they will teach. (p. 13)

Exploring terminology such as *identity*, *inverse*, *commutativity*, and *equivalence* is common practice in abstract algebra courses, which affords prospective secondary mathematics teachers the opportunity to develop deeper conceptual understandings of these concepts. In addition, Papick, Beem, Reys, and Reys (1999) suggested that abstract algebra enables students to experience "a rigorous examination of

A. L. Suominen (✉)
Department of Liberal Arts, Savannah College of Art and Design, Atlanta, GA, USA
e-mail: asuomine@scad.edu

N. H. Wasserman (ed.), *Connecting Abstract Algebra to Secondary Mathematics, for Secondary Mathematics Teachers*, Research in Mathematics Education,
https://doi.org/10.1007/978-3-319-99214-3_8

arithmetic properties in various algebraic structures [that] deepens the understanding of traditional arithmetic and accentuates the importance of axiomatic mathematics" (p. 306). The results of such a rigorous examination ideally provide prospective secondary mathematics teachers with opportunities to solidify their conceptual understandings of school arithmetic and the algebraic structures and properties found in the curriculum.

Prospective secondary mathematics teachers likely do not, however, access as many benefits from learning abstract algebra when mathematical connections between abstract algebra and secondary school mathematics are not recognized. In a report entitled *The Mathematical Education of Teachers (MET),* CBMS acknowledged, "Unfortunately, too many prospective high school teachers fail to understand connections between [abstract algebra and number theory] and the topics of school algebra" (2001, p. 40). Cofer (2015) confirmed this assertion when studying how prospective secondary mathematics teachers communicate the conceptual connections between abstract algebra and their teaching practices. She discovered that prospective teachers were able to identify very few connections, despite having just finished an abstract algebra course. Bukova-Güzel, Ugurel, Özgür, and Kula (2010) conducted a similar study in Turkey in which prospective secondary mathematics teachers were unable to identify many connections between their undergraduate courses and the secondary school mathematics curriculum. Ultimately, the failure to recognize such connections undermines the purpose of an abstract algebra course and thus represents a significant problem in collegiate mathematics education.

Although several research studies have focused on the teaching and learning of abstract algebra, little work has been done on specifically classifying the types of mathematical connections between abstract algebra and school mathematics. Given the importance of an undergraduate course in abstract algebra for teacher preparation programs, outlining these mathematical connections is important. This study attempts to examine which mathematical connections mathematicians and mathematics educators identify between abstract algebra and secondary school, and how they describe them.

Theoretical Perspective

Businskas (2008) and Singletary (2012) defined a *mathematical connection* as a relationship between ideas. A mathematical connection is "a true relationship between two mathematical ideas, A and B" (Businskas, 2008, p. 18) and "a relationship between a mathematical entity and another mathematical or nonmathematical entity" (Singletary, 2012, p. 10). In existing literature on mathematical connections, these relationships take three distinct forms: (1) mathematical connections as a characteristic of mathematics; (2) mathematical connections as an artifact of learning; and (3) mathematical connections as an active process of doing mathematics.

The first perspective, *mathematical connections as a characteristic of mathematics*, can be interpreted in a few different ways. For instance, Coxford (1995)

Table 8.1 Mathematical connections categories

Category	Description
Alternative representation	One concept is represented in different ways, such as symbolic, graphic, pictorial, or manipulative.
Comparison through common features	Two concepts share some features in common, which allows a comparison through the concepts being similar, exactly the same, or not the same.
Generalization	One concept is a generalization of another specific concept.
Hierarchical relationship	One concept is a component of or included in another concept. Since one concept is included or contained in the other concept, a hierarchical relationship exists between two concepts.
Real-world application	One concept is an example of another concept in the real-world (i.e., a concept refers to another concept outside the current mathematical context).

described mathematical connections as unifying themes (e.g., function and variable) within the discipline. Others (Skemp, 1987; Zazkis, 2000) have explained this perspective as explicit links between two specific concepts. Still, other researchers (Chappell & Strutchens, 2001; Hodgson, 1995a) have defined mathematical connections as equivalent representations across mathematical ideas, such as equivalent solving methods for systems of linear equations (substitution, elimination, etc.). Despite some slight differences, all of these researchers maintained the belief that mathematical connections exist and should be acknowledged throughout the discipline of mathematics.

Singletary (2012) identified eight ways to categorize these concept-to-concept mathematical connections. However, this list of eight connection types was then modified by the author based on a prior study about abstract algebra textbooks. The five categories of mathematical connections used in this study are given in Table 8.1. An alternative representation connection is when a concept is discussed in multiple ways. Specific types of groups using geometric transformations introduced alongside the formal definition would be an example of an alternative representation connection. A comparison through common features connection is when characteristics of two concepts are compared. For example, introducing a group, ring, or field by comparing their defining properties to those of familiar number systems and arithmetic operators. A generalization connection is when one concept is stated to be the generalization of another concept. An example of a generalization connection made in abstract algebra is the field of quotients being the generalization of fractions and operations on fractions. A hierarchical relationship connection is when one concept is a component of or contained in another concept. This type of connection is similar to, and yet subtly different from, a generalization. That is, a hierarchical relationship does not require one concept to be the general case. Finally, a real-world application connection is when a connection is made between an abstract algebra concept and another concept outside of mathematics.

A second perspective of connections found in the literature is *mathematical connections as an artifact of learning*. Businskas (2008) described this type of mathematical connection as "a process that occurs in the mind of the learner(s) and the connection is something that exists in the mind of the learner" (pp. 12–13). Mathematical connections, in light of this perspective, are a necessary component to learning because new mathematical ideas are connected to preexisting schemas or networks within the mind of the learner. Hiebert and Carpenter (1992) wrote, "A mathematical idea, procedure, or fact is understood thoroughly if it is linked to existing networks with stronger or more numerous connections" (p. 67). In light of this literature, mathematical connections are a necessary component of learning. Ultimately, Hazzan (1999) and Hiebert and Carpenter (1992) concluded that the teacher plays a pivotal role in helping students to construct these connections among mathematical ideas.

The final perspective, *mathematical connections as a mathematical process or activity*, blends the initial two perspectives in acknowledging that connections exist across the discipline and the learner should be involved in the activity of establishing these connections. That is, mathematical connection-making is not just a description of the (connected) nature of the relationship between two mathematical ideas, but also an active process in which one engages. Boaler (2002) asserted, "The act of observing relationships and drawing connections, whether between different functional representations or mathematical areas, is a key aspect of mathematical work, in itself, and should not only be thought of as a route to other knowledge" (p. 11). Thus, research from this perspective affirmed that the activity of making connections across mathematics is a significant aspect of doing mathematics.

Related Literature

Teaching and Learning of Abstract Algebra

Research on the teaching and learning of abstract algebra has indicated that the conceptual understanding of undergraduate students in abstract algebra is less than satisfactory (Dubinsky, Dautermann, Leron, & Zazkis, 1994; Hazzan & Leron, 1996). Leron and Dubinsky (1995) asserted that the teaching of abstract algebra is considered to be a failure by both professors and undergraduate students. As a result, many undergraduate and graduate students, including prospective teachers, struggle to grasp even the most fundamental concepts of this course (Dubinsky et al., 1994; Hazzan & Leron, 1996; Leron & Dubinsky, 1995). For many students, taking abstract algebra is the first time they experience such a high level of mathematical abstraction and formal proof. It is often the first tertiary course in which teachers expect students to "go beyond learning 'imitative behavior patterns' for mimicking the solution of a large number of variations on a small number of themes (problems)" (Dubinsky et al., 1994, p. 268). Nevertheless, it is widely acknowledged that abstract algebra is an essential part of undergraduate mathematical learning despite its high level of difficulty at the collegiate level (Gallian, 1990; Hazzan, 1999; Selden & Selden, 1987).

Other researchers (Dubinsky et al., 1994; Leron & Dubinsky, 1995) have shown undergraduate and graduate students encountering abstract algebra for the first time often struggle and fail to comprehend many of the fundamental concepts. Much of the research affirms students' difficulties in learning fundamental concepts in group theory (Asiala, Dubinsky, Mathews, Morics, & Oktaç, 1997; Larsen, 2004, 2009; Leron, Hazzan, & Zazkis, 1995), with little concentration on students' learning of other algebraic structures like rings or fields (Cook, 2012). Despite the overarching focus on discovery-based instructional approaches in the teaching of abstract algebra, none of the studies directly addressed the mathematics content connections between abstract algebra and school mathematics. Much of this research has endeavored to build on students' intuitions about mathematics without explicitly discussing a primary source of that mathematical knowledge—school mathematics. Even though abstract algebra is considered to be "a generalization of school algebra" (Findell, 2001), thus far, the explicit generalizations from school algebra to abstract algebra have not been elaborated thoroughly.

In response to an early awareness of the difficulties students were having in learning abstract algebra, several researchers have introduced alternative teaching methods. One of the earliest papers on teaching abstract algebra was from Pedersen (1972), in which she presented an activity to introduce the noncyclic 6-group. In this activity, students folded a strip of paper with 10 equilateral triangles. The students folded the paper in various ways to construct the noncyclic 6-group. Similarly, Huetinck (1996) and Larsen (2004, 2009) utilized an activity called SNAP in which students, first, rotated and translated equilateral triangles on an overhead transparency sheet and, second, used a nine-peg 3×3 square array board with three rubber bands. Both Huetinck and Larsen used this activity to help students understand how students reinvent the concepts of group and group isomorphism. In both activities, students were encouraged to build connections between their geometric knowledge and group theory. Similar to Larsen, Cook (2012) utilized the instructional design theory of Realistic Mathematics Education (RME) and guided reinvention to study the teaching and learning of rings, fields, and integral domains. Through discovery-based learning activities, the participants in these studies were able to identify similar features and structures to develop mathematical connections between concepts.

Gallian (1976) offered another approach to teaching abstract algebra by using computers. He first suggested using Fortran[1] to investigate finite groups. More

[1]Fortran is a programming language that was developed by IBM in the 1950s. This language has been especially useful for numeric computation and engineering applications.

recently, Leron and Dubinsky (1994), in response to the findings of Dubinsky et al. (1994), introduced key ideas related to group theory through the use of the programming language ISETL. This program allowed students to construct an algebraic environment or structure with a set of axioms and properties in order to work through specific examples and problems. Hodgson (1995b) reported successful learning results from having students use ISETL to learn abstract algebra. Similarly, Asiala et al. (1997) implemented the ACE teaching cycle (Activities, Class discussion, and Exercises), where students rotated in groups between classroom activities and computer activities using ISETL. Asiala et al. affirmed that through this teaching method students demonstrated a deep understanding of cosets, normality, and subgroups based on their performance on the two given tests and a final exam, as well as participants' responses during two sets of interviews.

Freedman (1983) offered yet another approach to teaching abstract algebra in detailing a unique lecture-based method that gradually required students to take an active part in the learning process through teaching. This three-stage teaching method had students (1) learn through traditional lecture, (2) complete a project and do some teaching, and (3) be solely responsible for preparing lectures. Freedman claimed the students were able to gain a strong understanding of the topics by explaining them to others in the class. Other researchers (Brown, 1990; Czerwinski, 1994; Leganza, 1995) have suggested the use of writing assignments in abstract algebra to enhance student involvement in learning. For instance, one of these assignments asked students to write about connections between a certain abstract algebra concept and real-world applications in order to make the abstract more concrete.

Types of Mathematical Connections in Abstract Algebra

Some of the earliest work that specifically identified mathematical connections between abstract algebra and secondary school mathematics was done by Usiskin. Usiskin (1974) examined specific concept-to-concept connections, such as the properties of addition and multiplication of real numbers compared to the structural properties of groups and fields. He wrote that "all properties of the reals follow from the complete field properties" (p. 279). In a second article, Usiskin (1975) highlighted similar mathematical connections but also discussed connections between the group structure and solving simple linear equations, the multiplicative group of invertible 2×2 matrices and solving systems of linear equations, and the additive and multiplicative groups with familiar number systems and groups of geometric transformations. Ultimately, Usiskin (2001) characterized these connections between concepts taught in abstract algebra as generalizations of school algebra.

The Conference Board of the Mathematical Sciences [CBMS] (2001, 2012), established by the American Mathematical Society (AMS) in partnership with the Mathematical Association of America (MAA), made recommendations regarding the mathematical knowledge necessary for teaching at all grade levels in *The Mathematical Education of Teachers (MET)* and *The Mathematical Education of Teachers II (MET II)*. In these reports, CBMS suggested ways to establish connections between undergraduate courses and secondary school mathematics, especially given that prospective secondary teachers were not currently recognizing many connections. Similar to Usiskin, the connections found in these reports were characterized as generalizations. For instance, abstract algebra was described in *MET* as the examination of "mathematical structures that are the foundation for number systems and algebraic operations" (p. 40). Thus, the algebraic structures are generalizations of familiar number systems and operators. Further, CBMS (2012) recommended the study of rings and fields as the underlying structures of operations with polynomials and rational functions. A focus on the mathematical concepts of *inverse* and *identity* was also suggested in *MET II*. The majority of the discussed connections implied that abstract algebra was a generalization of school algebra. A connection to school geometry was also mentioned when describing isometry and symmetry groups as generalizations of the geometry of transformations of regular polygons.

Difficulty of Connecting Secondary and Tertiary Mathematics

Bukova-Güzel et al. (2010) conducted a qualitative study that asked 36 students in the last year of their teacher preparation program, from four different universities in Turkey, to discuss the connections between their undergraduate mathematics courses and secondary teaching. The prospective secondary mathematics teachers were asked to answer four open-ended questions regarding their undergraduate content courses in Calculus, Analytic Geometry, Linear Algebra, Abstract Mathematics, Topology, and Differential Equations. The purpose of the study was threefold: (1) identify student teachers' opinions on whether their mathematics courses fully prepared them to teach; (2) determine whether these courses were necessary for their understanding and abilities to teach; and (3) ascertain the participants' perceptions of the quality of these courses. Overall, Bukova-Güzel et al. found that 83% of the prospective secondary mathematics teachers did not see connections between the undergraduate content courses they had taken and the secondary school mathematics curriculum. However, 25% of the participants did believe that first-year undergraduate courses, such as Calculus and Analytic Geometry, were coherent and related to the secondary curriculum. One participant's response stated:

> Since the courses are based on memorizing theorems and passing exams, it is really hard for us to apply even useful knowledge. At least on my own behalf, I was better at secondary school mathematics topics when I graduated from secondary school. (p. 2236)

Furthermore, 42% of the participants recommended designing new undergraduate courses that were directly related to secondary school mathematics. In fact, 56% of the prospective secondary mathematics teachers felt insufficiently prepared, mathematically, to teach. From these results, we see that the participants did not feel that their undergraduate content courses made connections to the secondary school mathematics curriculum.

Cofer (2015) confirmed these results when she considered how prospective secondary mathematics teachers in the USA communicated mathematical connections between abstract algebra and their teaching practices. In this interview study, five upper-level tertiary students were asked to give explanations about how abstract algebra concepts were connected to the school mathematics curriculum. The participants had recently completed coursework in abstract algebra the semester prior to the study. Cofer asked the participants questions about concepts related to division by zero and even and odd numbers. She inquired as to how a certain concept could be explained to a secondary student, and how the concept related to abstract algebra. Cofer discovered that the students were unable to relate abstract algebra concepts to the school mathematics curriculum, despite having just completed the abstract algebra course.

Cook (2012) hypothesized in his dissertation that the difficulty students experience in learning abstract algebra is due to the lack of established connections between undergraduate mathematics and school mathematics. He wrote that prospective teachers "do not build upon their elementary understandings of algebra, leaving them unable to communicate traces of any deep and unifying ideas that govern the subject" (p. xvi). Similarly, Cuoco (2001) noted in an article about secondary teacher preparation programs, "Most teachers see very little connection between the mathematics they study as undergraduates and the mathematics they teach. This is especially true in algebra, where abstract algebra is seen as a completely different subject from school algebra" (p. 169). To put it simply, students are not recognizing or establishing the mathematical connections between abstract algebra and high school mathematics (Usiskin, 1988). Ultimately, this inability to make these connections hinders students' learning of advanced mathematics, as well as potentially hurts future teachers' ability to teach school mathematics.

Methodology

This study was guided by two research questions: (1) What types of mathematical connections do mathematicians and mathematics educators identify between abstract algebra and secondary school, and how do they describe them? (2) What types of connections for teaching do mathematicians and mathematics educators identify between abstract algebra and school mathematics?

For this study, I used purposive case homogeneous sampling to choose participants who had expertise in abstract algebra (Patton, 2002; Roulston, 2010). To be more specific, mathematicians and mathematics educators were chosen if they had taught or were currently teaching an undergraduate abstract algebra course, or if they were involved in mathematical research in the area of abstract algebra. These participants ranged from pure mathematicians that have only published in pure mathematics journals, with little to no experience with the secondary school curriculum, to mathematics educators that have only published in mathematics education journals, with several years of secondary school teaching experience. In total, I interviewed 13 mathematicians and mathematics educators.

Each participant was involved in one interview that lasted between 40 and 70 min. A semistructured interview protocol consisted of a predetermined list of open-ended content questions and reflection questions based on question types given in Patton (2002), Taylor and Bogdan (1984), and Zazkis and Hazzan (1998). For instance, I began each interview by asking each participant about his or her background in teaching or researching abstract algebra and in teaching secondary school mathematics. I then asked how each participant defines a mathematical connection. Further, I asked each participant if he or she specifically made connections between abstract algebra and secondary school mathematics when teaching. Then, I asked them, more specifically, to detail these precise connections. Given the range of participants' experiences with the secondary school mathematics curriculum, I prepared in advance a list of some of the curriculum topics found in secondary school mathematics based on NGA and CCSS (2010). Each interview was audiotaped and then transcribed. A running list of identified mathematical connections was made and modified to include each explicit mathematical connection mentioned during each interview. I employed theoretical sampling throughout the interviews to ensure validity in this list of mathematical connections (Charmaz, 2000). In particular, I added to the connection list each mathematical connection explicitly mentioned by the participants. However, if two sequential participants disagreed with a connection, I removed the connection from the list.

After conducting all the interviews, I analyzed each interview transcript using an inductive and iterative coding process. I concentrated on different aspects of mathematical connections each time I examined the data. First, I concentrated on the *mathematical connections as a characteristic of mathematics*. Thus, my first round of data analysis consisted of me identifying each of the concept-by-concept connections mentioned by participants, so that I could organize a (reasonably) comprehensive list of mathematical connections. Next, I coded and analyzed the data thematically in light of the other perspectives of mathematical connections—*as an artifact of learning* and *as a mathematical activity*. Finally, I focused on emerging mathematical connections that were not previously identified in existing research literature. For instance, one theme, the historical development of abstract algebra, provided a lens for me to understand the types of connections the participants discussed.

Results

Mathematical Connections as a Characteristic of the Discipline

In this study, the participants discussed connections between abstract algebra and secondary school mathematics in ways that aligned with the previously established connection types shown in Table 8.1. In particular, the participants described connections that aligned with four of the five connection categories: comparison through common features, generalization, hierarchical relationship, and real-world applications. All the different connections identified are in Appendix 1. However, although the connections identified were the same, the participants emphasized the impact or usefulness of each type of connection differently. Most participants identified and talked about connections by comparing common features, whereas only one participant mentioned real-world applications.

Comparison Through Common Features

Recall that for this category two concepts are compared as being similar, exactly the same, or not the same due to common features. In their connection talk, the participants compared specific concepts, as well as compared mathematical structures. The mathematicians and mathematics educators in this study identified structure as being crucial to abstract algebra. In fact, many explicitly stated that abstract algebra is the study of structure. Naturally, these participants mentioned how the focus of the course would be on a comparison of these various structures. The ways in which these participants talked about structure, however, seemed to vary slightly. For instance, one participant described structure as, "When mathematicians say structure they generally mean in a set and some types of properties overlaid on that set." This participant then stated, "Thus, abstract algebra is about comparing structures and their properties." Another participant mentioned, "So one of the reasons to do abstract algebra is to give a structure to all those properties you learned."

As a result of their focus on structure, the majority of these mathematicians and mathematics educators, when teaching abstract algebra, explicitly concentrated on the development of students' abilities to recognize structural connections. They took the time to focus on these ideas, even at the expense of teaching other abstract algebra topics. Five participants emphasized that once a structure is known, questions can be asked about its properties to determine appropriate expectations and assumptions about similar structures. Overall, twelve of the thirteen mathematicians and mathematics educators discussed mathematical connections between abstract algebra and secondary school mathematics as a comparison of structural features.

One participant who emphasized structural similarities defined a mathematical connection as “when two contexts that look different on the surface have the same underlying structure.” For this participant, connections were more about abstract similarities than about studying surface features. Thus, he stressed that many problems in abstract algebra may appear to be different from other mathematics despite being quite similar to (or even the same as) problems that students have seen before. Another participant mentioned that students can think about and approach problems in completely different ways despite the content and underlying structure being the same. For instance, abstract algebra provides structure to school algebra, so that although the two may look different, the mathematical content is actually the same. A third participant shared similar sentiments:

> It is the way algebraists think, when we talk about structure we immediately start thinking about, “Oh is this something I’ve seen before?” ... Algebraists like to divide things up into chunks that behave similarly, so once you’ve started saying, “Does this behave like something I already know?” Oh well, let’s chunk it with those things.

For these participants, mathematical connections between abstract algebra and secondary school mathematics involved recognizing structural similarities, despite dealing with phenomena that initially appeared to be different.

Furthermore, the majority of the participants highlighted similarities relating number theory and solving polynomial equations to specific algebraic structures. In both instances, these participants primarily concentrated on the similarities between the operations and properties of the structures. They noted how these structures behave similarly and different across different domains. For instance, one participant discussed:

> When you translate a problem from one domain to another, that’s a connection, like constructing a regular polygon with a straightedge and compass turns out to be equivalent to finding the roots of the equation $x^{17} - 1$ and being able to express those in terms of radicals. So there’s structural similarities.

Another participant mentioned the structural parallels between the secondary school mathematics concepts of *function* and *domain* and the abstract algebra concepts of *operation* and *set*. This participant discussed how these concept pairs behave or act similarly despite seeming quite different at first.

Given the major focus placed on structural similarities by the mathematicians and mathematics educators in this study, it was not surprising that five of these participants underscored the importance of isomorphism in an abstract algebra course. One participant explained isomorphism as, “If you have these two systems and then somebody points out, ‘hey they are really the same, there’s just a different language here.’” In fact, this participant shared that one of the main reasons for studying abstract algebra is isomorphisms and the similarities of structures. Another participant stressed the amount of time he devoted to discussing *prototypes*, which he defined as a structure that new concepts can, and should be, compared to when teaching abstract algebra. Two examples of prototypes that he provided are the dihedral group of order 8 for non-abelian groups, and $\mathbb{Z}_n$ for infinite cyclic groups of order n. The other three participants who identified the importance of isomorphism described the structural similarities between the exponential or logarithmic functions and the real numbers. For instance, one participant discussed the connection in this way: the exponential and logarithmic rules, $e^x e^y = e^{x+y}$ and $\log ab = \log a + \log b$, "are both from the fact that the real numbers under addition is isomorphic to the positive real numbers under multiplication and you can use either e^x or $\log x$ as the mapping." This same participant emphasized the role of equivalence in problem-solving and explicitly detailed the various ways equivalence is presented in abstract algebra, such as two objects or structures being isomorphic.

Generalization

Four of the thirteen mathematicians and mathematics educators described mathematical connections as abstract algebra concepts being generalizations of secondary school mathematics concepts. An additional participant acknowledged the existence of generalization connections and provided one example but mentioned that he thinks about mathematical connections and teaches abstract algebra from another perspective.

Although four participants identified specific connections that were generalizations of school algebra, only one participant explicitly characterized abstract algebra as the generalization of school algebra. He defined mathematical connections as "seeing how new knowledge or general concepts in abstract algebra inform what students have learned previously or underpin what they learned previously." When discussing specific connections, he detailed how the ring and field properties are generalizations of the properties that are useful to solving polynomial equations. Similarly, another participant noted, "Each new abstract algebra concept is a generalization of number systems from the K-12 curriculum." He also elaborated on how ring and field properties are generalizations of the properties needed to solve polynomial equations. Another participant described how binary operators are generalizations of arithmetic operators.

In addition, one participant discussed the importance of teachers knowing the generalization relationship between polynomial rings and polynomials taught in secondary school mathematics. He explained:

> So you need a halfway decent understanding of them and you don't really understand them until you see the broader context of rings. For example, something as easy as a polynomial has at most n degree roots. Well, that's not always true and once you understand why that's not always true, it gives you better insight to what is going on and I think that it is really important for a teacher to know what is going on, not only to be able to teach it well but to be able to communicate. Once you embed these specific cases into the natural broader context I think it really aids the understanding.

This participant understood the secondary school mathematics concept *polynomial* as a specific case of the more general abstract algebra concept of a *polynomial ring*. In fact, he stressed the need for secondary teachers to possess deep understandings of the more general polynomial ring structure in order to better understand and teach polynomials in secondary school.

One of the four participants identified specific connections that were generalizations of school geometry. This participant described abstract algebra concepts such as *group* and *isomorphism* as generalizations of the representation of symmetries of regular polygons. He mentioned several activities used in his class to build on this idea. For example, he provided his students with opportunities to explore the symmetries of an equilateral triangle to enable students to generalize a set of rules regarding these symmetries to build the definition of a group. Additionally, he provided his students a mystery table and asked them if the table represented a group when teaching about isomorphism. After his students determined that the mystery table represented a group, they worked towards finding a correspondence between the table and their regular polygon symmetry tables. This participant hoped his students would notice that the two tables were the same, but with different symbols, so that they would generalize this idea of "sameness" to better grasp the definition of isomorphism.

Hierarchical Relationship

Two of the thirteen mathematicians and mathematics educators discussed hierarchical relationships between secondary school mathematics concepts and abstract algebra concepts by talking about specific secondary school mathematics concepts as being components of or included in abstract algebra concepts. One of these participants emphasized student knowledge about polynomials to understand polynomial rings and knowledge of functions to understand the bijections on a set for the definition of a group. This participant stated, "These topics provide a firm foundation to the material learned in abstract algebra." The other participant described the geometric concepts that students should learn in secondary school mathematics that will influence their ability to learn group theory. Some examples of these concepts include: rotation, reflection, axes, regular polygons, and degree measures. In addition, this participant affirmed that abstract algebra students should be very familiar with secondary school mathematics concepts, such as complex numbers, properties of real and rational numbers, the Euclidean algorithm, and least common multiples and greatest common divisors. He then detailed the connections between these topics and abstract algebra concepts. For instance, knowledge about least common multiples and greatest common divisors is necessary for students to understand LaGrange's theorem.

Real-World Application

Only one participant described mathematical connections between abstract algebra and secondary school mathematics in terms of real-world applications. He specifically discussed a connection between direct products taught in abstract algebra and molecules taught in secondary school chemistry. He explained:

> When you get to direct products, you can say that you glue those things together to get a much more complicated group by taking the basic component. Just like water is made out of hydrogen and oxygen, many groups are made out of $\mathbb{Z}_2$ cross $\mathbb{Z}_4$ cross a $\mathbb{Z}_8$ or something like that.

Further, he mentioned that students want to be taught this type of mathematical connection so that they can recognize how mathematical concepts are related to concepts outside of mathematics.

Other Perspectives of Mathematical Connections

While all thirteen mathematicians and mathematics educators evinced the perspective that connections exist across mathematics, the other two perspectives of mathematical connections—as an artifact of learning and as a mathematical activity—also emerged when the participants talked about connections.

Mathematical Connections as an Artifact of Learning

Three of the thirteen mathematicians and mathematics educators described mathematical connections as an artifact of learning. Businskas (2008) defined this type of mathematical connection as "a process that occurs in the mind of the learner(s) and the connection is something that exists in the mind of the learner" (pp. 12–13). These participants concentrated on the mental connections developed within the students' minds when talking about these mathematical connections.

One participant characterized these connections as "more a state of mind," so she did not think it was as important to explicitly state every connection to her abstract algebra students but rather to teach students how to make connections for themselves. She challenged her students' assumptions about previous knowledge and properties because students "aren't blank slates when they come in the classroom." She believed her students often took their algebraic knowledge for granted. Despite these beliefs about mathematical connections, she still stated that she makes a point to explicitly mention as many connections as possible when teaching. A second participant shared these sentiments by discussing mathematical connections as "process connections." She stressed the importance of having students think deeply about previously established concept definitions for secondary school mathematics to build mental connections between those ideas and abstract algebra concepts.

Consequently, both participants discussed the shared belief that students learn best when starting from something familiar and building on what they already know by challenging their minds with something new to extend the knowledge.

Another participant had a slightly different perspective on this type of mathematical connection. He focused his connection talk on students' development of mathematical habits of mind and the role mathematical connections play in this development. For instance, he believed students must first establish connections between concepts in their minds in order for them to be able to identify patterns or make conjectures. Thus, abstract algebra students cannot make conjectures about an unknown structure without first thinking about how a specific number system learned in school mathematics, for example, relates to the unknown structure. After that connection is made, they are free to investigate the reasons why the structures are similar.

Mathematical Connections as a Mathematical Activity

Six participants discussed the involvement of the learner in the activity of establishing or identifying these connections through proof writing. Several participants emphasized student learning of proof techniques over specific abstract algebra theorems when teaching abstract algebra. However, these participants also mentioned a few abstract algebra concepts that provide opportunities for enhanced proof learning. These concepts included inverses, cosets, normal subgroups, and ideals. One participant noted that the latter three concepts, in particular, introduce an extra qualifier that students need to pay attention to when developing proofs.

While elaborating on the importance of proof writing, one participant asserted that proofs can illustrate how much or little a student knows about a certain topic. She explained, "If you really understand that, then the proof is trivial. If they have no idea how to start it, then you know they don't have the right understanding." Ultimately, she concluded that the more mathematical connections a student knows or has developed, the more sophisticated of a proof that student can write and understand. Analogously, another participant mentioned how mathematics was taught in the Middle Ages to teach people to think clearly about the world; correspondingly, abstract algebra students are taught to think, read, and write proofs in order to help them think abstractly about mathematics. A third participant commented, "High school teachers should be comfortable with proofs, and abstract algebra is certainly one of the best courses for that because you have to think abstractly." Another participant acknowledged that throughout an abstract algebra course students learn several proof techniques or templates that teach them to think clearly and parse information. She further added that proofs require students to think about the role of definitions and how proofs build on these definitions by identifying connections to establish a sequence of logical arguments.

Emerging Mathematical Connections

Two additional perspectives emerged from the interview data: Connections to the Historical Roots of Group Theory and Connections for Teaching. These emerging mathematical connections are unique in that they were not discussed in previous connections research.

Emerging Mathematical Connections: Historical Roots of Group Theory

For many mathematicians and mathematics educators, the historical developments of algebraic structures (groups, rings, and fields) have shaped their understandings of them. In this study, the participants tended to highlight mathematical connections related to one of the four historical origins of abstract algebra: solving polynomial equations, generalizing solutions to number theory problems, satisfying preestablished axioms, and characterizing geometric transformations.

Four of the thirteen mathematicians and mathematics educators in this study explained mathematical connections between abstract algebra and secondary school mathematics in terms of solving equations. Three additional participants acknowledged that this perspective was important to abstract algebra but they did not elaborate on these connections. The four participants who did elaborate on these connections used descriptions similar to this one:

> I say to them, "Let's do solve equations." Then ask which properties they used to solve the equations, could you use fewer. Then the students came and presented on the board and say the following properties can be used for this one, these can be used for this one, etc. … If we want to solve any equation, which properties do we need, let's write a definition that captures that. That's how I introduce the definition of groups.

These participants then emphasized concept-to-concept connections such as inverse, identity, and operators. One participant pointed out that an inverse is first taught as a number. To be specific, an additive inverse is first introduced to middle school students as the negation of a number and a multiplicative inverse is introduced as the reciprocal of a number. High school students are then taught a more operational approach to inverse in learning how to find inverse functions and inverse matrices. In abstract algebra, students learn that inverses are necessary elements of sets (in, say, a group), whether that set is exemplified by solving equations or in the definition of an algebraic structure. After a discussion about inverse, identity, and operators, participants mentioned that these properties naturally come out through solving equations. Further, one participant noted his surprise that more abstract algebra students do not recognize the connection between the zero product property and integral domains with solving quadratic equations through factoring. The other three participants also mentioned this specific connection.

Four of the thirteen mathematicians and mathematics educators described mathematical connections related to number theory. Two additional participants acknowledged the importance of number theory to the historical development of abstract algebra, but they did not elaborate on these connections. The four participants discussed how the known number system can act as a foundation to facilitate discussion of the properties and operations associated with algebraic structures. For instance, these participants elaborated on the similarities between integers and polynomial rings. One of these participants mentioned the similar structural nature of base 10 expansions with polynomials, whereas three participants focused more on the parallels between integer long division and polynomial long division, factoring numbers and polynomials, and the Euclidean and other division algorithms. In addition, one participant explained structural connections between complex numbers and polynomials and how they both relate to the construction of splitting fields:

> A lot of high school students, if you watch them work, they will calculate the complex numbers as if i were x, and they work on them as if they are polynomials, and when they are all done they replace i^2 with -1. Well, that's actually a very deep idea. And we develop complex numbers from that point of view, building on what high school kids typically do. This is the idea that Kronecker used to construct splitting fields from polynomials. Basically, what you are doing is taking a polynomial of one variable with real numbers and reducing it modulo $x^2 + 1$, so that's the whole approach to complex numbers.

Ultimately, these four participants relied on students' background knowledge of number systems from secondary school mathematics to develop relevant abstract algebra concepts.

Only one participant discussed mathematical connections axiomatically. One additional participant mentioned this perspective, but only as a historical origin of abstract algebra. The former participant preferred to concentrate on the axioms, such as commutativity and associativity, that he believes students should have learned in secondary school mathematics when teaching abstract algebra. He then builds students' understandings of algebraic structures on these properties. Much of his talk about connections focused on the role commutativity plays in learning about abelian groups. Additionally, he discussed the notion of "losing" commutativity in studying certain structures, such as moving from dimension two complex numbers to dimension four quaternions. As a result, he emphasized logically deriving definitions and proving theorems from known axioms.

Four of the thirteen mathematicians and mathematics educators described mathematical connections between abstract algebra and secondary school geometry. Two additional participants acknowledged the existence of these connections but strongly preferred another perspective and, as a result, would not teach their students these connections. The four participants who emphasized geometric connections liked the visual nature of this approach to learning abstract algebra. These participants explicitly mentioned connections between geometric transformations, such as rotation and reflection to algebraic structures. One participant, who is representative of the four participants, described an approach to teaching abstract algebra:

> Basically, there's a long sequence of classes in the beginning that starts with symmetries on the equilateral triangle. The students reinvent their own symbols for those symmetries, then we talk about combining those symmetries. They develop a set of rules as to why that works. This leads into the axioms of a group.

This same participant reinforced these ideas by assigning his abstract algebra students a symmetry journal for homework in which students explored the symmetries of regular polygons not already examined in class. In addition, two participants discussed the mathematical connections of geometric transformations. One of these participants detailed the notions of identity and inverse in terms of these transformations. More specifically, she explained how an object reflected twice over an axis is, in fact, itself, so the action of two reflections acts as an identity and one reflection acts as an inverse. The other participant elaborated on using geometric transformations to explain the associative and commutative properties and parity. He noted:

> So I say look at your table of the symmetries of the square. They remember from the earlier work that a reflection and a reflection gives a rotation and a rotation and a rotation gives another rotation, if you mix them you get a reflection, so those act like evens and odds.

For these four participants, students' background knowledge of geometry from secondary school mathematics was pivotal to understanding abstract algebra concepts.

Emerging Mathematical Connections: Connections for Teaching

Three of the thirteen mathematicians and mathematics educators discussed another type of connection: mathematical connections for teaching. These participants emphasized that mathematical connections are important for prospective secondary school mathematics teachers to identify in order to enhance their future teaching. All three of the participants detailed how abstract algebra knowledge can be used for a teacher's preparation of lesson plans. For instance, one participant noted that secondary school mathematics teachers can employ what they know about a structure to help them plan a lesson. When a teacher is in the midst of extending an activity, he or she could quickly determine whether the extension is valid based on whether it fits within the structure. By knowing about the structure the students are working with, the teacher could also realize the questions she may need to ask students about the problem in order to extend student understanding and learning.

Another participant described how abstract algebra could inform lesson planning, such as using arithmetic of quadratic fields and rational points on conics to develop problems that work out nicely. Similarly, another participant explained:

> You are teaching them geometry and you want to give them some examples of triangles in a plane. It would be very nice if the side lengths of the triangle were integers, because it makes it much easier to work with, and by using abstract algebra it tells you how you can construct many examples.

He also mentioned that one of the things he likes to discuss when teaching abstract algebra is the impossibility of trisecting an angle, which comes up in secondary school geometry. He described another mathematical connection, "In high school you could take the affine group on the plane and construct an equilateral triangle with medians and apply the affine group, then you can see the properties are invariant of the affine group with midpoints." Ultimately, for this participant, to teach something effectively means the teacher needs to know why something is true, as well as be able to communicate why you need certain hypotheses or properties so that a structure does not break down. Consequently, to him, mathematical connections between abstract algebra and secondary school mathematics involve those beneficial to teaching effectively: important connections are those that both provide necessary background knowledge and improve a teacher's ability to communicate that knowledge.

Conclusions and Implications

In conclusion, many stakeholders and educational movements have emphasized the importance of recognizing mathematical connections between mathematical ideas to build students' understanding of mathematics. Although previous research has examined abstract algebra learning as well as mathematical connections from a variety of perspectives, this study provided a way to classify the mathematical connections between abstract algebra and secondary school mathematics. Identifying and characterizing connections between abstract algebra concepts and secondary school mathematics concepts offers abstract algebra instructors and faculty additional information that can be used to enhance undergraduate students' understandings of abstract algebra. In addition, this work provides a common vocabulary to discuss these mathematical connections.

Even though previous literature (CBMS, 2001, 2012; Usiskin, 2001) has characterized abstract algebra as the generalization of school algebra, this study revealed that mathematical connections between abstract algebra and secondary school mathematics are not simply generalizations. In fact, other connections were discussed in greater detail in this study. The participants primarily described specific concept connections between abstract algebra and secondary school mathematics using comparison through the common features. To be specific, twelve of the thirteen mathematicians and mathematics educators described mathematical connections through a comparison of structural features. These participants detailed ways that two structures could be compared and identified the particular structures they typically used while making these comparisons. For instance, the participants discussed the structural similarities between the properties defining group theory and the properties needed to solve a linear equation with one operator. These results are inconsistent with extant literature that has suggested the primary mathematical connection between abstract algebra and secondary school mathematics is generalization. The rationale for requiring prospective secondary mathematics teachers to take an abstract algebra course should then include abstract algebra as a further study of familiar mathematical ideas through studying structural comparisons, building upon previous mathematical concepts, and using alternative representations of algebraic concepts. In other words, a change must occur in the way in which we explain why abstract algebra is a required course for prospective secondary mathematics teachers.

Similarly, this study's results revealed that the mathematical connections described by the participants were linked to secondary school geometry nearly as often as secondary school algebra. Abstract algebra can no longer be considered simply as the generalization of school algebra, but rather it should be regarded as an extension of previous mathematical knowledge from algebra and geometry. Four participants prioritized connections to school algebra and four other participants prioritized connections to school geometry in their discussion about connections. These results were also inconsistent with extant literature, which suggests that the importance of abstract algebra lies in its mathematical connections to school algebra. Thus, abstract algebra provides prospective secondary mathematics teachers knowledge that is important to their understandings of school geometry, as well as their understandings of school algebra. For prospective teachers, it is especially important for abstract algebra instructors and faculty to explicitly discuss connections to both geometry and algebra, as future secondary teachers may teach either subject.

Another implication that can be drawn from the results of this study is that a variety of mathematical connections between abstract algebra and secondary school mathematics can be made, and these connections can be described in various ways. Not surprisingly, the participants identified various mathematical connections and prioritized certain connection types more than others. The mathematicians and mathematics educators' descriptions of connections reflected their own experiences with abstract algebra and secondary school mathematics. The pure mathematicians in this study, who had only published in pure mathematics journals and had little to no experience with the secondary school curriculum, made fewer connections than the participants with mathematics education experience, especially those having several years of secondary school teaching experience. To be more specific, the pure mathematicians discussed approximately five to eight mathematical connections, whereas the other participants discussed approximately ten or more connections. Further, not all of the participants described connections solely using concept-to-concept connections. Three participants described connections as an artifact of learning by concentrating on the mental connections developed within the students' minds between concepts when learning abstract algebra. And six participants discussed the involvement of the learner in a mathematical activity of establishing or identifying these connections through proof writing.

In addition, the mathematicians and mathematics educators differed according to their individual conceptualizations of group theory. This study revealed that individual conceptualizations of group theory influenced the participants' descriptions of the mathematical connections between abstract algebra and secondary school mathematics. To be specific, participants with views of abstract algebra based on axioms, solving equations, number theory, or geometry prioritized different sets of connections. Thus, abstract algebra instructors and faculty should be aware of how their individual conceptualizations of abstract algebra influence their teaching. An instructor or faculty member who focuses on certain connections or connection types over others can limit the understandings of his or her students by not providing them with opportunities to identify a different set of connections. For instance, one who favors the axiomatic approach to group theory may fail to discuss the geometric connections between group theory and secondary school mathematics. As a result, students, and especially prospective secondary mathematics teachers, will not reap the full potential from their abstract algebra course if they do not see the connections to secondary school geometry. Therefore, abstract algebra instructors and faculty should consciously consider their individual conceptualizations of abstract algebra and how it affects their students' learning.

Given the emphasis placed on the importance of recognizing mathematical connections between mathematical ideas to build students' understanding of mathematics, and the requirement of most secondary mathematics teacher education programs to include coursework in abstract algebra, identifying and characterizing connections between abstract algebra concepts and secondary school mathematics concepts becomes increasingly critical. It is not sufficient to simply make a list of mathematical connections but rather one must consider the ways in which these connections are discussed. As this study has shown, the participants' own individual conceptualizations of abstract algebra and their experiences with the secondary curriculum greatly influenced which mathematical connections were identified and the ways in which they were described. It is my hope that instructors and faculty reevaluate the way abstract algebra is being taught to undergraduate students, especially prospective secondary mathematics teachers, by examining the approach(es) used to elicit students making meaningful mathematical connections to school mathematics.

Appendix 1: Mathematical Connections List After Interviews

Abstract algebra concept	Secondary school mathematics concept
Algebraic structures (group, ring, integral domain, field) and their properties	Function and domain; identity; inverse; number systems and known operators; solving linear equations
Binary operator	Arithmetic operators and number systems; domain; function; function composition; function transformations
Commutative ring theory (localization)	Fractions
Compass/geometric constructions	Geometry concepts including: Points, lines, circles, regular n-gons, angles, intersection, and trisection
Congruence	Solving linear equations
Cyclic group	Division algorithm; greatest common divisor; imaginary unit; rotations and periodicity
Direct product	Cartesian plane and ordered pairs; matrices for area and volume
Equivalence	Equal sign; inequality; similarity; solving equations
Equivalence classes	Decimal expansions; equivalent fractions; linear functions
Equivalence relation	Congruence; inequality; similarity; symmetry
Extension field/splitting field	Complex numbers; domain; roots of a polynomial
Fundamental theorem of algebra	Roots of a polynomial
Galois theory	Radicals; roots of polynomial equations
Groups and specific types of groups	Function composition; geometric transformations and symmetries
Homomorphism/isomorphism	Equality; function; infinity and finitely infinite; invariance; mapping
Ideal	Number systems; subset
Inverse	Multiplicative reciprocal; negative numbers
Irreducible polynomial	Factoring polynomials
Kernel	Nullspace of a matrix
Lagrange's theorem	Euclidean algorithm; greatest common factor; least common multiple
Nilpotent	Geometric series and convergence
Permutation group; product of cycle decomposition	Function and function composition; permutation; symmetry

Polynomial ring	Operations with polynomials and polynomial long division; polynomial vocabulary (degree, coefficients, roots, etc.); power series
Quotient group/quotient field	Equivalent fractions; fractions and operations with fractions;
Quaternions	Complex numbers
Sign rule in a ring	Product of two negative numbers is positive
Subgroup	Subsets
Unary operators	Negation; trigonometric functions
Unit	Invertible matrices
Zero divisors	Geometric reflections and rotations; solve quadratic equations by factoring

References

Asiala, M., Dubinsky, E., Mathews, D. M., Morics, S., & Oktaç, A. (1997). Development of students' understanding of cosets, normality, and quotient groups. *Journal of Mathematical Behavior, 16*(3), 241–309.

Boaler, J. (2002). Exploring the nature of mathematical activity: Using theory, research and "working hypotheses" to broaden conceptions of mathematics knowing. *Educational Studies in Mathematics, 51*, 3–21.

Brown, A. (1990). Writing to learn and communicate mathematics: An assignment in abstract algebra. In A. Sterrett (Ed.), *Using writing to teach mathematics* (pp. 131–133). Washington, DC: Mathematical Association of America.

Bukova-Güzel, E., Ugurel, I., Özgür, Z., & Kula, S. (2010). The review of undergraduate courses aimed at developing subject matter knowledge by mathematics student teachers. *Procedia: Social and Behavioral Sciences, 2*, 2233–2238.

Businskas, A. M. (2008). Conversations about connections: How secondary mathematics teachers conceptualize and contend with mathematical connections. (Unpublished doctoral dissertation). Burnaby, BC, Canada: Simon Fraser University.

Chappell, M. F., & Strutchens, M. E. (2001). Creating connections: Promoting algebraic thinking with concrete models. *Mathematics Teaching in the Middle School, 7*(1), 20–25.

Charmaz, K. (2000). Grounded theory: Objectivist and constructivist methods. In N. K. Denzin & Y. S. Lincoln (Eds.), *Handbook of qualitative research* (2nd ed., pp. 509–535). Thousand Oaks, CA: Sage.

Cofer, T. (2015). Mathematical explanatory strategies employed by prospective secondary teachers. *International Journal of Research in Undergraduate Mathematics Education, 1*(1), 63–90.

Conference Board of the Mathematical Sciences. (2001). *The mathematical education of teachers (Issues in mathematics education)* (Vol. 11). Providence, RI: American Mathematical Society.

Conference Board of the Mathematical Sciences. (2012). *The mathematical education of teachers II (Issues in mathematics education)* (Vol. 17). Providence, RI: American Mathematical Society.

Cook, J. P. (2012). A guided reinvention of ring, integral domain, and field. (Doctoral dissertation, University of Oklahoma, Norman, Oklahoma). Available from ProQuest Dissertations and Theses database. (UMI No. 3517320).

Council of Chief State School Officers. (2010). *Common core state standards for mathematics.* Washington, DC: National Governors Association Center for Best Practices Author.

Coxford, A. F. (1995). The case for connections. In P. A. House & A. F. Coxford (Eds.), *Connecting mathematics across the curriculum* (pp. 3–12). Reston, VA: National Council for Teachers of Mathematics.
Cuoco, A. (2001). Mathematics for teaching. *Notices of the American Mathematical Society, 48*(2), 168–174.
Czerwinski, R. (1994). A writing assignment in abstract algebra. *Primus, 4*, 117–124.
Dubinsky, E., Dautermann, J., Leron, U., & Zazkis, R. (1994). On learning fundamental concepts of group theory. *Educational Studies in Mathematics, 27*, 267–305.
Findell, B. (2001). Learning and understanding in abstract algebra. Unpublished doctoral dissertation. Durham, NH: University of New Hampshire.
Freedman, H. (1983). A way of teaching abstract algebra. *American Mathematical Monthly, 90*(9), 641–644.
Gallian, J. A. (1976). Computers in group theory. *Mathematics Magazine, 49*(1), 69–73.
Gallian, J. A. (1990). *Contemporary abstract algebra* (2nd ed.). Lexington, MA: D. C. Heath.
Hazzan, O. (1999). Reducing abstraction level when learning abstract algebra concepts. *Educational Studies in Mathematics, 40*(1), 71–90.
Hazzan, O., & Leron, U. (1996). Students' use and misuse of mathematical theorems: The case of Lagrange's theorem. *For the Learning of Mathematics, 16*(1), 23–26.
Hiebert, J., & Carpenter, T. P. (1992). Learning and teaching with understanding. In D. A. Grouws (Ed.), *Handbook of research on mathematics teaching and learning* (pp. 127–146). New York, NY: Macmillan.
Hodgson, T. R. (1995a). Connections as problem-solving tools. In P. A. House & A. F. Coxford (Eds.), *Connecting mathematics across the curriculum* (pp. 13–21). Reston, VA: National Council of Teachers of Mathematics.
Hodgson, T. R. (1995b). Reflections on the use of technology in the mathematics classroom. *Primus, 5*, 178–191.
Huetinck, L. (1996). Group theory: It's a SNAP. *Mathematics Teacher, 89*(4), 342–346.
Larsen, S. (2004). Supporting the guided reinvention of the concepts of group and isomorphism: A developmental research project. (Unpublished doctoral dissertation). Phoenix, AZ: Arizona State University.
Larsen, S. (2009). Reinventing the concepts of group and isomorphism: The case of Jessica and Sandra. *Journal of Mathematical Behavior, 28*(2–3), 119–137.
Leganza, K. (1995). Writing assignments in an abstract algebra course. *Humanistic Mathematics Network Journal, 11*, 29–32.
Leitzel, J. R. C. (Ed.). (1991). *A call for change: Recommendations for the mathematical preparation of teachers of mathematics (MAA Reports No. 3)*. Washington, DC: Mathematical Association of America.
Leron, U., & Dubinsky, E. (1994). *Learning abstract algebra with ISETL*. New York, NY: Springer-Verlag.
Leron, U., & Dubinsky, E. (1995). An abstract algebra story. *American Mathematical Monthly, 102*(3), 227–242.
Leron, U., Hazzan, O., & Zazkis, R. (1995). Learning group isomorphism: A crossroads of many concepts. *Educational Studies in Mathematics, 29*, 153–174.
National Council of Teachers of Mathematics. (2000). *Principles and standards for school mathematics*. Reston, VA: Author.
Papick, I., Beem, J., Reys, B., & Reys, R. (1999). Impact of the Missouri Middle Mathematics Project on the preparation of prospective middle school teachers. *Journal of Mathematics Teacher Education, 2*, 301–310.
Patton, M. Q. (2002). *Qualitative research and evaluation methods* (3rd ed.). Thousand Oaks, CA: Sage.
Pedersen, J. J. (1972). Sneaking up on a group. *Two-Year College Mathematics Journal, 3*(1), 9–12.
Roulston, K. (2010). *Reflective interviewing: A guide to theory and practice*. Thousand Oaks, CA: Sage.

Selden, A. & Selden, J. (1987). Errors and misconceptions in college level theorem proving. Proceedings of the Second International Seminar on Misconceptions in Science and Mathematics, (Vol. 3, pp. 457–470). Ithaca, NY: Cornell University.

Singletary, L. M. (2012). Mathematical connections made in practice: An examination of teachers' beliefs and practices. (Unpublished dissertation). Athens, GA: University of Georgia.

Skemp, R. R. (1987). *The psychology of learning mathematics*. Harmondsworth: Penguin.

Tall, D., & Vinner, S. (1981). Concept image and concept definition in mathematics with particular reference to limits and continuity. *Educational Studies in Mathematics, 12*, 151–169.

Taylor, S. J., & Bogdan, R. (1984). *Introduction to qualitative research methods: The search for meanings*. New York, NY: Wiley.

Usiskin, Z. (1974). Some corresponding properties of real numbers and implications for teaching. *Educational Studies in Mathematics, 5*, 279–290.

Usiskin, Z. (1975). Applications of groups and isomorphic groups to topics in the standard curriculum, grades 9–11. *Mathematics Teacher, 68*(3), 99–106 235–246.

Usiskin, Z. (1988). Conceptions of school algebra and uses of variables. In A. F. Coxford (Ed.), *The ideas of algebra, K–12: NCTM 1988 yearbook* (pp. 8–19). Reston, VA: National Council of Teachers of Mathematics.

Usiskin, Z. (2001). Teachers' mathematics: A collection of content deserving to be a field. *The Mathematics Educator, 6*(1), 86–98.

Zazkis, R. (2000). Factors, divisors, and multiples: Exploring the web of students' connections. *CBMS Issues in Mathematics Education, 8*, 210–238.

Zazkis, R., & Hazzan, O. (1998). Interviewing in mathematics education research: Choosing the questions. *Journal of Mathematical Behavior, 17*(4), 429–439.

Chapter 9
Making Mathematical Connections Between Abstract Algebra and Secondary Mathematics Explicit: Implications for Curriculum, Research, and Faculty Professional Development

James A. Mendoza Álvarez and Diana White

Introduction

This commentary chapter offers a reflection on the four chapters in this section, which focus on connecting abstract algebra to secondary mathematics. The chapters are synergistic and complementary, providing specific examples emerging from an archaeological look at solving quadratic equations (McCallum, Chap. 5), connections from a curriculum for college-track students from over 45 years ago (Smith, Chap. 6), connections taught in a capstone course for preservice secondary mathematics teachers (PSMTs) (Bass, Chap. 7), and connections identified by mathematicians and mathematics educators in a recent qualitative research study (Suominen, Chap. 8). Overall, the chapters in this section provide interesting examples for connecting abstract algebra to secondary mathematics for PSMTs, offer important information about how a selected group of mathematicians and mathematics educators describe these connections, expose professional development needs of faculty seeking to make these desired connections in their undergraduate courses, and generate a call for ideas for further development of curriculum and for further research on how connections are made in the college classroom, how to assess whether PSMTs make the connections, and how to determine the impact on mathematics teaching and learning in the secondary schools. Underlying the ideas in these chapters rests the importance of partnerships between mathematicians and mathematics educators.

J. A. M. Álvarez (✉)
Department of Mathematics, The University of Texas at Arlington, Arlington, TX, USA
e-mail: james.alvarez@uta.edu

D. White
Department of Mathematical and Statistical Sciences, University of Colorado Denver, Denver, CO, USA

N. H. Wasserman (ed.), *Connecting Abstract Algebra to Secondary Mathematics, for Secondary Mathematics Teachers*, Research in Mathematics Education,
https://doi.org/10.1007/978-3-319-99214-3_9

In his chapter, Bass (Chap. 7) sheds light on the challenges confronted in making explicit connections between abstract algebra and secondary mathematics for PSMTs by describing the seeming paradox that while the "power of mathematical abstraction is its generality, thus having the potential to conceptually unify many apparently distinct mathematical contexts" (p. 127), this comes at a cost for a novice learner's sense-making. Furthermore, he describes college courses in abstract algebra as "remote from the needs of K-12 teachers" (p. 128) and that the generality and basic examples in these courses rarely treat the rich algebraic structures present in school mathematics in sufficient detail. This contrast, which includes the potential utility of a course in abstract algebra, coupled with the potential costs—both from a learners' perspective and from a teachers' perspective—help set the stage for our commentary.

In this commentary chapter, we have organized our reflections around: (1) learning goals for PSMTs, which, from these chapters, we characterize as promoting structuralist thinking; (2) curricular materials for PSMTs; and (3) an examination of these efforts, leading to calls both for further research and for addressing the professional development needs of faculty. We integrate various examples from the four chapters as we elaborate on each of these topics.

Learning Goals: Promoting Structuralist Thinking

McCallum's (Chap. 5) "Excavating School Mathematics" uses the metaphor that the school curriculum may be viewed as an "archeological record of the history of mathematics and of previous efforts at reforming school instruction" (p. 87). He reaches to historical examples and redirects them to ways in which a focus on algebraic structure may illuminate mathematical understanding, mathematics teaching, and school curriculum standards.

His case for understanding algebra from Klein's higher standpoint of abstract algebra relies on promoting structures as an organizing principle for (school) mathematics, the significance of which was fully understood under the influence of Noether in the 1920s (Hausberger, 2018). Notably, McCallum's perspective on structure, as related to abstract algebra, aligns with the views of 12 of the 13 mathematicians in Suominen's study (Chap. 8); this structural emphasis might be regarded as slightly different from viewing abstract algebra as a generalization of school algebra.

Hence, promoting PSMT preparation that develops *structuralist thinking* in future teachers establishes a desirable target. Hausberger (2018) characterizes structuralist thinking "in terms of classes of objects, relationships between these classes and (structural) stability of properties under operations on structures" (p. 83). However, difficulties related to undergraduate students' learning of abstract algebra and to their transition to advanced mathematics, as described in Suominen (Chap. 8) and Bass (Chap. 7), need to be considered in both developing structuralist thinking in future teachers and the effectiveness of translating this to the school curriculum.

The specific connections McCallum provides offer tangible examples from which to imagine the practical ways in which understanding structure may influence how secondary mathematics teachers think about, say, connections between completing the square and the quadratic equation; or, say, the failure of PEMDAS to generalize to algebraic thinking. Given this learning goal of developing structuralist thinking in PSMTs, courses in abstract algebra may better serve future mathematics teachers by incorporating more examples like these or by including tasks that require PSMTs to examine school curricula and determine how structuralist thinking may enhance a prospective teacher's mathematical knowledge for teaching and how a structuralist teaching approach may impact the learning of school-age students.

Indeed, the Secondary School Curriculum Mathematics Improvement Study (SSMCIS) material described by Smith (Chap. 6)—which was in fact intended for school-age students (middle and high school)—may be one example of a type of structuralist teaching approach, since each year's materials explicitly refer to "mappings that 'preserve structure'." A comparison of this material with modern resources, as well as the examples provided by McCallum, Bass, and Suominen (Chaps. 5, 7, and 8, respectively, of this volume), may uncover important ways in which attending to structure can be addressed in school mathematics.

McCallum (Chap. 5) also explains how the Common Core State Standards in Mathematics (CCSSM) attempt to incorporate a structuralist perspective in their design. The hope remains that secondary mathematics teachers who have internalized the "higher standpoint" of abstract algebra, and developed structuralist thinking for themselves, may be better equipped to interpret the CCSSM domain and cluster headings that focus on the structure. However, more research is needed to support this hope and to see how this, in turn, would help teachers teach the concepts in a manner that reflects structuralist thinking.

For university faculty who have PSMTs in their abstract algebra courses, an awareness of Verret's "*desynchretisation* of knowledge"—evident in mass university education (as cited in Kondratieva & Winsløw, 2018)—may give insight into how elements of knowledge that were originally combined and united become separated for efficiency or economy of exposition. Such disaggregation is likely undesirable when considering the relationship between abstract algebra and secondary school mathematics, from the perspective of PSMTs. More work to synchronize the topics in abstract algebra courses to school algebra for PSMTs may enable them to fundamentally rethink school mathematics topics, such as the purpose of factoring as a procedure for solving equations. McCallum suggests that this staple of school algebra be recast as, or reunited with, the notion of seeing equivalence between forms and conveying how the factored form of the expression immediately reveals important information about the quantity. Making connections between abstract algebra and school mathematics (i.e., the learning goal of structuralist thinking), at its core, is about uniting (not disaggregating) teachers' mathematical knowledge.

Although Usiskin (as cited in Suominen, Chap. 8), and the Conference Board for the Mathematical Sciences' (CBMS) *Mathematical Education of Teachers* reports (MET and MET II) (CBMS, 2001, 2012) characterize abstract algebra as a generalization of school algebra, Suominen's research reveals that mathematicians and mathematics educators do not solely hold this view. Their prevalent views about structure suggest primary connections between abstract algebra and secondary school mathematics more akin to the structuralist ideas in McCallum (Chap. 5) and Smith (Chap. 6). As mentioned above—and supported by Suominen's conclusions—it is not sufficient to create a list of connections. More work is needed to document how the connections are discussed in the abstract algebra classroom and how they translate to knowledge about school mathematics and teaching.

Providing Curricular Materials

The MET II report (CBMS, 2012) calls for preparing a new generation of teachers with "a coherent view of the structure of mathematics in order to develop reasoning skills in their students" (p. 56). It recommends spending time in undergraduate courses "looking back" at the content of K-12 mathematics as an important and worthy part of a mathematics major that prepares future teachers (p. 54). The original MET report (CBMS, 2001) also promotes examining core ideas in high school mathematics from an advanced standpoint. In addition, the Mathematical Association of America Committee on the Undergraduate Program in Mathematics' (CUPM) 2015 *CUPM Curriculum Guide to Majors in the Mathematical Sciences* recommended that PSMTs not only explore advanced mathematics, but also examine and engage with school mathematics content from an advanced perspective.

Substantial curricular improvement in the undergraduate preparation of PSMTs requires a cultural shift within the university mathematics community and significant attention to the professional development of the faculty teaching or designing these courses. The MET report (CBMS, 2001) calls for the development of new undergraduate capstone courses, as well as a major redesign of core mathematics major courses, but little is documented about successful attempts to redesign core courses in response to the MET report. That is, in terms of curricular materials for such courses for PSMTs, there is a dearth of available resources. In addition, the MET II report (CBMS, 2012) reiterates the need for a direct focus on the content of high school mathematics from an advanced perspective *throughout* a preservice teacher's preparation, not just in a capstone experience.

Smith (Chap. 6) gives an overview of the SSMCIS material developed over 45 years ago with the goal of presenting a unified treatment of mathematics through six courses designed for school-age students, which were linked to students' interests and abilities, rather than grade-level ties. This is an example of a curricular resource, albeit one for school-age students. A key component in the SSMCIS development appears to derive from the rich partnership between mathematicians and mathematics educators. Partnerships and collaborations between mathematicians, mathematics educators, and mathematics education researchers remain critically important today; undoubtedly, we see a need for increasing opportunities to establish productive partnerships between mathematicians and mathematics educators.

In his chapter, Smith points out that abstract algebra only entered the undergraduate curriculum in the 1960s and that student-friendly undergraduate textbooks did not exist when these collaborations and curriculum development began. Yet we regard the materials—long out of print—as a potentially useful resource in connecting concepts in abstract algebra courses to secondary mathematics for PSMTs. As a curriculum, Smith asserts that SSMCIS materials serve as an existence proof of sorts that creation of this type of material is possible. Specifically, he states that the SSMCIS material "demonstrates that it is possible to incorporate sophisticated mathematical content into high school and junior high school curricula" (p. 103). More concrete examples, similar to those in McCallum (Chap. 5), may arise from a careful revisiting of these materials from the perspective of connecting abstract algebra to school mathematics for teachers. With this in mind, PSMTs (not just secondary students) may benefit from close examination of examples that connect their undergraduate mathematics knowledge to topics from school algebra, such as the example given regarding connecting "A Field of 2×2 Matrices" to the complex number system.

The SSMCIS material provides an important perspective for thinking about ways to connect abstract algebra to secondary mathematics for secondary mathematics teachers. Since the materials were written with college-bound junior high and high school students in mind, adaptations *of* the material, or inspirations *from* the material, could produce tasks or curricular materials for PSMTs as a means to help them conceptualize a more unified and structuralist approach to school mathematics. Few would argue that the profile of college-bound students has changed significantly over the last 45 years, and thus the importance of considering the design of the materials for this select population is critical. The research literature on the learning of abstract algebra (e.g., Suominen, Chap. 8) reveals the difficulties students face in this course and, from the personal experience of one of the authors of this chapter, the "basic idea of an isomorphism of groups" is still quite challenging for many undergraduates, despite having had an entire course introducing them to mathematical proofs. The manner in which these ideas were addressed in the SSMCIS Courses 7 and 8 may inform curriculum designers or research design related to this topic.

More explicit in terms of curricular materials for PSMTs, Bass (Chap. 7) offers a description of a capstone course he designed for PSMTs. The course aims to make curricular connections across mathematical topics and presents novel problem-solving formats to prompt connection-making. He also offers suggestions related to pedagogy that would be central to this treatment for future teachers, such as collaboration and group work and an emphasis on mathematical exposition, explanation, and justification.

Although this capstone course does not solely focus on abstract algebra, the topics in algebra presented provide useful examples for how connections across mathematical topics for PSMTs can be approached. The problem formats offer a promising design for prompting students to make connections. The idea of sorting a diverse group of problems by grouping those that are essentially the same provides PSMTs important practice at seeing structural connections between seemingly unrelated problems. The other format presents students with a set of problems that share a common mathematical structure. Students must identify and explain the common structure, as well as show how it arises in each of the problems. These problem formats provide opportunities for PSMTs to engage in making connections; however, it is unclear how these connections develop unique mathematical understandings for teaching. Abstract algebra courses, and the materials in this capstone course, to some extent, begin with a grounding of mathematical topics at the advanced level and connections are then made vertically to school mathematics, either explicitly or implicitly. In contrast, Epperson and Rhoads (2015) describe "high-yield mathematical tasks for secondary teachers" that are grounded first in the work of secondary school mathematics and teaching, and, by connecting to the advanced level, aim to build deeper and more flexible understandings of the secondary concepts, and reinforce mathematical habits of mind. More examples of tasks for PSMTs that emanate from school mathematics and create a need to understand structure and connections to higher level mathematics need to be developed.

Research and Implications

Suominen's chapter (Chap. 8) reports on a study of 13 mathematicians and mathematics educators in which she conducted semi-structured interviews to probe the mathematical connections and connections for teaching that they identified between abstract algebra and secondary school mathematics. She adopted *mathematical connections* as described in Businskas (2008) and Singletary (2012) to inform the list of mathematical connections that arose from her interview data, and used theoretical sampling throughout the interviews to support the validity of this list. Like others, she asserts that the importance of a course in abstract algebra for prospective secondary mathematics teachers can "help them explain and unite concepts found in school mathematics" (p. 149). However, it is recognized that more research is needed to validate this claim; also, she acknowledges that PSMTs probably do not derive benefits from learning abstract algebra in the absence of explicit mathematical connections between abstract algebra and secondary school mathematics and cites the MET report (CBMS, 2001) and Cofer (2015) to support the latter.

Her work sets out to possibly list and describe mathematical connections between abstract algebra and secondary school mathematics as indicated by her participant pool since "little work has been done on specifically classifying the types of mathematical connections between abstract algebra and school mathematics" (p. 150). Her findings support the notion of structure (i.e., developing structuralist thinking) as a critical mathematical connection identified by 12 of the 13 participants. Specific examples are offered, such as the connection between exponential and logarithmic rules, $e^x e^y = e^{x+y}$ and $\log ab = \log a + \log b$, which follows from the fact that the real numbers under addition is isomorphic to the positive real numbers under multiplication. This also relates to an example in the chapter by Smith (Chap. 6) that describes an example for the SSMCIS material Course 4 in which students are asked, "Which arithmetic problem would you prefer to solve: (a) 2.40 • 2.75 or (b) .3802 + .4393?" (p. 110) The prevalent use of technology today may render the latter less relevant and may require some rethinking to adapt the question in a way that accomplishes the goal of focusing on structure.

Directions for Research

Suominen's work outlines *mathematical connections* between abstract algebra and secondary mathematics. This is important work; yet further studies that examine not just lists of mathematical connections, but how PSMTs might develop knowledge of such mathematical connections, are needed and important. In addition, the data on *teaching connections* between abstract algebra and school mathematics does not provide strong evidence that Suominen's group of participants were making substantive teaching connections. There is a vague description about how secondary mathematics teachers can employ what they know about structure to help them plan a lesson, but no specific example is given and the assertion seems mostly hypothetical in nature. Other teaching connections mentioned convey that the knowledge gained in abstract algebra helps teachers come up with examples in their teaching and helps them decide or know whether something is true. Again, without more examples from the classroom and interviews with secondary mathematics teachers, these connections, while the hope is that they are true, lack sufficient evidence (e.g., Zazkis & Leikin, 2010). More research and scholarship in this area is needed. This speaks to the need for empirical studies that might investigate how knowledge of such mathematical connections might translate into teaching connections. In addition, and equally important, the professional development needs of faculty teaching these courses must be considered, since there appear to be too few resources [we anticipate this volume to serve as at least one starting point] and little opportunity to make these connections before being tasked to teach a course that explicitly attends to the needs of PSMTs. We address this more explicitly in the subsequent section.

McCallum states, "A mathematician brings to the examination of school mathematics a set of structural sensibilities born of the accumulated architecture of the subject itself" (p. 100). This statement suggests that exposing secondary mathematics teachers to archaeological or historical artifacts related to school mathematics topics in algebra may help them understand the purpose and structure of the topics they teach. Thus, further curricular development and research investigating the degree to which incorporating courses about the history of mathematics into teacher education are productive is another line of inquiry that may help us investigate ways to teach abstract algebra courses to PSMTs that are meaningfully connected to their future professional work. Notably, we regard all of this future work as requiring that mathematicians, mathematics educators, and mathematics education researchers collaborate in order to (1) characterize structural sensibilities, (2) develop curriculum that attends to these with respect to school mathematics, and (3) create or apply theoretical frameworks to study the impact on teacher knowledge, teacher teaching, and student learning.

Professional Development for Faculty

To achieve and support substantive improvements in courses for PSMTs, faculty teaching or designing these courses must have access to professional development experiences that address how to incorporate connections of advanced mathematical content to school mathematics, how to design and implement tasks that elicit these connections, and how to include active learning strategies for engaging students in making these connections—such as the two problem formats advocated for by Bass (Chap. 7).

Most faculty have not had opportunities to think deeply about the connections between advanced undergraduate mathematics (e.g., abstract algebra) and school mathematics, or how they would inject this into their teaching. This relates to Souminen's (Chap. 8) finding in her study that the pure mathematicians who had only published in mathematics journals identified fewer connections than mathematics educators, who often have some secondary teaching experience. Since graduate-level abstract algebra courses designed for mathematics students generally focus on a streamlined approach to the essential content, we see this narrow focus as potentially explicating Souminen's findings. As such, in graduate mathematics studies, overt connections to school mathematics are rarely, if at all, discussed; in cases where the connections arise explicitly, it may be seen as an interesting afterthought, but rarely assessed in any manner. As educators, we recognize that students focus on topics that will be assessed and the rest often "gets lost." As such, many mathematicians graduate from their doctoral programs with a key gap in the mathematical content knowledge needed for effective teaching of abstract algebra courses that make connections to school mathematics for PSMTs.

For example, mathematicians may take numerous graduate courses in abstract algebra, even earning their doctorate in some subfield of it, without making an observation that the four group axioms are required to complete the steps needed to solve a linear equation of the form $x + a = b$, where a, b are integers—an example Wasserman (2014) used to describe vertical content knowledge for school mathematics teachers. Observe the following:

$$\begin{aligned}
x + 7 &= 3 \\
x + 7 + (-7) &= 3 + (-7) && \text{(equivalence, not a group axiom)} \\
x + [7 + (-7)] &= 3 + (-7) && \text{(associativity)} \\
x + 0 &= 3 + (-7) && \text{(inverse)} \\
x &= 3 + (-7) && \text{(identity)} \\
x &= -4 && \text{(closure)}
\end{aligned}$$

This simple illustration shows how an elementary topic in secondary school mathematics (equation solving) gives rise precisely to the four group axioms. That is, to solve it, the operation and the elements of the set must satisfy these properties. However, while accessible to students in an abstract algebra course, this is rarely explicitly discussed or connected for undergraduate and graduate students.

Some graduate programs, especially in departments that emphasize teacher training and teacher professional development, provide opportunities for graduate students to assist with professional development programs for teachers, and to teach mathematics courses for future elementary teachers. These opportunities are ripe with what are often new mathematical approaches and new perspectives on school mathematics for these graduate-student mathematicians. Incorporating such experiences may be a low-hanging fruit for future faculty. However, the profession must also call for more coordinated professional development targeted toward faculty with responsibility for teaching and designing mathematics courses for PSMTs that highlight explicit links to school mathematics.

As a profession, we have, and still do, struggle to shape and reshape the undergraduate mathematics major from one focused primarily on preparing students for graduate school in mathematics to one that positions students for a much wider variety of careers (including not only mathematical research but also secondary mathematics teaching, etc.). Courses and books focused on connections to high school mathematics have been developed, including ones specifically aimed at abstract algebra (e.g., Cuoco & Rotman, 2013). Yet such textbooks may look foreign to many mathematicians, as they may not adhere to a standard "groups, rings, fields" approach.

Mathematicians, like teachers, tend to teach how they were taught. The preparation of mathematicians to conduct mathematical research, or the minimal experiences they receive in the basics of teaching, do not prepare them with the specialized knowledge of teaching mathematics that is useful for teaching an abstract algebra course to undergraduate students. Research indicates that professional development for faculty may be most effective when the issues related to effective pedagogy or task design emanate from a faculty member's sense of content expertise. For example, for mathematics faculty involved in a professional development seminar focused on teaching concepts of proof, Blanton and Stylianou (2009) drew upon the depth of faculty content knowledge to create content-centered discussions to "motivate issues concerning curriculum, mathematical tasks, and classroom practice" (p. 81). They built a community of practice (Lave & Wenger, 1991; Wenger, 1998) grounded in discussions of mathematical content to highlight issues of practice. Evidence from the K-12 arena supports the idea that professional development is more effective in prompting improvements and supporting innovations when it is "situated in practice" (Smith 2001 as cited in Blanton & Stylianou, 2009). Hence, as we consider this pressing need for professional development of faculty teaching such courses to PSMTs, we suggest finding ways both to leverage their content expertise and to develop meaningful communities of practice.

Conclusion

The chapters by McCallum, Smith, Bass, and Suominen converge on emphasizing structure as a way to connect abstract algebra to school mathematics for secondary mathematics teachers. This aligns with the beliefs of other mathematics educators and mathematicians that argue that recognizing form should be emphasized in school algebra (e.g., Wheeler, 1996; Pimm, 1995). Kieran (2007) asserts that algebra teachers' pedagogical content knowledge, especially for beginning teachers, lacks awareness of students' misconceptions and teaching strategies to address them. The challenge we see ahead for the field rests in connecting what PSMTs understand about algebraic structures to the actions they take in the classroom (or in lesson planning) and the ways in which this enhances their students' understanding of algebra.

Good examples have been offered in these chapters, which convey the idea of connecting abstract algebra to school mathematics for secondary teachers with an emphasis on structural thinking. Yet it is clear that the mathematics and mathematics education community must create more meaningful examples that derive not only from the advanced mathematics, but also from the school mathematics (cf. Bass, Chap. 7; Epperson & Rhoads, 2015). Research on the impact of the curriculum development efforts on PSMTs' mathematics learning, and how this links to their work as teachers of algebra, remains a critical component to tie together these efforts.

As noted, to make a broad impact on the mathematical education of PSMTs, faculty who embed curricular materials centered on structuralist thinking into their abstract algebra courses require access to professional development that supports their implementation efforts. There is much across these chapters for a mathematician or mathematics educator to ponder. Several challenges have been outlined with respect to PSMT learning, research-based curriculum development, and professional development for faculty teaching abstract algebra for PSMTs. The structural sensibilities we strive to create for PSMTs are intimately connected to a coherent educational effort that attends to the learning goals (i.e., meaningful connections between abstract algebra and school mathematics), the curricular goals (i.e., creating materials that address learning outcomes), and examination of these efforts from a research perspective.

References

Blanton, M. L., & Stylianou, D. A. (2009). Interpreting a community of practice perspective in discipline specific professional development in higher education. *Innovative Higher Education, 34*(2), 79–92.

Cofer, T. (2015). Mathematical explanatory strategies employed by prospective secondary teachers. *International Journal of Research in Undergraduate Mathematics Education, 1*(1), 63–90.

Conference Board of the Mathematical Sciences. (2001). *The mathematical education of teachers*. Providence RI and Washington DC: American Mathematical Society and Mathematical Association of America.
Conference Board of the Mathematical Sciences. (2012). *The mathematical education of teachers II*. Providence RI and Washington DC: American Mathematical Society and Mathematical Association of America.
Cuoco, A., & Rotman, J. J. (2013). *Learning modern algebra: From early attempts to prove Fermat's last theorem*. Washington, DC: Mathematical Association of America.
Epperson, J. A. M., & Rhoads, K. (2015). Choosing high-yield tasks for the mathematical development of practicing secondary teachers. *Journal of Mathematics Education at Teachers College, 6*(1), 37–44.
Hausberger, T. (2018). Structuralist praxeologies as a research program on the teaching and learning of abstract algebra. *International Journal of Research in Undergradraduate Mathematics Education, 4*(1), 74–93.
Kieran, C. (2007). Learning and teaching algebra at the middle school through college levels: Building meaning for symbols and their manipulation. In F. K. Lester Jr. (Ed.), *Second handbook of research on mathematics teaching and learning* (Vol. 2, pp. 707–762). Reston, VA: National Council of Teachers of Mathematics (NCTM).
Kondratieva, M., & Winsløw, C. (2018). Klein's plan B in the early teaching of analysis: Two theoretical cases of exploring mathematical links. *International Journal of Research in Undergraduate Mathematics Education, 4*(1), 119–138.
Lave, J., & Wenger, E. (1991). *Situated learning: Legitimate peripheral participation*. Cambridge: Cambridge University Press.
Mathematical Association of America. (2015). *2015 CUPM curriculum guide to majors in the mathematical sciences*. Washington, DC: Mathematical Association of America Author.
Pimm, D. (1995). *Symbols and meanings in school mathematics*. London: Routledge.
Wasserman, N. (2014). Introducing algebraic structures through solving equations: Vertical content knowledge for K-12 mathematics teachers. *PRIMUS, 24*(3), 191–214.
Wenger, E. (1998). *Communities of practice: Learning, meaning, and identity*. Cambridge: Cambridge University Press.
Wheeler, D. (1996). Backwards and forwards: Reflections on different approaches to algebra. In N. Bernarz, C. Kieran, & L. Lee (Eds.), *Mathematics education library, Approaches to algebra* (Vol. 18, pp. 317–325). Dordrecht, The Netherlands: Springer.
Zazkis, R., & Leikin, R. (2010). Advanced mathematical knowledge in teaching practice: Perceptions of secondary mathematics teachers. *Mathematical Thinking and Learning, 12*(4), 263–281.

Part III
Exploring Approaches to Secondary Teacher Education: Engaging in Abstract Algebra in Relation to Developing Disciplinary Practices

Chapter 10
Making Connections from the Secondary Classroom to the Abstract Algebra Course: A Mathematical Activity Approach

Rose Mary Zbiek and M. Kathleen Heid

The very relevance of Abstract Algebra,[1] the theoretical foundation of School Algebra, to the algebra taught in secondary mathematics classrooms is regularly questioned by secondary teachers (e.g., Zazkis & Leikin, 2010). Prospective teachers ask, "What good is what I am learning?" and practicing teachers ask, "Does what I have learned in courses like Abstract Algebra have anything to do with what I am teaching?" This disconnect between School Algebra and Abstract Algebra, manifested in a compartmentalization of representations of the same essential algebraic concept in the two settings, has been identified even in some students who were highly successful in their Abstract Algebra course (Wasserman, 2017). Perhaps the extent to which prospective and current secondary mathematics teachers do not see a connection between Abstract Algebra and School Algebra is a signal of work that needs to be done to improve students' experiences in both arenas.

An Alternative Approach to Connecting School Algebra and Abstract Algebra

Several practitioners and scholars have pursued development and implementation work to understand and disrupt the traditional teaching of School Algebra and

[1]We capitalize Abstract Algebra and School Algebra not to refer to particular courses, but to highlight and juxtapose the two mathematical settings.

R. M. Zbiek (✉) · M. K. Heid
Department of Curriculum and Instruction, The Pennsylvania State University, University Park, PA, USA
e-mail: rzbiek@psu.edu

N. H. Wasserman (ed.), *Connecting Abstract Algebra to Secondary Mathematics, for Secondary Mathematics Teachers*, Research in Mathematics Education,
https://doi.org/10.1007/978-3-319-99214-3_10

Abstract Algebra. Mathematicians and mathematics educators, like practicing and prospective teachers, often question the relationship between School Algebra and Abstract Algebra, as they study or teach Abstract Algebra and look for direct connections between the content of Abstract Algebra and the content of School Algebra. A different approach to seeing the contribution of Abstract Algebra to the preparation of secondary mathematics teachers reverses the direction of the questioning. Rather than beginning with Abstract Algebra, we begin with School Mathematics, specifically with the lived experiences of teachers, as we examine the work of preservice teachers and current teachers to locate opportunities to identify what Abstract Algebra experiences could inform teachers in these moments. Since this approach begins in the work of teaching, questions of how the examples are relevant to teaching secondary school mathematics are implicitly addressed.

We also challenge questions about the usefulness of Abstract Algebra to secondary mathematics teachers by focusing on mathematical activity, or practices, rather than the content details of particular courses. Our Mathematical Activity approach is consistent with recent efforts to extend school mathematics beyond a focus on content comprised of such things as procedures and concepts to a focus in the spirit of recent attention to practices (Common Core State Standards Initiative, 2010), processes (Zbiek, Heid, & Blume, 2012), and habits of mind (Cuoco, Goldenberg, & Mark, 1996). The approach also resonates with Cook's (2014) approach to teaching abstract algebra through mathematical activities, such as the engagement of students in the mathematical activity of defining a structure as a means of learning about it. Since mathematical activities are touted as important across multiple areas, it might seem odd to limit their discussion to School Algebra and Abstract Algebra. Rather than serving to limit attention, the focus on algebra reflected our selection of a mathematical area that has great impact. School Algebra is widely taught in secondary schools and it is often taught by beginning school mathematics teachers, and Abstract Algebra is commonly required in teacher preparation programs. Making the practices explicit in these two scenarios could affect a large number of school students and prospective teachers, and help them establish early in their careers that school mathematics and collegiate mathematics are connected in useful ways.

The Mathematical Activity connection between the two educational settings suggests the need to revisit and extend these efforts. In particular, what would happen if Abstract Algebra were taught with an attempt to foreground the mathematical activities? And, what would School Algebra look like if it assumed the mathematical activities should be explicit? In this chapter, we argue for that approach and explore what it offers not only for improving connections between School Algebra and Abstract Algebra, but also for challenging the ways in which mathematics is taught in these two settings. Our description of Mathematical Activity and examples of classroom events follow from portions of our earlier work with the Situations Project (cf., Heid, Wilson, & Blume, 2015), which we briefly describe in the following section.

The Situations Project: Starting with the Work of Teachers

The MAC-MTL–CPTM[2] Situations Project, or Situations Project, started with mathematical opportunities that arose in the context of secondary school mathematics teaching—including, but not limited to, School Algebra settings. Each mathematical opportunity grew into a Situation that includes several components, each with a specific purpose.

Descriptions of these mathematical opportunities are called Prompts. For each Prompt, the MUST Project team developed a collection of mathematical understandings on which teachers might draw in the given setting. Expanded descriptions of these strands of mathematical understanding were developed and evolved into what are called Mathematical Foci. Discussions of connections and extensions of the mathematical ideas for each set of Foci are called Commentaries. The set consisting of the Prompt, related Mathematical Foci, and various Commentaries constituted what the MUST Project team called a Situation. Almost 50 Situations were fully developed, and these Situations were analyzed to develop a framework: Mathematical Understanding for Secondary Teaching.

The collection of Mathematical Foci across all MUST Situations constitute a reservoir of descriptions of mathematical knowledge that teachers could productively use in the context of teaching school mathematics. The attention to secondary mathematics coupled with actual classroom events we had witnessed as the starting point distinguished the MUST project from other efforts to understand what secondary mathematics teachers know or could use. An extensive analysis of those descriptions (see Heid et al., 2015) generated a triune framework to capture the breadth and nature of mathematics for teaching secondary school. The framework accounts concurrently for types of mathematical understanding (Mathematical Proficiency), types of mathematical actions (Mathematical Activity), and contexts in which classroom practice calls for a teacher's mathematical knowledge (Mathematical Contexts of Teaching). Although mathematical understanding for teaching includes all three perspectives, it is the second of these perspectives that provides the opportunity that we take in this chapter to focus on Mathematical Activity, rather than on particular content topics.

Strands of Mathematical Activity

The MUST team synthesized the Mathematical Foci of the Situations as three strands of the Mathematical Activity perspective: *Mathematical Noticing*, *Mathe-*

[2]The MAC-MTL–CPTM Situations Project was an NSF-funded collaboration between the Mid-Atlantic Center for Mathematics Teaching and Learning and the Center for Proficiency in Teaching Mathematics involving mathematics education faculty and graduate students at The Pennsylvania State University (MAC-MTL) and the University of Georgia (CPTM). The project produced the Mathematical Understandings for Secondary Teaching (MUST) framework.

matical Reasoning, and *Mathematical Creating*. Each of these strands is further explicated in terms of particular types of mathematical work in which teachers and School Algebra students engage.

Mathematical Noticing involves *noticing structures of mathematical systems*, *noticing symbolic form*, and *noticing the form of a mathematical argument*. Noticing these mathematical entities also involves recognizing the actions that can appropriately be applied to those structures and forms. At the secondary level, students are regularly asked to decide on and perform symbolic manipulation, requiring them to recognize the structure of symbolic forms (e.g., expressions or equations) and to generate equivalent forms. Recognizing structures and symbolic forms requires being aware of the similarities and differences among structures and forms, and using this awareness to choose wisely among available strategies and actions. Secondary teachers need to help their students focus on making and using these similarities and differences. Mathematical noticing also appears in the form of *connecting within and outside mathematics* as secondary school students learn to interpret algebraic expressions, with meaning in mathematical contexts (e.g., features of graphs from literal expressions) and to the world as they know it (e.g., meaning of rates of changes as unit costs). Secondary teachers need to develop a broad repertoire of connections among facets of the mathematics they are teaching and between the mathematics they teach and the mathematics in the world around them.

The *Mathematical Reasoning* strand is not limited to *justifying/proving* for the production of mathematical arguments. It also includes *reasoning when conjecturing and generalizing*, as in the generation and testing of mathematical claims, and *constraining and extending* systems and objects. At the secondary level, students generate conjectures about mathematical relationships and test their generalizations with examples, counterexamples, and informal arguments. Secondary students might also be exposed to systematic deductive arguments to support mathematical claims as they encounter proofs in addition to less formal justifications. Secondary teachers need proficiency in a broad range of proof forms, allowing them to vary the level of formality of their proofs without compromising the integrity of the argument. Extending and constraining systems and objects at the secondary level includes, for example, students' reasoning that when they extend their prior experience with real numbers to other number systems, expanding or limiting domains of functions is required. When familiar mathematical systems and objects are constrained or extended, students need to recognize the need for modifications of the familiar entities (e.g., when matrices are multiplied, the "zero-product rule" does not apply). Secondary teachers need to model, and develop in their students, the ability to constrain and extend mathematical systems and objects in ways that account for related modifications.

The *Mathematical Creating* strand involves *representing*, *defining*, and *modifying/transforming/manipulating*. Engaging in mathematical creating requires generating new ways to explain mathematical objects and transforming familiar representations of mathematical objects into alternative forms. Secondary mathematics teachers engage in mathematical creating in their daily work as they assess and challenge students' thinking by developing examples, counterexamples, and alternative representations to explain a mathematical idea. As they assess students' descriptions of mathematical ideas, secondary teachers need to identify, articulate, evaluate, create, and draw on sets of defining attributes for mathematical objects. Secondary mathematics teachers need to be adept at transforming symbolic forms and other mathematical representations to create the forms that will give them the information they seek.

In the following section, we use seven MUST Prompts as examples to argue that the Mathematical Activity perspective reveals mathematical actions that cut across School Algebra and Abstract Algebra. They are activities in which professors and Abstract Algebra students, as well as School Algebra teachers and students, engage frequently. We focus on examples that are especially relevant in connecting the mathematical activity of Abstract Algebra to the mathematical activity of and for School Algebra. For each example, we note a Prompt to connect directly to School Algebra as lived by teachers in their practice and state explicitly what the mathematical activity is in School Algebra and/or in Abstract Algebra. We also discuss how the example is faithful to both School Algebra and Abstract Algebra. Although prompts in this chapter are from the Situations Project, the analysis of the examples, with its focus on Mathematical Activity in Abstract Algebra, is new work.

Examples from the Work of Teaching

If School Algebra connects to Abstract Algebra through mathematical activity, examples from the work of teaching secondary school mathematics should exemplify both mathematical activity and connections to Abstract Algebra. The collection in this section of seven examples related to the MAC-MTL–CPTM Situations Project are evidence of the connection between School Algebra and Abstract Algebra through mathematical activity. The examples are loosely organized by the mathematical activity they evince—Noticing, Creating, and Reasoning—and to a lesser extent, by the sophistication of the School Algebra content.

Each example begins with a few introductory words and a statement or description of a Prompt that captures an incident from the work of teaching school mathematics. After descriptions of related School Algebra events and Abstract Algebra events that involve the same mathematical activities, the connections between School Algebra and Abstract Algebra are highlighted in terms of mathematical activities.

Example 1: Graphing Quadratic Functions with Mathematical Noticing, Specifically Noticing Symbolic Form

School algebra includes the study of a variety of properties of each of several different types of functions. The challenge of teaching algebra for understanding is to help students make sense of the properties and not simply memorize them. A property of quadratic functions is the topic of the prompt from Situation 21, Graphing Quadratic Functions.

> When preparing a lesson on graphing quadratic functions, a student teacher found that the textbook for the class claimed that $x = \frac{-b}{2a}$ was the equation for the line of symmetry of a parabola $y = ax^2 + bx + c$. The student teacher wondered how this equation was derived.
> (Situation 21, MUST, p. 257)

In School Algebra, teachers know that $x = k$ is the equation of the vertical line of symmetry. Thinking about the equation of the parabola in terms of $f(x) = ax^2 + bx + c$, teachers can notice that symmetry requires $f(k - 1) = f(k + 1)$ for some real number, k. They can also see $f(k - 1)$ and $f(k + 1)$ as two different names for the same number, though they do not know—and do not need to determine—a specific value for this number. Rewriting $f(k - 1) = f(k + 1)$ as $a(k - 1)^2 + b(k - 1) + c = a(k + 1)^2 + b(k + 1) + c$ requires interpreting both $k - 1$ and $k + 1$ as single, real numbers.

To see parallel mathematical activity in Abstract Algebra, one can turn to cosets. If H is a coset with elements a and b, then $aH = bH$. This idea is similar to the School Algebra use of $f(k - 1) = f(k + 1)$. In both scenarios, two different expressions name the same thing—the same coset or the same function value. In each case, the two expressions look somewhat similar, but may cause students to pause. There could be something unsettling about setting $f(k - 1)$ equal to $f(k + 1)$ when the former involves a difference and the latter involves a sum. In the Abstract Algebra case, aH and bH look similar, but might cause one to have that moment of thinking, inappropriately, that a and b would then have to be the same element.

Discussion of cosets can also require thinking about a symbol string as a single object, paralleling the way in which the School Algebra setting requires interpreting both $k - 1$ and $k + 1$ as single real numbers. In a discussion of cosets, for a and b in coset H, we might rewrite b as ah for some h in H and know hH is another name for H. As a result, $bH = ahH = aH$. In this example, ahH needs to be interpreted as ah applied to H and as a applied to coset hH, which is H. Interpreting a string of symbols ($ahH, hH, k - 1, k + 1$) both as a relationship among two or more things and as one thing is critical to following the argument about cosets and the equality of the two function values.

The mathematical activity highlighted in this example is one of *Mathematical Noticing* with a focus on *noticing symbolic form*. Teachers need to recognize the symbolic forms as real numbers, as well as sums or differences in School Algebra, and as cosets, as well as combinations of cosets and elements thereof in Abstract Algebra. They also need to look at the symbolic forms and know what rules or properties could be applied, such as the properties used in writing bH, then ahH, and then aH. Abstract Algebra, in this case, does not provide additional insights into the specific concepts involved in the function argument, but the experience does challenge a student's ability to parse and interpret strings of symbols.

When teaching about cosets, Abstract Algebra instructors can foster students' abilities to notice symbolic form by focusing their students on parsing symbolic expressions in several different ways, for example interpreting ahH as ah applied to H and as a applied to coset hH. This work with parsing expressions could consider the polysemic and homonymous roles of symbols in Abstract Algebra, paralleling Zazkis and Kontorovich's (2016) elaboration on the polysemy and homonymous nature of symbols in function translation tasks. Each of hH, and ahH needs to be associated with the single term "coset." Yet ahH must be understood both as a coset and as a applied to coset hH, with the meaning implied by its context.

As Zazkis, Liljedahl, and Gadowsky (2003) observed, "–3" was a surface level cue for translation of $f(x) = x^2$ to lower values for $f(x) = x^2 - 3$, but the surface level cue was misleading when translating $f(x) = x^2$ to obtain $f(x) = (x - 3)^2$. In the Abstract Algebra discussion, we note how italicized characters function as surface level cues that a, h, and ah are elements. However, italicized characters as a surface level cue can become an obstacle when ahH needs to be viewed as a coset, rather than an element. As the existence of the function translation example suggests, focusing on the meaning of symbolic expressions only when studying cosets, however, is not sufficient support for developing a tendency to notice symbolic form. Attention to interpreting symbolic forms that permeates the study of Abstract Algebra, however, has the potential to help students develop this tendency.

Example 2: Connecting Factoring with the Quadratic Formula with Mathematical Noticing, Specifically Connecting within Mathematics

The strength of mathematical ideas sometimes lies in how broadly these ideas can be represented. Exploration of several mathematical representations for the same mathematical structure can shed light on the properties of those structures. In the following example, a student is looking for connections between two processes—as will be shown, those processes can be connected through alternate representations.

A teacher who had completed a unit on factoring quadratic polynomials began a unit on the quadratic formula. One student asked whether there was a direct connection between factoring quadratic polynomials and the completing the square process used to establish the quadratic formula.
(Adapted from MUST, p. 263)

A major portion of students' time in School Algebra is historically spent on factoring quadratic polynomials. One of the ways that teachers establish the quadratic formula is through a process called "completing the square." To make the process more concrete, teachers form an analogy between the factors of a quadratic and the area of related rectangles, with side lengths corresponding to the factors of the quadratic. A rectangle with sides of length $x + 3$ and $x + 5$ is shown to be comprised of an $x \times x$ square, eight $x \times 1$ rectangles, and fifteen 1×1 squares, as shown on the left side of Fig. 10.1. The process of completing the square can be represented as a rearrangement of these squares by moving the pieces labeled 1, 2, 3, and 4 to create the arrangement on the right side of Fig. 10.1, which can be interpreted as $(x + 4)^2 - 1$. The equivalence of the areas supports the claim that $(x + 3)(x + 5)$ is equivalent to $(x + 4)^2 - 1$.

Representing relationships among polynomials, as done in School Algebra by using area as a proxy for products, helps to concretize mathematical relationships. With a "hands-on" representation of a mathematical idea, students can reason about the mathematical idea. For instance, students familiar with the area representation for products of polynomials can envision the sometimes mysterious (to students) process of completing the square.

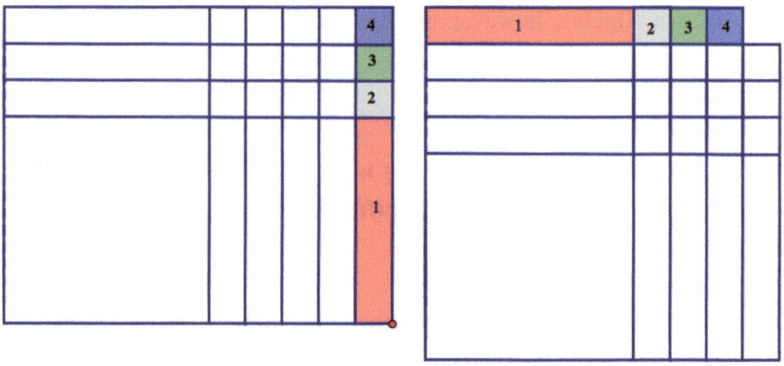

Fig. 10.1 Regions represented by components of $(x + 3)(x + 5)$, shown on the left of the figure, are rearranged to show its equivalence to $(x + 4)^2 - 1$, whose area model is shown on the right of the figure

In Abstract Algebra, there are also opportunities for concrete representations of mathematical constructs. One popular example is the group of permutations of a set of order three (with composition as its operation). The group can be exemplified as the set of mappings of an equilateral triangle to itself. Rotations of 120°, 240° and 360°, and reflections over each of the medians, correspond to the members of this permutation group. The elements of this group can be embodied in the six placements of an equilateral triangular card, whose front and back are distinguishable, such that each placement aligns the card with the card's original position. The properties of this set of placements can be explored through (acceptable) physical manipulations of the triangular card. Another example that can concretize ideas in Abstract Algebra is in the context of cyclic groups. Students can gain insight into the order of an element in cyclic groups by physically rotating about its center a marked block in the shape of a regular polygon. A sequence of physical turns is a composition of rotations. A sequence of turns through a given angle that places the block in its original position corresponds to the identity, and the number of turns in this sequence is the order of the element. Physical objects that represent abstract elements and a physical operation, defined on those elements, that captures the properties of algebraic structures can open up those ideas to exploration.

The mathematical activity of *Mathematical Noticing*, as particularized in *connecting within and outside mathematics*, is central to both the example from School Algebra and the examples from Abstract Algebra. Work in Abstract Algebra that involves students in exploration of tangible representations of mathematical structures opens the Abstract Algebra curriculum to "hands-on" explorations of abstract mathematical ideas. Students engaged in these activities have an enhanced opportunity to notice that different manifestations of the same mathematical structure have the same characteristics. Just as the exploration of different representations of a mathematical structure afford students the opportunities to notice the characteristics of the structure, it also may provide a venue that can foster inaccurate characterizations. As teachers and students explore various exemplars of mathematical structures, they may notice the nuances, the affordances, and the limitations of each of these representations. As they see different manifestations of an abstract mathematical idea, they may develop dispositions to look for mathematical structures elsewhere in mathematics, as well as in real-world relationships.

A caution about developing this disposition is that the mapping of real-world phenomena to mathematical structures is tricky at best. Students (and sometimes teachers) may have a tendency to make analogies that are not based in mathematical structures (Shimizu, 2013). For example, teachers may explain the product of signed numbers by referring to "good" (+) and "bad" (−) people moving in (+) or out of a city (−), with "bad" people moving out of a city being a "good" thing (+). The analogy falls apart in that there is no reason to attach this motion to multiplication, rather than to addition of signed numbers. Analogies like this do not inherit the structure of a mathematical system; rather, they serve only as a reminder of a rule. Teachers might develop a collection of analogies to use as handy classroom explanations, but not be able to unpack the analogies, as Zazkis et al. (2003) report. Unprobed analogies arguably might be a sign that the teachers experienced courses in which they accepted outcomes from others without sufficiently engaging in the mathematical activity that led others to those outcomes. Abstract Algebra instructors may be able to mitigate the use of amathematical analogies by having students explore not only what characteristics a representation suggests about a mathematical system but also why that characteristic follows and how it is connected to other characteristics of the system.

Example 3: Inverse Trigonometric Functions with Mathematical Noticing, Specifically Noticing Structure of a Mathematical System

Teachers of School Algebra commonly witness students applying inappropriate processes in required algebraic calculations. The processes they apply are often familiar routines applied to an inappropriate mathematical object. Students may "solve" an expression by setting it equal to 0 and factoring, or they may "simplify" an equation by cross-multiplying. This misapplication of a learned routine is a sign that students are not distinguishing the nature of the mathematical objects with which they are working. One such instance of this confusion was manifest in a high school trigonometry class.

> High school students were encountering trig functions for the first time. They started calling sine, cosine, and tangent the *three basic trig functions*. Some of the students, when asked to identify "the inverse trig functions," identified them as secant, cosecant, and cotangent. In this case, their teacher recognized that students needed to reconsider what is meant by an inverse.
> (Adapted from Situation 16, MUST, pp. 217–222)

In this example, the students are confusing the inverse of a function under the operation of composition of functions with the multiplicative inverse. This behavior is not unusual. In School Algebra, students are introduced to the concept of inverse in many different settings. They encounter additive inverses, multiplicative inverses, inverses of algebraic transformations, and inverses of other functions. Generally, the context in which they are working can cue students regarding which inverses to apply in a given setting. Work on solving linear equations can cue the application of additive and multiplicative inverses. Work on more complicated equations can cue the application of inverses of a range of other functions. As systems of linear equations are encountered, the applications of inverses of matrices can be cued, and the problematic nature of the nonexistence of inverses may arise. The teacher of the students in this situation knew the relevance of the concept of inverse in their comments. She wanted students to understand the concept of inverse in general and its application to functions in particular.

In their Abstract Algebra and linear algebra courses, students study mathematical objects and operations that can be performed on those objects. They study inverses, both specific to an algebraic structure and more general. As an example, they encounter inverses related to the set of $n \times n$ matrices together with the operation of multiplication of matrices. They learn that the identity element, I, is the matrix that, when multiplied by any matrix, A, in the set, gives the matrix, A, as the result, and they learn that only certain matrices are "invertible." That is, if, for a matrix, A, in a set, M, of $n \times n$ matrices, there is a matrix B in M that has the property that $A * B = B * A = I$, then B is called the inverse of A. As students study more mathematical structures (e.g., rings, groups, fields), they encounter inverses related to those structures. Students learn to think about inverses as comprised of three components: a set of mathematical objects, an operation to be performed on elements of that set, and the identity element for that operation on that set.

The mathematical activity that is involved in work with inverses in both School Algebra and Abstract Algebra is that of *Mathematical Noticing*—of *noticing the structure of a mathematical system*. The term, inverse, was a familiar one to the teachers and students, but the students were not taking into account all of the relevant components. The teacher needed not simply to recognize the presence of "inverse" in the conversation, but to attend to the group under consideration—functions with composition, rather than real numbers with multiplication. This kind of thinking may not be immediately acquired in Abstract Algebra courses, as evidenced by the work of the participants in Wasserman's (2017) study. Such evidence underscores the need to engage Abstract Algebra students directly in noticing the structure of a mathematical system. Knowing essential features of the mathematical structure of an inverse (the set, the operation on members of the set, and the identity element for that operation on that set) helped the teacher notice that the students had failed to recognize the appropriate operation (composition) for inverse trigonometric functions. The teacher's Abstract Algebra experience helped her think about how the different ways that the term inverse is used in School Algebra are all examples of the abstract mathematical entity of inverse. Having noted the use of inverse and knowing the multiple manifestations of inverse across different mathematical systems, she could reason about the mapping between the mathematical entities in the definition of inverse and the components of the entities her students studied in a School Algebra setting.

Abstract Algebra instructors can identify terms that are common content in School Algebra (e.g., associativity, closure, commutativity, distributivity, divisor, factor, function, identity, inverse, power, divisor), have students recall definitions of those terms from School Algebra, and work with students to expand or extend those definitions to apply to other mathematical systems and sets.

Example 4: Summing the Natural Numbers with Mathematical Creating, Specifically Representing and Manipulating

Although much of the day-to-day work of School Algebra may appear to students as practicing learned routines for manipulating strings of symbols, the art of School Algebra requires having the insight to write algebraic expressions purposefully in forms that reveal structures and relationships that were not prominent in the original expressions. One Situation in which this restructuring arose is in the context of finding the sum of the first n natural numbers (one version of this problem was developed in MUST, pp. 121–133).

> A high school algebra class was focused on finding a formula for the sum of the first n natural numbers. One student thought he remembered that the formula involves n and $n + 1$, (or was it $n - 1$). Another recalled $n(n + 1)/2$ but was uncertain that her recollection was correct. A third asked "How can we figure it out for sure?"
> (Situation 4, MUST, p. 121)

This task can be approached in myriad ways that are accessible to students in School Algebra classes. Many of these strategies involve the purposeful creation of a rearrangement of geometric representations of the sum. For instance, one can use the idea of starting with a "staircase" of unit squares and completing a rectangular array of unit squares, as illustrated in the case of $n = 5$ shown in Fig. 10.2. An upside down staircase of grey unit squares (representing the sum of the first 4 natural numbers) completes a 5×5 rectangular array of unit squares from a staircase of unit squares that represents the sum of the first 5 natural numbers. This approach to the problem requires generating a visual image of something that has not yet been present in the classroom discussion.

The students' production of the staircase figure corresponds to their creation of a symbolic representation. The figure suggests that the sum of the first n natural numbers, S_n, added to the sum of the first $n - 1$ natural numbers is

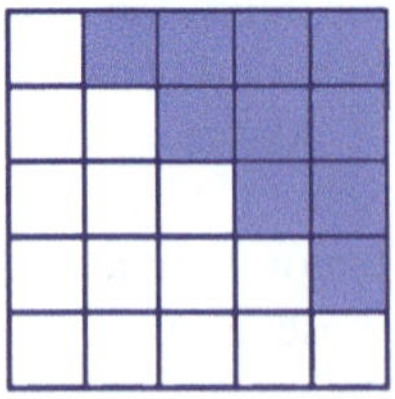

Fig. 10.2 Two staircases, one 5 units wide and the other 4 units wide are juxtaposed to form a 5×5 square

n^2. That is, $(1 + 2 + \ldots + n) + (1 + 2 + \ldots + (n - 1)) = n^2$, or $S_n + (1 + 2 + \ldots + (n - 1)) = n^2$. Adding n to each member of this equation results in $S_n + (1 + 2 + \ldots + (n - 1) + n) = n^2 + n$, or $2S_n = n^2 + n$, yielding $S_n = (n^2 + n)/2$, or $S_n = (n(n + 1))/2$. Two clever rearrangements (one geometric and one algebraic) reveal a formula for the sum of the first n natural numbers. There are additional clever rearrangements of symbolic expressions that, if created by the students, would yield the same formula. For example, writing the sum of the natural numbers from 1 to n in two different ways paves the way to finding a closed-form formula for the sum. This is illustrated in the following sequence of symbolic statements:

$$S_n = 1 + 2 + \cdots + (n - 1) + n$$

$$S_n = n + (n - 1) + \cdots + 2 + 1$$

Adding the left and right members of these two equations yields:

$$2S_n = (1 + n) + (2 + (n - 1)) + \cdots + ((n - 1) + 2) + (n + 1)\text{, or}$$

n pairs, each of which has a sum of $1 + n$.

So $2S_n = n(n + 1)$, and $S_n = n(n + 1)/2$.

Noticing that it can be productive to represent the sum and to do it in different ways that "align" nicely (e.g., 1 aligns with n, and 2 aligns with $n - 1$) is critical.

This need for creative rearrangement comes to the fore in Abstract Algebra in instances such as stating and proving claims about relationships involving inverse elements, such as "the inverse of a product is the product of the inverses in reversed order." For example, the justification of the claim that $(ab)^{-1} = b^{-1}a^{-1}$ readily includes symbolic work in this spirit:

$$(ab)(ab)^{-1} = e$$

$$a^{-1}(ab)(ab)^{-1} = a^{-1}e$$

$$\left(a^{-1}a\right) b(ab)^{-1} = a^{-1}$$

$$b(ab)^{-1} = a^{-1}$$

$$b^{-1}\left(b(ab)^{-1}\right) = b^{-1}a^{-1}$$

$$\left(b^{-1}b\right)(ab)^{-1} = b^{-1}a^{-1}$$

$$(ab)^{-1} = b^{-1}a^{-1}$$

This work relies on a series of critical decisions: realizing that a product can be used to represent the identity (such as $a^{-1}a$); representing the identity in different ways (such as $(ab)(ab)^{-1}$ and e); and "aligning" b^{-1} and a^{-1} with b and a, respectively.

The mathematical activity that underpins both the School Algebra and the Abstract Algebra examples is *Mathematical Creating* in the context of *representing and manipulating representations.* The point is not routine generation and mindless manipulation but rather purposeful choice of multiple symbolic representations of the same thing and planful transformations of those representations. In the case of School Algebra, the sum of the first n positive integers was represented in several different ways, and those ways were creatively and purposefully juxtaposed. In the case of the Abstract Algebra example, the identity was expressed in different ways and then used creatively and purposefully in the production of equivalent equations. The essence of this work is to create and use equivalent expressions, such as $a^{-1}(ab)(ab)^{-1}$ and $(a^{-1}a)b(ab)^{-1}$, while manipulating a string of symbols.

As Abstract Algebra instructors and students create and present certain proofs, they can identify places at which the form of a mathematical entity changes and how those changes accentuate different information. School algebra teachers can do the same, producing different forms of quadratic equations, such as $f(x) = ax^2 + bx + c$ (revealing that the graph of $y = f(x)$ will have a y-intercept of $(0, c)$) and $g(x) = a(x - h)^2 + k$ (revealing a vertex of (h, k) and a minimum or maximum value of g).

Example 5: Powers with Mathematical Reasoning, Specifically Extending and Constraining

Seemingly innocent questions from School Algebra students can lead to nontrivial leaps in mathematical complexity. Such questions can easily arise when students are working with natural numbers or integers and then ask about rational numbers. A move from positive integer powers of 2 to a rational power of 2 is the basis of the prompt from Situation 11, Powers.

> During an Algebra 1 lesson on exponents, the teacher asked the students to calculate positive integer powers of 2. A student asked the teacher, "We've found 2^2 and 2^3. What about $2^{2.5}$?"
> (Situation 11, MUST, p. 185)

The student might be asking merely about a more complicated number than 2 and 3 as the exponent, and simply happened to say 2.5. Perhaps the first number that came to mind for the new exponent was the arithmetic mean of 2 and 3. Regardless of the student's motivation, the question about 2.5 as an exponent of 2 can be interpreted, from the teacher's perspective, as a question of whether the domain of the exponent, b, in 2^b can be expanded from positive integers to rational numbers. To view this situation as the extension of a domain requires being aware of sets and supersets and draws on awareness of what it means to extend a set.

In Abstract Algebra, noting closure for an operation on a set leads to the conclusion that, under that operation, a positive integer power of any element of S is an element in S. If S is the set of integer powers of 2, then 2^2 and 2^3 are in S. Unaware of these structures, the student questions whether $2^{2.5}$ is an element of S. Since S does not include $2^{2.5}$, the move is to extend S to $S\left(\sqrt{2}\right)$ by adjoining the square root of 2. This move to a superstructure of the set of all positive integer powers of 2 parallels the extension of the domain of the exponent in the Prompt from positive integers to rational numbers.

The teacher, and arguably not the student, knows about structures. The teacher who notices a situation in which students suggest what might be viewed as an extension of a set might first ask the student why the issue matters. For example, why did the student choose to ask about 2.5? It could be that the student was suggesting the extension of the set of allowable powers to all non-negative real numbers. The student response might suggest that the student or class could build the extended domain and test claims about the new superstructure. This kind of activity challenges the assumption of some Abstract Algebra students that the structures they study in college mathematics are not relevant to secondary school mathematics.

The Mathematical Activity in reacting to the School Algebra Powers situation is *Mathematical Reasoning*, specifically *extending and constraining* structures. An underlying Abstract Algebra practice is the adjoining of a needed element to extend an algebra to create a new structure, as in moving from Q, to $Q\left(\sqrt{2}\right)$ or to $Q\left(\sqrt{3}\right)$, and eventually to $Q\left(\sqrt{2}, \sqrt{3}\right)$. While engaging students in exploring an algebraic entity with a restricted domain for its parameters, an Abstract Algebra instructor can extend that exploration by having students explore the feasibility of extending the domain of the parameter.

Example 6: Zero-Product Property with Mathematical Reasoning when Generalizing

At the heart of School Algebra is the refinement of student skills with symbolic transformations of symbolic expressions and equations. The following Prompt demonstrates how those skills may go astray.

A student in an Algebra 1 class wrote the following solution to a homework problem on the whiteboard:

$$x^2 - 4x - 5 = 7$$
$$(x - 5)(x + 1) = 7$$
$$x - 5 = 7 \text{ or } x + 1 = 7$$
$$x = 12 \text{ or } x = 6$$

A different student commented that 6 was a solution to the equation because $6^2 - 4(6) - 5 = 7$ but 12 was not a solution because $12^2 - 4(12) - 5 \neq 7$. (Adapted from Situation 17, MUST, pp. 223)

Symbolic routines are the core of School Algebra, and students often work to refine their performance of symbolic routines by replicating them step-by-step. The steady methodical nature of this replication may foster a habit of executing the steps without thinking about why each step makes sense. Failure to account for the conditions that permit a particular symbolic transformation can result in overgeneralization. The example of such an overgeneralization in the Prompt illustrates a student applying a corruption of the zero-product property ($a{\cdot}b = 0 \Rightarrow a = 0$ or $b = 0$) that does not account for the condition that the property applies only to products that have a value of 0. The student assumes that ($a{\cdot}b = c \Rightarrow a = c$ or $b = c$), no matter the value of c. Had the student considered the meaning of the routine, he could have recognized that the product of two numbers being 7 does not mean that one of the numbers must be 7, and that there is an infinite number of pairs of real number factors whose product is 7. This further consideration would have brought to the student's attention the uniqueness of 0 and, therefore, the importance of the zero-product property.

In Abstract Algebra, students develop the understanding that potential generalizations must be tested. For example, they may be tempted to treat the commutativity of multiplication—a property that applies in so many of the algebraic systems they encounter—as if it applies in all algebraic systems. In Abstract Algebra, they learn that groups in which multiplication is commutative are special enough to be designated with a special name, "abelian." The existence of the label can serve to prompt students to note that some groups are not abelian. One way in which secondary school students can test the claim that an operation is commutative is to look for counterexamples. A quintessential counterexample for a generalization that all groups are abelian is the set of 2×2 matrices under multiplication, as illustrated in the following calculation:

$$\begin{bmatrix} 1 & 2 \\ 3 & 4 \end{bmatrix} \times \begin{bmatrix} 0 & 1 \\ 1 & 1 \end{bmatrix} = \begin{bmatrix} 2 & 3 \\ 4 & 7 \end{bmatrix} \text{ and } \begin{bmatrix} 0 & 1 \\ 1 & 1 \end{bmatrix} \times \begin{bmatrix} 1 & 2 \\ 3 & 4 \end{bmatrix} = \begin{bmatrix} 3 & 4 \\ 4 & 6 \end{bmatrix}, \text{ but } \begin{bmatrix} 2 & 3 \\ 4 & 7 \end{bmatrix} \neq \begin{bmatrix} 3 & 4 \\ 4 & 6 \end{bmatrix}.$$

In this particular example, students' specific experience with integral domains in Abstract Algebra might have included the comparison of $\mathbb{Z}_6$, in which neither 3_6 nor 4_6 is 0_6 but $3_6 \times 4_6 = 0_6$, with the real numbers under multiplication, in which the product of any two nonzero elements is nonzero. Such experiences alert prospective teachers to the importance of questioning the nature of factors, including whether one of the factors in $(x - 5)(x + 1) = 7$ has to be 7.

The mathematical activity illustrated in this example is that of *Mathematical Reasoning when Generalizing*. An essential component of generalizing is determining the extent of the domain to which a set of properties applies. The domain of the generalization for c in $(a{\cdot}b = c \Rightarrow a = c$ or $b = c)$ is the set of values of c that make the generalization true; that is, $c = 0$. Without sensitivity to such limitations, overgeneralization is likely. School Algebra teachers are frequent witnesses to student errors that stem from unwitting overgeneralizations. Their experience in Abstract Algebra with mathematical structures that have various collections and configurations of properties helps them learn to distinguish between structures with different set of properties (e.g., abelian groups and nonabelian groups). The distinctions alert Abstract Algebra students to the need to test potential attribution of properties to an expanded set of systems, or, in the case of the zero-product property, to a set of examples that fails to account for an essential condition. Constant attention is needed in an Abstract Algebra class on the domain of applicability of properties students may have otherwise considered as universal.

Example 7: Circumscribing Polygons with Noticing Structure of Mathematical Systems and Noticing Symbolic Form

General statements presented as axioms, postulates, or theorems are expected in School Algebra and Abstract Algebra. Teachers can be prepared to engage students with the mathematics of these statements as generalizations. More challenging situations arise when students pose questions about classes of objects. The prompt from Situation 34, Circumscribing Polygons, presents a generalization for the teacher to unpack.

> In a geometry class, after a discussion about circumscribing circles about triangles, a student asked, "Can you circumscribe a circle about any polygon?" (Situation 34, MUST, p. 293)

The link between Abstract Algebra and this School Geometry problem might seem out of place in a chapter that, otherwise, argues that the mathematical activity of Abstract Algebra informs how teachers might think about School Algebra. However, this example does illustrate important mathematical activity involving classes and subclasses of objects in geometry, just as classes and subclasses matter in School Algebra. Examples of circumscribed triangles lead to a student's desire to extend the class of objects to which a generalization applies. The simple answer to the question of whether all polygons can be circumscribed is "no." The mathematically active teacher capitalizes on the opportunity to reason about classes of objects. The extension to all regular polygons seems natural. An example of a square and a non-circumscribable quadrilateral can quickly address the student's question with a response such as, "No, you cannot circumscribe an arbitrary polygon." The artful teacher can steer discussion to a question of which classes of quadrilaterals can be circumscribed.

A school mathematics curriculum helps students move from natural numbers through integers and rational numbers to real numbers and then imaginary and complex numbers. The various types of numbers seemingly fade into the background in School Algebra. On various occasions students consider which type of numbers is involved and decide which procedures or results are reasonable. For example, students learn to solve different types of equations, and they observe how different methods are more or less efficient, or even effective, depending upon the classes of numbers to which coefficients and solution values belong. Real-world contexts further contribute to what types of numbers are meaningful as numerical answers.

In Abstract Algebra, attention to classes and subset–superset relationships are unavoidable. The classes of objects to which operations are applied underpin the type of algebraic structure that is formed. For example, consider the differences between $(\mathbb{Z}_5, +_5)$, $(\mathbb{Z}\backslash\{0\}_5, +_5)$, and $(\mathbb{Z}\backslash\{0\}_5, \bullet_5)$. Consideration of the subset–superset relationship, and the operations involved, can be overlooked by Abstract Algebra students. (We also note the potential usefulness of the subscripts to highlight the correct operations for college students who [initially] are strongly wedded to real numbers and their operations.) The simple elimination of one element from $\mathbb{Z}_5$ with the same additive operation yields $(\mathbb{Z}\backslash\{0\}_5, +_5)$, a proposed structure devoid of closure and not a group. When paired with a change in the operation, the move from $\mathbb{Z}_5$ to $\mathbb{Z}_{5^*} = \mathbb{Z}\backslash\{0\}_5$ yields $(\mathbb{Z}_{5^*}, \bullet_5)$, which is a group different from $(\mathbb{Z}_5, +_5)$. The significance of these relationships seems critical in considering whether a set with two operations forms a field. The subtle notion of omitting the additive identity to form the algebra for the multiplicative relationship is elusive to Abstract Algebra students who readily claim that "real numbers with addition and multiplication form a field" without unpacking the nuance of the statement. These observations suggest the need for explicit attention to symbolic noticing in Abstract Algebra courses in the context of sets and operations, such as those involved in $(\mathbb{R}, +, \cdot)$ as well as in $(\mathbb{Z}_5, +_5, \bullet_5)$.

Mathematical Noticing in the form of *noticing structure of mathematical systems* and some *noticing symbolic form* is the mathematical activity present across the School Geometry, School Algebra, and Abstract Algebra settings discussed in this example. Inescapable attention to classes of objects in Abstract Algebra occurs as the class of objects to which operations are applied is extended, suppressed, and changed. The Abstract Algebra experience might encourage attention to the differences among classes of objects that seemingly develop over time in school mathematics.

Summary

These seven examples include all of the three types of mathematical activity: *Mathematical Noticing*, *Mathematical Creating*, and *Mathematical Reasoning*. A close look at the details of the activity suggests there might be a slight emphasis on some specific activities, such as *noticing symbolic form* and *noticing structure* of a mathematical system. The prevalence of some particular subcategories of activities in this chapter might reflect the nature of Abstract Algebra. The MUST Framework was designed to cross areas of mathematics, so some categories of activities might not be as prevalent as others simply as a result of the application of MUST exclusively to the mathematical strand of algebra. Some mathematical activities that are important in other strands of mathematics might not be foregrounded in algebra.

Given the attention to symbolic work in School Algebra, perhaps it is not surprising that *noticing symbolic form* and *noticing structure* are readily present in the examples. While these mathematical activities are common, they are far from trivial. For example, School Algebra teachers often lament their students' use of "cancellation" in inappropriate venues, such as the following: $\frac{\cancel{(x-3)}+7}{\cancel{(x-3)}} = 7$. Abstract Algebra instructors can observe seemingly parallel moves, such as $a^{-1}bab = \cancel{a^{-1}}b\cancel{a}b = b^2$.

These erroneous moves might be an issue of underdeveloped symbolic noticing. Students might become sufficiently savvy to notice the multiple instances of a common expression (e.g., $x - 3$) or the presence of an element and its inverse (e.g., a^{-1} and a in $a^{-1}bab$), but they seemingly do not attend to the context that supplies the larger structure or they might be led astray by inappropriate visual cues. Engagement in a broader range of mathematical activities, as in Cook's (2014) use of the defining as an activity for students in a teaching experiment with fields that drew attention to student development of the concepts of zero divisor and unit, has potential for pushing students to develop robust mathematical understandings.

The seven examples from the work of teachers capture events that are not uncommon in School Algebra settings, and the corresponding Abstract Algebra content is similarly present in college courses. This commonality suggests the prevalence of Mathematical Activity in each of these contexts. It further suggests that experience with Mathematical Activity could be important to prospective and current mathematics teachers as a way to see how their college mathematics experience connects with their school teaching work. It also prepares teachers to engage their students in activity that is mathematical rather than actions that mimic mathematical activity.

Concluding Thoughts

The seven examples from the MUST Situation Prompts illustrate multiple ways in which types of mathematical activity are present in both School Algebra and Abstract Algebra. This body of evidence suggests that a benefit of Abstract Algebra for secondary mathematics teachers is not simply the content knowledge they gain; it is also the experience they have with mathematical activities. In fact, one might consider what it means to become more agile in mathematics for others, as well as for teachers. Command of procedures and concepts is not sufficient; one must also become agile in mathematical activity that draws on, extends, and combines known content to form new outcomes.

The call to mathematical activity for teachers is also a call for students to engage with mathematics in particular ways that include the ideas of practices (Common Core State Standards Initiative, 2010), processes (Zbiek et al., 2012), and habits of mind (Cuoco et al., 1996). Perhaps the extent to which prospective and current secondary mathematics teachers do not see a connection between Abstract Algebra and School Algebra is a signal of work that needs to be done to improve students' experiences in both arenas. Several practitioners and scholars have pursued development and implementation work to understand and disrupt the traditional teaching of School Algebra and Abstract Algebra. The Mathematical Activity connection between the two educational settings suggests the need to revisit and extend these efforts. In particular, what would happen if Abstract Algebra were taught with an attempt to foreground the mathematical activities? And, what would School Algebra look like if it assumed that mathematical activities should be explicit?

The teaching of Abstract Algebra and School Algebra, with attention to mathematical activity, is the potential means for foregrounding activity and/or making activity explicit. It might not be a question of should it be *either* content *or* mathematical activity, but a need for *both* content *and* mathematical activity.

References

Common Core State Standards Initiative. (2010). *Common core state standards for mathematics*. Washington, DC: National Governors Association Center for Best Practices and the Council of Chief State School Officers.

Cook, J. P. (2014). The emergence of algebraic structure: Students come to understand units and zero-divisors. *International Journal of Mathematical Education in Science and Technology, 45*(3), 349–359.

Cuoco, A., Goldenberg, E. P., & Mark, J. (1996). Habits of mind: An organizing principle for mathematics curricula. *Journal of Mathematical Behavior, 15*(4), 375–402.

Heid, M. K., Wilson, P. S., with Blume, G. W. (Eds.). (2015). *Mathematical understanding for secondary teaching: A framework and classroom-based situations*. Charlotte, NC: IAP.

Shimizu, J. K. (2013). *The nature of secondary mathematics teachers' efforts to make ideas of school algebra accessible* (Doctoral Dissertation). The Pennsylvania State University.

Wasserman, N. H. (2017). Making sense of abstract algebra: Exploring secondary teachers' understandings of inverse functions in relation to its group structure. *Mathematical Thinking and Learning, 19*(3), 181–201.

Zazkis, R., & Kontorovich, I. (2016). A curious case of superscript (−1): Prospective secondary mathematics teachers explain. *Journal of Mathematical Behavior, 43*, 98–110.

Zazkis, R., & Leikin, R. (2010). Advanced mathematical knowledge in teaching practice: Perceptions of secondary mathematics teachers. *Mathematical Thinking and Learning, 12*(4), 263–281.

Zazkis, R., Liljedahl, P., & Gadowsky, K. (2003). Students' conceptions of function translation: Obstacles, intuitions and rerouting. *Journal of Mathematical Behavior, 22*(4), 437–450.

Zbiek, R. M., Heid, M. K., & Blume, G. W. (2012). Seeing mathematics through processes and actions: Investigating teachers' mathematical knowledge and secondary school classroom opportunities for students. Paper presented at the 12th International Congress on Mathematical Education (ICME-12), Seoul, Republic of Korea.

Chapter 11
Learning Mathematical Practices to Connect Abstract Algebra to High School Algebra

Erin E. Baldinger

Abstract algebra is often a required course in mathematics teacher preparation. However, it is also a course that many teachers find unrelated to the content that they end up teaching. So it is critically important both to identify the connections between abstract algebra and school mathematics that exist, and also to highlight those connections for future teachers. In this way, the course becomes more than just a requirement; instead, it becomes a course taken with purpose and one that supports the teaching of high school mathematics.

Researchers have explored different ways in which abstract algebra might be connected to the school curriculum. In this chapter I explore the role of mathematical practices in connecting abstract algebra to school mathematics. Mathematical practices are ways in which people engage in the work of doing mathematics. Practices include work such as making conjectures, generalizing, attending to precision, and representing mathematical thinking. Practices are, by nature, related to multiple content areas and are used at all levels of mathematics. Because of these features, they are a natural place to investigate the existence and potential importance of connections. I investigate how completing an abstract algebra course

Author Note: An earlier version of this paper was presented at the North American Chapter of the International Group for the Psychology of Mathematics Education Annual Meeting, in November 2015, in East Lansing, MI. The research for this paper was conducted as part of a dissertation study at Stanford University, and was supported in part by a Stanford Graduate School of Education Dissertation Support Grant.

E. E. Baldinger (✉)
Department of Curriculum and Instruction, College of Education and Human Development, University of Minnesota, Saint Paul, MN, USA
e-mail: eebaldinger@umn.edu

N. H. Wasserman (ed.), *Connecting Abstract Algebra to Secondary Mathematics, for Secondary Mathematics Teachers*, Research in Mathematics Education,
https://doi.org/10.1007/978-3-319-99214-3_11

designed for teachers supported participants in learning to engage in mathematical practices themselves, and how these practices contributed to making connections between abstract algebra and mathematics teaching.

Literature Overview

Connecting Abstract Algebra and School Mathematics

Abstract algebra courses tend to focus on structures such as groups, rings, and fields, and the relationships between them. However, these structures themselves are not explicit objects of study in the school curriculum, even though they underlie much of school mathematics (Katz & Barton, 2007; Novotná & Hoch, 2008; Usiskin, 1988). This can make it seem to future teachers, who are often required to take abstract algebra, that the mathematics they are learning is not relevant to their work as teachers. Despite this apparent disconnect, there are numerous connections that exist between abstract algebra and school mathematics. I consider connections related to both mathematical content and ways of thinking about and engaging with that content (Murray, Baldinger, Wasserman, Broderick, & White, 2017). Content connections might include the importance of inverse and identity in both abstract algebra and school mathematics. Another connection might be recognizing that the real number system is a field. In addition to these content connections, abstract algebra and school mathematics are connected by mathematical practices—ways of engaging in or doing mathematics. Practices include conjecturing, justification, and attention to precision (National Governors Association Center for Best Practices & Council of Chief State School Officers, 2010). Even if the content is not the same, the work of doing mathematics and the habits of mind entailed when doing mathematics (Cuoco, Goldenberg, & Mark, 1996) serve as critical connections across different levels of mathematics.

Neither content nor practice connections are necessarily readily apparent to future teachers. Teachers struggle to see the relevance of their university mathematics courses (Goulding, Hatch, & Rodd, 2003). Additionally, professors teaching courses such as abstract algebra need support to make connections explicit (Hodge, Gerberry, Moss, & Staples, 2010). Identifying connections helps both professors and future teachers with the question of how courses like abstract algebra are relevant for teacher preparation. Drawing on connections can help make school mathematics more authentic to the discipline, and more engaging for students. If the work of mathematics includes making conjectures, generalizing, and proving, then that ought to be the work of students. This work is much more compelling than rote work such as following algorithms without understanding.

Beyond having a basic awareness of how content and practices might connect abstract algebra with school mathematics, teachers ought to be able to draw on these connections in their teaching practice (Wasserman, 2016). A deeper understanding of the connections between abstract algebra and school mathematics might influence the language teachers use, the tasks they select, or the habits of mind they aim to support their students to develop (Murray et al., 2017). Incorporating a focus on mathematical practices in the curriculum is a key element of teaching mathematics. It follows that mathematical practices can represent an important connection between abstract algebra and school mathematics.

Mathematical Practices

People engage in mathematics through mathematical practices. Mathematical practices are the ways in which people *do* mathematics. This might include activities such as generalizing, proving, making conjectures, and representing mathematical ideas. Naming these practices has resulted from attention to the work of mathematicians (Moschkovich, 2013). "Mathematical practices are not singular, monolithic, or homogeneous. Mathematical practices include multiple forms ranging over a spectrum of practices such as academic, workplace, playground, street selling, home, and so on" (Moschkovich, 2013, p. 264).

There have been numerous approaches to name and describe mathematical practices. Moschkovich (2013) categorizes these approaches by the method of study: autobiogrpahical, ethnographic, philosophical, historical, and cognitive. Autobiographical approaches (e.g., Bass, 2011; Cuoco et al., 1996) name mathematical "habits of mind." According to Moschkovich (2013), Burton (1999) used self-reports from mathematicians to describe the multiple ways they engage in mathematics, while Lakatos (1976) and Pólya (1957) took a more "philosophical persecptive" to describing the work of doing mathematics. The variety of approaches to investigating mathematical practices contributes to the conclusion that mathematical practices are not uniform.

Over the past few decades in the USA, mathematical practices have emerged as central features of the school curriculum. Incorporating mathematical practices into standards for student learning was a major emphasis of the National Council of Teachers of Mathematics (NCTM) standards (1989, 2000) with their foregrounding of "mathematical processes." Later, mathematical practices were codified in the Common Core Standards (2010), adopted by more than 40 states, as the ways in which students should be supported to engage in mathematics. Due to the importance of mathematical practices in doing mathematics, and the incorporation of mathematical practices into the school curriculum, a focus on teacher engagement in them has become a particularly salient issue for current research. In fact, "conceptions of teacher knowledge have seldom considered the kinds of mathematical practices that are central to teaching. For example, rarely do teachers have opportunities to learn about notions of definitions, generalization, or mathematical reasoning" (Ball, 2003, p. 21). Teachers' engagement in mathematical practices matters because the ways in which teachers engage in practices may affect how they teach their students to engage in practices. Through engagement in mathematical practices, teachers also begin to develop a sense of how students might engage in them. Practices are a key mechanism by which teachers and their students learn mathematics.

Teacher Engagement in Mathematical Practices

Much of the existing literature on teacher engagement in mathematical practices emphasizes proof. Stylianides and Ball (2008) argue that teachers need to understand not just the structure and features of proof, but also the situations that support learning about proof. Research on teachers and proof has investigated teacher beliefs about proof and its role in the classroom (Knuth, 2002a, 2002b; Staples, Bartlo, & Thanheiser, 2012) and teacher knowledge related to the analysis of specific types of proof (e.g., Stylianides, Stylianides, & Philippou, 2007; Tabach et al., 2011). However, few studies investigate how teachers engage in proof themselves (for an exception, see Tabach et al., 2011). Additionally, research related to teachers' beliefs about proofs tends to be situated within the context of school mathematics content, while research about teachers' engagement in proof tends to be situated in the context of college-level mathematics. This highlights an important opportunity to explore more directly the connections between proof in a course such as abstract algebra and proof in school mathematics.

Equally important is research that investigates practices beyond looking only at proof. Mathematical practices rarely occur independently of one another, making it critical to look at them in concert with each other and within a variety of mathematical content domains. For example, Larsen and Zandieh (2007) argue that defining plays an important role in how students engage in more formal mathematical reasoning. How might teachers engage in the practice of defining, and how might it support their engagement in other mathematical practices? Investigating teacher engagement in sets of mathematical practices helps describe the varied ways in which teachers do mathematics. This suggests multiple possible connections between abstract algebra and school mathematics via different mathematical practices.

Conceptualizing Teacher Learning

One of the ways abstract algebra can provide relevant support for future teachers is through giving them the opportunity to engage in mathematics as learners themselves. As such, abstract algebra courses can be sites for teachers to learn to engage in mathematical practices. I take a situated view of teacher learning. Lave and Wenger (1991) define learning as a change in participation in a community of practice. Additionally, "new learning always involves the application and extension of previous knowledge" (Simon, 1994, p. 78). Evidence of learning can come from changes in the way teachers interact with one another in the context of solving a mathematics problem, and it can also come from changes in the way they individually reason about a mathematics task (Cobb & Bowers, 1999). Opportunities to learn can come from experiencing dissonance, interacting with peers, and exploring mathematical ideas from a variety of perspectives (Silver, Clark, Ghousseini, Charalambous, & Sealy, 2007). This implies that learning can be investigated by examining changes in participation in the community of practice following a learning opportunity.

Applying this view of learning to future teachers' learning to engage in mathematical practices requires making precise what that change in participation might entail. Moschkovich (2004) proposes the notion of appropriation as a way to understand learning mathematical practices. She writes,

> Appropriation involves joint productive activity, a shared focus of attention, and shared meanings (Rogoff, 1990). Appropriation also involves taking what someone else produces during joint activity for one's own use in subsequent productive activity while using new meanings for words, new perspectives, and new goals and actions. (Moschkovich, 2004, p. 51)

Appropriation, in this view, contrasts simple imitation of procedures. For example, learning to engage in a mathematical practice, such as making conjectures, would not be the result of a teacher simply imitating the conjectures made by a professor. Instead, the teacher would need to try out making conjectures alongside the professor. Together, they would create meaning around the mathematics under consideration. Then, the teacher would be able to take on the practice of making conjectures in different settings.

Using appropriation as a way to understand how teachers learn to engage in mathematical practices helps clarify the nature of learning opportunities. First, appropriation requires taking what someone else has done and transforming it. This implies that learning opportunities might include seeing examples of mathematical practice engagement as part of joint work and attention. For example, the professor might model the practice of making conjectures while working together with teachers on a particular mathematics task. Second, appropriation requires joint productive activity. In other words, teachers must have experiences engaging in mathematical practices themselves. These opportunities would provide participants with an opportunity to potentially transform their use of mathematical practices.

This chapter explores the role that an abstract algebra course might play in supporting future teachers to learn to engage in mathematical practices and describes how this learning can act as a connection between abstract algebra and high school mathematics. I consider the following question: What did participants in an abstract algebra course for teachers learn about engaging in mathematical practices? I describe elements of the course that provided learning opportunities around mathematical practices and discuss how connections to teaching mathematics emerged.

Table 11.1 Participants' mathematics and teaching backgrounds (all names are pseudonyms)

Name	Mathematics background	Teaching background
Daniel	Engineering major, career in engineering and business	Tutoring
Laura	Mathematics major	Paraprofessional
Sam	Mathematics major; associates degree in engineering	Substitute teacher
Tim	Mathematics and physics major	A few education courses

Methods

This study considers the learning of participants in an abstract algebra course designed specifically for preservice teachers. I captured engagement in mathematical practices at the beginning and end of the course through task-based interviews. Using a multiple case approach (Miles, Huberman, & Saldaña, 2013), I describe participant learning related to engagement in mathematical practices. Finally, participant learning is connected to relevant opportunities to learn in the abstract algebra course.

Participants

This study examines future teachers' learning in a preparation program that was purposively sampled (Miles et al., 2013) because it had a strong focus on the mathematical content preparation of future teachers. During the year of the study, four future teachers were enrolled in this program and all agreed to participate in the study (see Table 11.1). All participants had either a mathematics degree or a degree in a related field (science or engineering), preparation that is consistent with that of other secondary mathematics teachers (Graham, Li, & Buck, 2000). They had an array of different experiences in teaching, ranging from a few education courses to substitute teaching.

Abstract Algebra Course Designed for Teachers

The preparation program emphasized rigorous content preparation as part of learning to teach. In addition to methods courses and general education courses, participants took four content courses: abstract algebra, applied mathematics, analysis, and geometry. At the time of the study, participants were taking abstract algebra, applied mathematics, a mathematics methods course, and two general education courses. They also tutored middle school students once a week. This study focuses on participation in the abstract algebra course, a 10-week summer course. It met once a week for 3 h per session. I observed all ten classes, took detailed field notes, and collected all handouts and homework assignments.

The abstract algebra course focused on proving the fundamental theorem of algebra. The professor chose a proof that was "algebraic in flavor," while also acknowledging that this theorem does not have a fully algebraic proof. On the first day of class, he explained to students,

> The topic of this class is algebra, and one of the challenges is that everyone comes in with a very different background. [. . .] I've selected a topic in algebra, the fundamental theorem of algebra, which is a key topic that impacts high school math and middle school math. We're actually going to prove it this quarter. There are lots of different proofs. I've selected the most difficult and exciting.

He went on to explain that the proof might not be the "easiest way" to prove the fundamental theorem, but it would expose the students to "a huge list of topics" in algebra. The class discussed the equivalence of many different statements of the fundamental theorem, and eventually proved the statement, "Every degree n polynomial ($n \geq 1$) with real coefficients has at least one complex root."

Each week the class explored key ideas that would support the proof, such as proof by induction, complex numbers, the fundamental theorem of arithmetic, and the pigeonhole principle. In this respect, the content covered in this course was very different from a traditional course focusing on groups, rings, and fields. The professor explained that he chose this focus for the course based on his own experience spending time in schools and trying to find mathematical content that would be particularly relevant for future teachers. The course explicitly attended to mathematical practices. The professor explained that he planned to emphasize problem-solving, proof, communication, and precision. The course was also structured around providing sustained opportunities for group work and problem-solving, so it differed substantially from more common lecture style courses (e.g., Speer, Smith III, & Horvath, 2010).

Even with this nontraditional approach to abstract algebra, the course still included opportunities for participants to engage with ideas from abstract algebra. For example, in one exercise, the participants were asked to compare and contrast multiplying two polynomials in $\mathbb{Z}[x]$, $\mathbb{Z}_7[x]$, and $\mathbb{Z}_8[x]$. This problem elicited conversation around the idea of zero divisors. In a different class, participants investigated permutations as an operation in order to develop a definition of symmetric polynomials. They built on this discussion to define a symmetric polynomial generated by a given monomial. Other topics included proving the equivalence of different statements of the fundamental theorem and exploring the necessity of field extensions.

Data Sources

At the beginning and end of the abstract algebra course, participants completed in-depth task-based interviews (Ginsburg, 1981; Goldin, 1997), where they solved secondary level algebra problems and reflected on their learning. Observation records provide evidence for opportunities to learn mathematical practices in the abstract algebra course (Baldinger, 2014).

Interview Design

The pre- and post-interviews were designed around solving a secondary level mathematics task focusing on topics in algebra. The problems were chosen to be accessible to participants with a range of mathematical proficiencies, to be challenging, to have multiple possible solution strategies, and to be nonfamiliar; that is, even though they dealt with secondary level content, they were not problems participants were likely to have seen or completed prior to the interview.

The pre-interview task was selected from among the TEDS-M study released items (TEDS-M International Study Center, 2010). The task selected addresses linear functions, which was not a topic explored in depth during the abstract algebra course, and it requires proof. A correct proof could use algebraic or visual strategies. The post-interview task was selected to be similar, but not identical, to the pre-task, in order to avoid the possibility that participants would remember their previous solutions rather than approaching the post-task as a new problem. The post-task maintained some of the key features of the pre-task while allowing others to vary (see Table 11.2). The content remained about linear functions and the task required proof. The wording of the problem was intentionally similar to the pre-task. The post-task allows for techniques similar to the pre-task, such as looking at specific cases and using a graphical approach.

Interview Protocol

Participants were asked to think aloud as they solved the problem (Ericsson & Simon, 1980; Taylor & Dionne, 2000), and they engaged in "free problem-solving" (Goldin, 1997). No hints were given; the only prompts were to remind the participant to keep talking. There was no time limit for solving the tasks. Participants did not receive feedback on their responses (Goldin, 1997). Possible correct responses are included in Appendices 1 and 2. Participants' solutions were both oral and written due to the think-aloud structure. All interviews were audio-recorded and video-recorded.

Table 11.2 Interview mathematics tasks and key features

	Pre-task	Post-task
Task features	High school level content	High school level content
	Linear functions	Linear functions
	Prove a general statement is true	Prove a general statement is true
	Focus on the sum of two functions at a particular point	Focus on the x- and y-intercepts of a particular line
Task statement	Prove the following statement: If the graphs of linear functions $f(x) = ax + b$ and $g(x) = cx + d$ intersect at a point P on the x-axis, the graph of their sum function $(f + g)(x)$ must also go through P. (TEDS-M International Study Center, 2010, p. 14)	Take a point (p, q) on the Cartesian plane. Reverse the coordinates to obtain a second point (q, p). Prove that on the line between these two points, the x-intercept and the y-intercept are the sum of the coordinates.

Data Analysis

Analysis proceeded in several phases: (1) focused memos considering engagement in mathematical practices, (2) case development by task, and (3) pre–post comparison. The first two phases focused on individual engagement in mathematical practices; the third phase focused on patterns across participants and time points. To prepare the data for analysis I transcribed each interview, using audio, video, and written work to develop as accurate a record as possible.

Capturing Engagement in Mathematical Practices

I began the memo writing process by describing instances from each interview where participants engaged in mathematical practices, and I used the Common Core Standards for Mathematical Practice (2010) as a guide. To do this, I decomposed each of the eight mathematical practice standards into smaller components. For example, I considered "making a conjecture," "attending to the givens in a problem," and "making sense of a problem" as potentially separate practices all connected to Mathematical Practice 1 in the standards document. At the same time, many of these smaller practices had the potential to overlap. For example, "attending to the givens in a problem" might have many similarities to "using stated assumptions in constructing arguments," which, according to the organization of the Common Core Standards would be part of Mathematical Practice 3. Because of this, the decomposition of the standards document did not serve as a traditional a priori coding scheme. Instead, the decomposition provided a lens through which I could begin to explore what mathematics a participant was doing during the interview. Using this decomposition of the mathematical practice standards to guide the memo writing process also helped me identify practices *not* engaged in by a participant.

In the memos, I included notes about practices even when there might be an error in the mathematics (e.g., a participant might incorrectly justify an argument, but is still engaging in justification).

Case Development

To develop each case, I first summarized the participant's general approach to the problem and their engagement in mathematical practices. I identified nuances in practice engagement (e.g., attention to precision at one stage of problem-solving but not at another). This allowed me to make assertions about each participant's engagement in mathematical practices based on the transcripts, including potential missed opportunities for practice engagement, or where *not* engaging in a practice (e.g., connecting representations, attention to precision) may have impeded a solution. For each assertion, I searched for disconfirming evidence to ensure that the assertion accurately captured the participant's practice engagement. For example, Daniel made several errors during the post-task that might have suggested a difficulty attending to precision, but later in the interview he corrected those errors by carefully connecting representations, thus providing disconfirming evidence and suggesting a more nuanced picture of how Daniel attended to precision. This phase of analysis concluded by identifying emerging themes to look for in other cases.

Pre-post Analysis

I investigated participant learning by looking across the pre- and post-tasks. Analysis began at the group level through identifying similarities and differences in the way participants engaged in mathematical practices across time points. Next, I focused on changes in engagement in mathematical practices for each participant. Taking the perspective that learning is evidenced by changes in how people participate in a given practice, I used these comparisons to explore how, if at all, participants engaged in practices in more nuanced, complex, flexible, or expert ways over time. In other words, the comparisons provided evidence of appropriation of mathematical practices. Appropriation could be evident in different ways. For instance, participants might exhibit more flexible use of a practice (e.g., applying a practice in a novel context). Participants might also demonstrate more expert use of a practice (e.g., engaging in the practice in ways more similar to how a mathematician might engage in the practice).

Findings

In this section, I show how participant engagement in mathematical practices changed from the pre-task to the post-task. Each participant showed a variety of shifts, not just in terms of which practices they engaged in, but also how they engaged in some of the same practices. I present each participant as a case foregrounding a particular theme that emerged during data analysis. I then explore some of the opportunities to learn from the abstract algebra course that relate to the learning demonstrated by participants. Finally, I present participants' reflections on the potential influence the abstract algebra course might have on their teaching.

Learning to Use a Special Case to Make Sense of a Problem

Laura began the teacher preparation program with a strong commitment to being a mathematics teacher, and a desire to find new ways to teach, but also expressed her lack of confidence about her own mathematical ability. During both interviews, this lack of confidence seemed to manifest itself through Laura's reluctance to spend much time engaging in the tasks. However, one notable difference between the pre-task and the post-task was Laura's more expert way of using a special case to help her make sense of the problem.

For both tasks, Laura engaged in the practice of choosing a special case. She selected example lines in the pre-task, and example points in the post-task. Solving the pre-task, she said, "I'm just going to put real numbers in this for a minute" and selected the lines $y = 3x + 2$ and $y = 4x + 6$. Laura found the point where the two lines intersected algebraically (at $x = -4$), found the sum of the two functions ($y = 7x + 8$), and based on that, conjectured that the statement she was to prove must be false, rather than true. After writing down her examples, she paused, and eventually said,

> I'm going to say no, it does not, because, how do I prove this though? That's the part I'm having difficulty with. The proving part. I'm not exactly sure how to prove that aside from using real numbers to prove it. Hmm. Yeah. Go on.

With that, Laura declined to continue working on the task. She did not attempt a proof by counterexample, for instance, nor did she engage in practices such as connecting representations that might have enabled her to notice that the example lines she chose did not match the condition of the problem, as they did not intersect at a point $(P, 0)$. Had she continued to work on the problem, Laura likely would have engaged in additional mathematical practices.

On the post-task, Laura still expressed doubt in her own mathematical ability, and began the problem in more or less the same way. She drew a graph for herself, plotted a point and said, "So if I take (p, q), okay, I'm just going to give this [*labels the point* $(1, 2)$], and (q, p) [*draws another point and labels it* $(2, 1)$]." She continued on with this special case, using slope-intercept form to find the y-intercept in her example, and while she wrote out the algebra, she said, "Okay. I'm just solving. That doesn't make sense, does it? Yeah. Okay. So I'm just checking my work. I tend to doubt a lot. Okay [*writes:* $y = -x + 3$]." Then she moved on to identifying the x-intercept using the definition of slope and counting out points on the graph. She concluded, "So the x-intercept is the sum and the y-intercept is the sum."

Laura's engagement in the practice of using a special case to make sense of a problem was more expert in the post-task than the pre-task. She selected an example that attended to the conditions of the problem and convinced herself that the statement to be proven was true. Additionally, rather than stopping as she had in the pre-task, Laura summarized the work she had done to reach her conclusions. After describing her work on the special case, she said,

> I don't think that I proved it though. I proved it for $(1, 2)$ and $(2, 1)$, but I didn't prove it for p and q. I know that. I'm able to recognize that. I'm just not sure how I would turn this into a proof of the (p, q) and (q, p).

At this point, she again stopped working on the task and chose to move on with the interview. However, her concluding statement here shows how learning to more expertly engage in the practice of selecting a special case helped her better make sense of the problem and wonder about what a formal proof might entail. Laura's engagement in this practice shows evidence of learning, and also enabled her to demonstrate her ability to engage in other mathematical practices, such as connecting representations and evaluating her conclusions.

Learning to Generalize from a Special Case

Sam began the teacher preparation program with a strong commitment to becoming a mathematics teacher. He brought a good work ethic and a commitment to perseverance that was evident in his problem-solving during the interviews. In both cases, Sam spent a great deal of time working through the problems, wanting to really make sense of the underlying mathematics and be confident in his final answer. In addition to these commonalities across the two interviews, Sam showed changes in the ways in which he engaged in mathematical practices. One notable change was the way in which Sam was able to generalize from a special case.

In both cases, Sam made sense of the problem by considering a particular example. In the pre-task, Sam first thought through the problem by drawing a graph and using the functions $f(x) = ax + b$ and $g(x) = cx + d$. He said,

> So then if you add them together like you're not... I'm thinking you just close them together, but how do I show that they also intersect at a point P? By just using algebra. So, I'm thinking this is like the, it's like a geometry problem. ... I mean so if you have $f(x)$, $(f + g)(x)$, you have a function... I mean I guess I maybe could, I could put in numbers. I could try to do it that way and see where it gets me.

As examples, he chose the functions $f(x) = -2x + 2$ and $g(x) = 2x - 2$. When he summed the functions, he found that $(f + g)(x) = 0$ for all values of x, not just at a single point P. Sam wanted to use this example as a basis for trying to understand the relationship in the problem statement. However, he did not find his example helpful, saying, "the function goes here which means you're on the x-axis which would also equal 0, that's not proving anything, if I put in numbers."

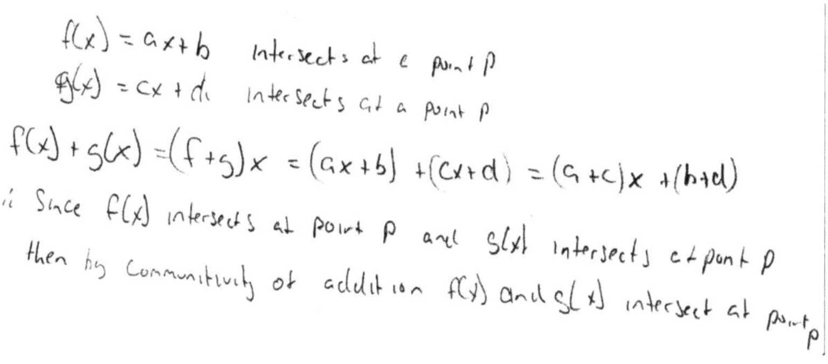

Fig. 11.1 Sam's generalization in the pre-task

Fig. 11.2 Sam's special case and the beginning of his generalization in the post-task

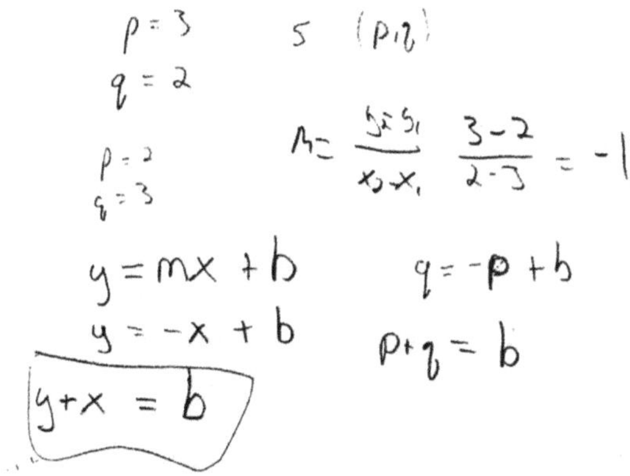

As Sam talked through his thinking about how the special case would generalize, he began treating the problem as a logic problem (see Fig. 11.1). He reasoned,

> By commutativity of addition, then, since p and q then since p is true and since f is true and g is true, then f and g are true. It's logic. I'm trying to come up with. I did the, just think, it's by commutativity of addition, right? By commutativity of addition, then f and g are true. [*Long pause*] That's it. I mean, if one does and the other one does. ... I know if p is true and f is true then p and f are true, or if p is true, f is true, and g is true, then f and g are true. Like I know that from logic, and that's how I solve this.

On the post-task, Sam also considered a particular case. He chose the point (3, 2) as an example and worked through the algebraic details of the problem for this case (see Fig. 11.2). He then worked to generalize this process to fully prove the statement.

As he generalized, he tried to use the same steps algebraically as he had done with his example (see Fig. 11.3). He said,

> I have – line between these two points – I can find the slope of this line – I don't know if that's gonna do anything. So, I know the slope of this line is -1. $y = -x + b$. So, in this equation, $y + x = b$. So my y-intercept is $y + x$.

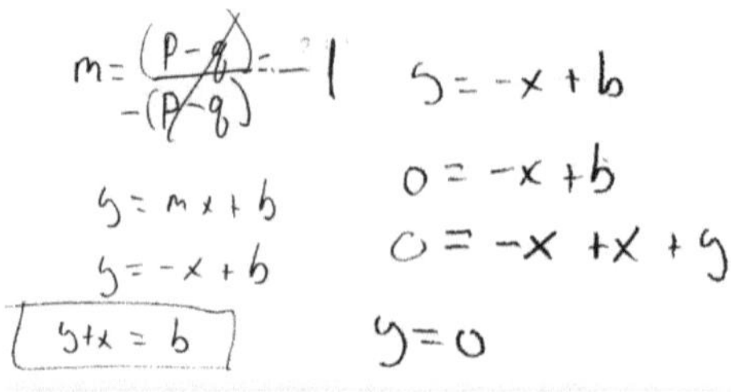

Fig. 11.3 Sam's general algebraic representation for the post-task

The way Sam generalized from his special case changed from the pre-task to the post-task. In the pre-task he felt the special case did not prove anything, while in the post-task he recognized how to use the structure of the algebra to generalize his thinking. This attention to the structure within the algebra highlights the more expert approach to generalizing from a special case that Sam took in the post-task.

In addition, the examples Sam chose in the two instances were not equivalent. In the pre-task, the example was in some ways too narrow—the sum of the two functions was zero everywhere, not just at a single point, whereas the example in the post-task illustrated the structure of the general proof more accurately. I argue that this represents learning to engage more expertly in the practice of generalizing a special case because in the pre-task, Sam did not attempt to investigate the structure of his special case. Had he done so, he might have discovered the additional assumption he had unintentionally embedded in his example through his engagement in other mathematical practices. However, it might alternately be considered a case of learning to choose better special cases or learning to attend to the conditions in a problem. It is also possible that if Sam had chosen a different example in the pre-task, he would have demonstrated the same level of expertise at both time points. These possibilities point to how mathematical practices are dependent upon one another. This case illustrates not just the learning of a single practice, but the potential for that practice to support engagement in others. I contend that Sam learned to more expertly generalize from a special case because in the pre-task, Sam did not look for any underlying structure in his example, nor did he re-evaluate the chosen example in light of the problem statement. In the post-task, Sam recognized that the structure of the solution ought to be illustrated through his example, and so he sought out that structure and built on it. Sam's demonstration of increased expertise in generalizing from a special case was supported by his engagement in other mathematical practices, such as developing a plan for his solution, monitoring his own progress, and critiquing his own reasoning. Sam engaged in numerous mathematical practices in a coordinated way to work through the problem.

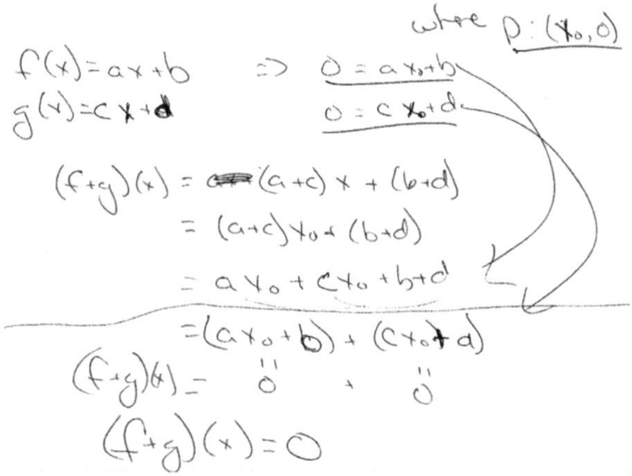

Fig. 11.4 Tim's written algebraic work on the pre-task

Learning to Distinguish Formal and Informal Proof

Tim entered the preparation program right after completing his undergraduate degree and was very eager and excited to become a teacher. For both tasks, Tim worked algebraically and illustrated his ideas by drawing graphs. He solved both tasks very quickly and was convinced of his answers.

One way that Tim changed his engagement in mathematical practices was in distinguishing formal and informal proof. In the pre-task, Tim defined a point P as $(x_0, 0)$, and used arrows to show substituting that point into the sum function (see Fig. 11.4). His notation here was informal. Additionally, Tim wrote $(f + g)(x) = 0$, rather than $(f + g)(x_0) = 0$, though he said, "And so let's see their sum. ... And then if I plug x-naught in ..." This mismatch between words and notation also suggests an informal approach to proof. Alternately, it might be interpreted as making a small typo that Tim may or may not have discovered had he continued working on the problem. Tim did not discuss any ideas related to formal and informal proof during the pre-task.

After solving the post-task, Tim commented that even though he had convinced himself that the problem statement was true, what he had written was not a formal proof. He said, "This – I mean, this is a terrible proof, but this is me – this is like definite scratch work." When prompted, Tim showed what a more formal proof would look like. His formal proof included writing out the reasoning that justified different steps of the proof (see Appendix 3). He was careful about notation and organized the formal proof in a way that he might share it with others. Tim's unprompted comment that he had written a "terrible proof" contrasts with his work on the pre-task, when he simply accepted his initial writing as sufficient without examining whether or not it might be formal proof. This provides evidence that Tim learned about the distinction between informal and formal proofs and was able to identify that distinction in his own proof writing.

Learning to Attend to Precision Through Defining Variables

Daniel entered the preparation program after a long career in engineering and business, excited about starting a new career. He was particularly interested in the possibility of teaching computer science. He approached both tasks by using visuals to make sense of the statements and working through the proofs algebraically. He regularly paused to make sure he was on the right track, and he moved between his algebraic and graphical representations to check his work.

One distinction between Daniel's pre-task and post-task solutions was his attention to precision in defining variables. In the pre-task, Daniel said, "So they intersect at a point P on the x-axis and so, starting with the first equation $f(x) = ax + b$, that means that at this point P, $aP + b = 0$. For the second one, we got a similar statement that $cP + d = 0$." Later, he wrote $(P, 0)$ on his paper, implying he had chosen P to represent a specific value, rather than P representing a free variable. Daniel went into more detail about this issue on the post-task. He said, "The general equation of the line that I'm going to use is $y - y_1$, which is the letters for subscripts are expected to be actual points. I'm not using the right term. The letters without subscripts are the variables." He again mentioned the distinction between free and fixed variables during his work on the post-task when he said, "The ps and qs are the x and y values for these two points."

This detail may seem small in the context of the overall proof, but it represents an important aspect of working with algebraic notation. It also is clear evidence of change in how Daniel engaged in this mathematical practice. Something he left implicit on the pre-task became explicit in the post-task, showing that Daniel not only learned how to more precisely define variables, but also that he saw it as an important aspect of describing his solution.

Common Themes

The engagement in mathematical practices by each participant highlights different themes in how engagement in practices might change over time. There were also common themes that emerged across participants. Across tasks, all participants used multiple representations to access and make sense of the problem, and they also worked to connect those representations. All four participants showed more expert practice in connecting representations on the post-task than they did on the pre-task. For example, on the pre-task, Daniel connected representations only to evaluate his completed proof, while on the post-task, he connected representations throughout to help shape his argument.

The participants also showed more expert practice in attending to precision in the post-task. In some cases, precision emerged in the context of more formal proof writing; in other cases, it emerged in the process of generalizing from a special case. This attention to precision carried over into the ways in which participants communicated their solutions. Tim's more formal proof on the post-task is the strongest evidence of this, and all of the participants showed more expert practice in communicating their solutions.

Connection to Opportunities to Learn

Each of the cases described above illustrates an example of learning to engage in mathematical practices that is related to opportunities for learning that occurred in the abstract algebra course. I highlight moments from the course where the class worked collectively on reasoning based on generic examples, distinguishing formal and informal proof, and attending to precision. Finally, I describe structures in the course that had the potential to support learning to engage in mathematical practices.

Generic Examples

One technique emphasized in the course was to use examples as a way to explore a more general problem. For example, the professor used an example to illustrate the structure of the proof of the rational root theorem. Using the equation $x^4 + 2x^3 + 4x^2 + 8x + 16 = 0$, the class set out to prove that any solution to this equation, t, must be irrational, using proof by contradiction. Together, with the professor, the class carefully went through the algebra, assuming initially that t was rational, and identifying all possible rational roots for the equation. Daniel questioned what they were really proving, and the professor responded,

> I wanted to prove the rational roots theorem but present it as an example. [...] That's one teaching tactic I have for a challenging proof. I try to come up with a simple example where all the reasoning for the general case is right there. A generic example. You'd see exactly this in a more general context. It's a sneaky way to lower the conceptual difficulty of a challenging proof, but you worked through all the important details.

With this, the professor emphasized how using strategically selected examples could illustrate the structure of a proof before proceeding with full generality. In this instance, the professor modeled engagement in a mathematical practice, a necessary part of appropriation.

Another instance of highlighting the usefulness of generic examples occurred when the class was proving the fundamental theorem of algebra. The professor said,

> This is something like a generic example. [...] A generic example illustrates a line of reasoning that generalizes. So, what we'll see in this example encapsulates a construction that will allow us to do the induction step. [... This will] lower the conceptual difficulty by making it more concrete, then it will be easier to think about full generality.

He then used an example to illustrate the approach he would be taking to prove the fundamental theorem of algebra by induction.

These two instances of using generic examples to illustrate the structure of a proof provided valuable opportunities for participants to gain insight into how to approach algebraic proofs. Using generic examples in class mirrors the approach Sam took on his post-task. He lowered the conceptual difficulty of the problem with a concrete example and then he generalized that argument. Similarly, Laura may have drawn on these instances in her work on the post-task as she recognized the necessity of generalizing from her example.

Participants also had the opportunity to engage in this practice themselves as they solved problems during class. One such instance occurred during an activity introducing the function e_2, defined as follows: "for every n in $\mathbb{N}$, the value $e_2(n)$ is 0 if n is odd, and otherwise, is the exponent of 2 in the prime factorization of n. For example, $e_2(2011) = 0$; $e_2(80) = e_2(2^4 5^1) = 4$." Participants generated data to fill in a table for $n = 1$ to $n = 16$, with columns for n, $\binom{n}{2}$, $e_2(n)$, and $e_2\left(\binom{n}{2}\right)$. Then they were tasked with making conjectures about any patterns they observed. In subsequent classes, participants proved their conjectures. This activity exposed participants to an unfamiliar function, e_2, and required them to write their own proofs. It also served as a gateway into abstract algebra content such as the degree of polynomials. The e_2 function was used later as part of the strategy for proving the fundamental theorem of algebra.

During the conjecturing stage, Laura initially conjectured that if $e_2(n) = 1$, then $e_2\left(\binom{n}{2}\right) = 0$. Through discussion with her classmates, Laura added a second conjecture, that if $e_2(n) = 0$, then $e_2\left(\binom{n}{2}\right) = e_2(n - 1)$. These conjectures illustrate how Laura was using examples to help her make sense of a problem. In that same activity, Sam made three related conjectures. His conjectures also used specific examples to begin to make sense of the larger structure of the task. Sam conjectured: (1) starting at $n = 2$, every fourth e_2 value is 1; (2) starting at $n = 3$, every second e_2 value is 0; and (3) starting at $n = 8$, every sixteenth e_2 value is 3. This string of conjectures would eventually be consolidated by the class as they generalized their descriptions of the patterns they observed.

Formal and Informal Proof

The emphasis on proof and proof technique permeated the course, and Tim's distinction between scratch work and formal proof in the post-task was reflective of a discussion in the abstract algebra class. During the first week, the professor said,

> At the beginning [of working on a problem] I don't know how to [write a proof], so start with what you want and try to get somewhere. This is sometimes called the "forwards-backwards" method. We can try to get to something we're confident in, so this is actually good scratch work for problem solving because it gives us a foothold into writing down the proof. You're likely to gain insight into *how* to solve problem [through scratch work] – it's not to derive something that's true.

To further emphasize the point, the professor then showed participants the "proof" that $0 = 1$. He pointed out how one of the steps in the "proof" cannot be reversed and emphasized that working "on both sides of the equation" was very valuable as scratch work but would not be sufficient for a formal proof. Tim very directly drew on this distinction in his work on the post-task. This focus was also supplemented by the professor's emphasis on attending to precision.

Participants also had multiple opportunities throughout the course to engage in transforming their informal work into formal proofs. As one example, participants were assigned weekly homework sets, which they were required to submit using LaTeX. This requirement meant that participants would work out their solutions on paper, and then make them more formal as they typed them up. Furthermore, they had the opportunity to revise each submission. Each revision showed participants improving the language and precision in their proofs.

Participants also engaged in proving during class time. For example, participants were learning about symmetric polynomials by way of transpositions and permutations. The class started with a warm-up considering the polynomial $ab + cd$, investigating what happens to the polynomial under different transpositions (selecting two variables and switching them), and determining whether they had found all the different possible polynomials these transpositions could produce. After conjecturing they had found them all, Sam said that they would need to provide justification. Together, the participants began to formalize their thinking about whether or not they had closure. This discussion led to defining symmetric polynomials, and, again, participants collectively engaged in formalizing their ideas about how they could prove whether or not a given polynomial was symmetric.

Defining Variables

Another aspect of attending to precision came up on the first day of class. The professor assigned a reading that addressed the difference between free and fixed variables. In class, participants explored this distinction in the context of an induction proof. The professor elaborated, "This [n] is a free variable, it represents any positive integer. Over here, N is our fixed variable; we're using it very differently than n. It's a number that's fixed, not changing, but it is arbitrary." This distinction was attended to throughout the course, for example, during the proof of the rational root theorem. Participants often noted this distinction for themselves during problem-solving or homework. Daniel's case illustrated learning around attending to this distinction.

Structures Supporting Learning to Engage in Mathematical Practices

The episodes described above were common in the course. The professor regularly engaged participants in problem-solving activities, and explicitly called attention to aspects of doing mathematics that are often left implicit. The class included sustained periods of group work, sometimes lasting upwards of an hour. Participants regularly made conjectures as they investigated new ideas. In other words, the course provided the two key features needed for appropriation of mathematical practices: modeling by the professor and the opportunities for participants to engage in the practices themselves (Moschkovich, 2004, 2013). Additionally, the class was explicitly focused on content that the professor considered to be the most relevant to teaching school mathematics.

Participants reflected on these course structures in interviews and expressed that they felt they had learned more about proof, justification, and attention to precision through the course. For example, Tim described how he valued the opportunity to explore mathematical ideas and make mistakes as part of his learning, saying, "I can take something, look at it and actually recognize it—'ooh, maybe induction would be good for this.' I'll give it a shot—maybe fail, but that's the fun in it." He also reflected, "I learned to write proofs better, I believe, and I've developed just a way of going into a problem and figuring out 'what do I know, where do I want to go, how can I fill in this space between?'" Laura talked about learning to make conjectures. She said, "Even conjecturing I had never experienced before. And so that was interesting." She reflected on her conjectures during the e_2 activity, which became central to the ongoing class discussion, saying, "I felt so good after that. I'm usually very good at finding patterns. That's just where I am I guess. So, I'm glad I was able to give something." All four participants recognized this course as being unlike previous mathematics courses they had taken.

Connecting Abstract Algebra and Teaching Mathematics

Each participant had the chance to reflect on how they imagined their participation in this abstract algebra course might influence their teaching. Sam found the course unrelated to teaching, but Daniel, Laura, and Tim all saw connections between the course and teaching. Daniel talked about mathematical thinking and problem-solving. He said,

> We are supposed to both be students and stand outside ourselves and look at what it is like to be a student. And try to appreciate that, because when you're teaching in a classroom, you know, you don't get that opportunity. So, the reminder of how hard it is and the reminder of how frustrating it is and things like that, that's very valuable.

He saw a connection in gaining insight into how students feel when learning new mathematics, and he appreciated how the course helped him understand the content at a deeper level.

Tim commented on how the content of abstract algebra would "trickle down" into his teaching. He said,

> I entered talking with [Sam], and he didn't want to [take abstract algebra] because he didn't think we should. I thought, "yeah, we probably don't need it, that much abstract algebra you know, we're just not going to be going beyond calculus really." And I just wanted – I

> was really open to it, because I just love doing math because it's really fun. But now that I went through it, I feel like it – all of it just trickles down as you go higher. All that goes down and trickles into all the earlier things you did, and I really have a greater appreciation for that now.

Tim was able to see strong connections to his own teaching beyond the particular content of abstract algebra, which he had initially viewed as not being connected to school mathematics.

Laura reflected on how the teaching approach in the abstract algebra course would be particularly influential in her own teaching. She said,

> [The professor's] continued use of posing questions and having us work together and tying everything up at the end are all things that I hope I'm able to do in the classroom. In terms of the math, definitely the thinking part of it, you know. If I was able to better my mathematical thinking just in those activities that we did, I'm sure that that's going to transfer to my students as well. So, if I can foster the same kind of culture in the classroom where asking questions is okay, and where I'm modeling. And actually, not only modeling, but actually making it part of my class, that I want students to work together and ask questions and come up with ideas and justify them and all those habits of mind.

Laura describes the way in which mathematical practices, such as making conjectures, asking questions, and justifying, serve as connections between abstract algebra and school mathematics.

All four participants reflected on the value of engaging in mathematical practices during the course. Even though Sam did not think the course would influence his teaching, he knew the course had influenced the way he solved problems. Daniel, Laura, and Tim talked about different aspects of the course in their reflections but grounded their reflections in the value of *doing* mathematics. They saw that they were doing mathematics themselves and expressed the desire for their students to similarly be doing mathematics.

Discussion

Through their participation in the abstract algebra course, participants learned to engage in a variety of mathematical practices. This study design does not allow for generalizations to larger populations of future teachers or to other abstract algebra courses. Though the analysis does not explore alternative learning opportunities (e.g., other concurrent courses), the close link between the types of mathematical practices used in the course and evident in participants' interviews, coupled with participants' reflections on their own learning, provides reasonable evidence for a connection. Additionally, the findings may be limited due to potential differences in the difficulty levels between the two tasks. However, both tasks provided opportunities for engaging in mathematical practices regardless of difficulty level, or whether or not a participant arrived at a correct answer. Acknowledging these limitations, I argue that engagement in mathematical practices is a key connection between abstract algebra and school mathematics.

At the beginning of the post-interview, Daniel commented on how he felt like the tasks I was asking him to solve were the only questions anybody had asked him to check if he actually understood high school mathematics. He said,

> I notice that your survey is, to date, the only or the best question measure of whether I actually know anything about high school math. This seems very whacko to me, in light of all the statements that are made about how important it is to know something about the math. I mean, it's fine, you know. In [the applied math] class, I learned some math. In [the abstract algebra] class, I learned some math. There's some overlap, but I'm very surprised that we had – that nobody's asked, you know, before I even got here, do you actually know anything about circles, you know.

This highlights a common disconnect between school mathematics and college level courses.

Despite this apparent disconnect, all the participants were able to draw on the learning opportunities in the course and apply them, not just on course homework assignments, but in the context of tasks explicitly situated in school level algebra. For example, the abstract algebra class talked about free and fixed variables in the context of an induction proof. Daniel applied these ideas in an entirely different setting. Similarly, Laura, Sam, and Tim capitalized on learning opportunities situated in abstract algebra while solving problems addressing secondary content.

In course activities and in the interviews, all participants engaged in practices in an interconnected way. Engaging in proving was more than just formal proof, it required making conjectures, connecting representations, attending to precision, and many other mathematical practices. The abstract algebra course made explicit the work of engaging in mathematics and how different practices support one another. Participants' engagement in mathematical practices during the interviews highlights how they can serve as a connection across levels.

The engagement in mathematical practices identified here suggests that abstract algebra courses for teachers can provide valuable, relevant learning opportunities focused on mathematical practices that will be applicable in the context of teaching. By giving participants the space to engage in mathematics themselves, the abstract algebra course provided them the opportunity to develop more expert engagement in multiple mathematical practices. The participants all capitalized on these opportunities and were able to leverage that newly developed expertise in the context of school mathematics.

Implications

Mathematical practices are an important place to consider connections between abstract algebra and school mathematics. The participants in this study were all able to draw on their experiences in an abstract algebra course and apply the mathematical practices they learned there in the context of school algebra. Of course, the abstract algebra course was unique; it had many features not commonly found in most abstract algebra courses. Moreover, the themes identified for learning mathematical practices may be entirely local to these four participants. These themes are not intended to imply that participants are fully expert in any practice; rather, they have become more expert than they previously had been and will continue to grow as they engage in more mathematics. The learning around mathematical practices exhibited by these participants provides an image of how one type of connection between abstract algebra and school mathematics can be made explicit in an abstract algebra course for teachers.

All teachers enter teacher preparation programs with differential strengths and areas for growth in their engagement in mathematical practices. In this case, despite these differences, all four participants began to more fully appropriate mathematical practices. Furthermore, these changes in engagement in mathematical practices were related to opportunities to learn in the abstract algebra course. This suggests the value of mathematics content courses as sites for participants to more fully develop their engagement in mathematical practices. Finally, the portability of the practices across content levels demonstrates a perhaps unexpected way that abstract algebra content can be made more relevant for secondary teaching.

One of the ways in which participants demonstrated more expert engagement in mathematical practices was through engaging with multiple practices in an interconnected way. This interconnectedness of the practices suggests the value of learning about practices in conjunction with one another. The implication for teacher preparation is to provide multiple opportunities for pre-service teachers to engage in a variety of practices, rather than focusing more narrowly on a single practice (e.g., formal proof). Similarly, there is value in looking more holistically at practice engagement in research rather than focusing on a single practice.

Abstract algebra is commonly a required course in teacher preparation. In addition to exploring how content can serve as a connection to school mathematics, it is important to consider the place of mathematical practices in this course. In the USA, the Common Core (2010) standards of mathematical practice provide a practical motivation for understanding how practices are learned and how they can be taught. It is reasonable to imagine that if teachers have not had opportunities to engage in mathematical practices themselves, it will be difficult for them to create opportunities for their students to engage in mathematical practices. Learning to engage in mathematical practices can be seen as a first step toward learning to teach others to do so, creating an imperative to more fully understand teacher learning around mathematical practices. Moreover, having the opportunity to learn mathematics through doing mathematics in an abstract algebra course can help make that course more relevant to the work of teaching.

Appendix 1: Possible Algebra Pre-task Solution

Prove the following statement:
If the graphs of linear functions $f(x) = ax + b$ and $g(x) = cx + d$ intersect at a point P on the x-axis, the graph of their sum function $(f + g)(x)$ must also go through P.

We are given that $f(x) = ax + b$ and $g(x) = cx + d$. These two functions intersect at a point P on the x-axis. Define P as $(p, 0)$.

The fact that the two functions intersect at point P implies that both contain $(p, 0)$. In other words, $f(p) = 0$ and $g(p) = 0$.

To determine whether the sum function also passes through point P, we must consider $(f + g)(x)$ and see if the value of the sum function at p is 0.

$$\begin{aligned}(f+g)(p) &= f(p) + g(p)\\ &= 0 + 0\\ &= 0\end{aligned}$$

Since the sum function is defined as the sum of the value of the functions f and g at every point in the domain, we can calculate the value of the sum function at p as shown above. The fact that f and g contain $(p, 0)$ guarantees that their sum also contains $(p, 0)$. ■

Appendix 2: Possible Algebra Post-task Solution

Take a point (p, q) on the Cartesian plane. Reverse the coordinates to obtain a second point (q, p). Prove that on the line between these two points, the x-intercept and the y-intercept are the sum of the coordinates.

Consider the points (p, q) and (q, p) for $p \neq q$. Then the equation of the line between the two points is:

$$
\begin{aligned}
y - q &= \tfrac{p-q}{q-p}\,(x - p) \\
y - q &= \tfrac{p-q}{-(p-q)}\,(x - p) \\
y - q &= -1\,(x - p) \\
y - q &= -x + p \\
y &= -x + p + q
\end{aligned}
$$

The y-intercept occurs when $x = 0$. Plugging in 0 for x gives:

$$
\begin{aligned}
y &= -0 + p + q \\
&= p + q
\end{aligned}
$$

And so the y-intercept is the sum of the coordinates p and q.

Similarly, the x-intercept occurs when $y = 0$. Plugging in 0 for y gives:

$$
\begin{aligned}
0 &= -x + p + q \\
x &= p + q
\end{aligned}
$$

And so the x-intercept is the sum of the coordinates p and q.

If $p = q$, then (p, q) and (q, p) are the same point, and an infinite number of lines could pass through it. Thus the proposition stated in the problem holds only for $p \neq q$. ■

Appendix 3: Tim's Scratch Work and Formal Proof, Algebra Post-task

Scratch work

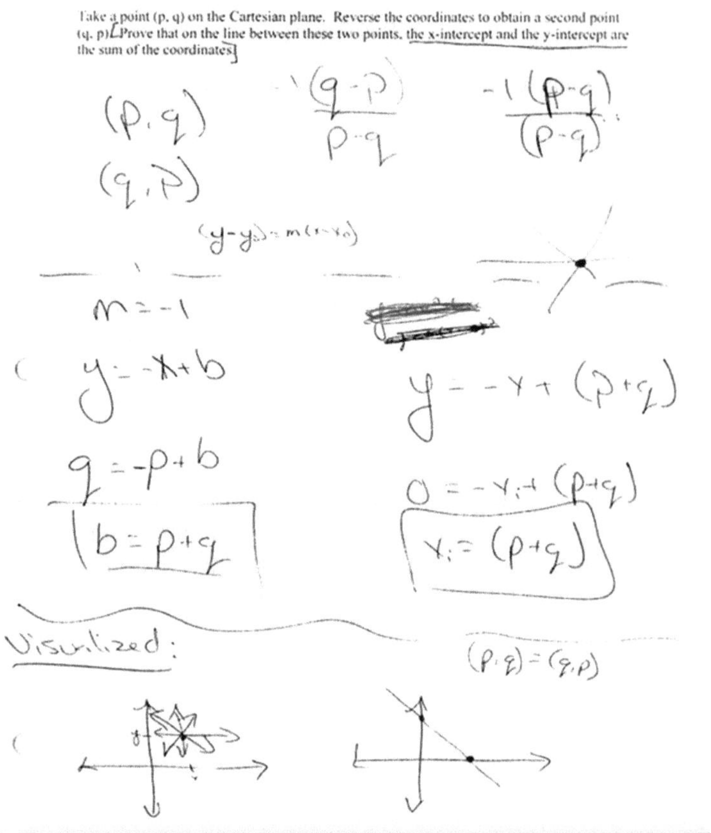

Formal work

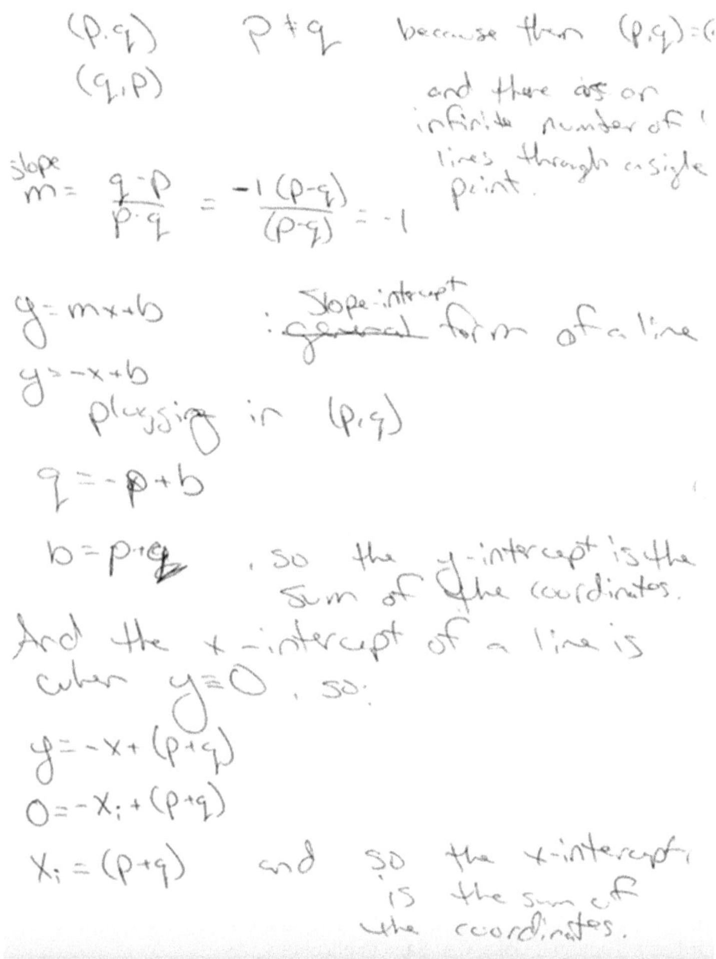
(p,q) p ≠ q because then (p,q)=(
(q,p) and there are an infinite number of lines through a single point.

slope $m = \frac{q-p}{p-q} = \frac{-1(p-q)}{(p-q)} = -1$

$y = mx+b$: Slope-intercept ~~general~~ form of a line
$y = -x+b$
plugging in (p,q)
$q = -p+b$
$b = p+q$, so the y-intercept is the sum of the coordinates.
And the x-intercept of a line is when y=0, so:
$y = -x+(p+q)$
$0 = -x_i+(p+q)$
$x_i = (p+q)$ and so the x-intercept is the sum of the coordinates.

References

Baldinger, E. E. (2014). Opportunities to develop mathematical knowledge for teaching in a math content course (Unpublished doctoral dissertation). Stanford, CA: Stanford University.

Ball, D. L. (Ed.). (2003). Teaching and learning mathematical practices. In *Mathematical proficiency for all students: Toward a strategic research and development program in mathematics education* (pp. 29–41). Santa Monica, CA: RAND.

Bass, H. (2011). A vignette of doing mathematics: A meta-cognitive tour of the production of some elementary mathematics. *Montana Mathematics Enthusiast, 8*(1&2), 3–34.

Burton, L. (1999). The practices of mathematicians: What do they tell us about coming to know mathematics? *Educational Studies in Mathematics, 37*(2), 121–143. https://doi.org/10.2307/3483312

Cobb, P., & Bowers, J. (1999). Cognitive and situated learning perspectives in theory and practice. *Educational Researcher, 28*(2), 4–15.

Cuoco, A., Goldenberg, E. P., & Mark, J. (1996). Habits of mind: An organizing principle for mathematics curricula. *Journal of Mathematical Behavior, 15*(4), 375–402. https://doi.org/10.1016/S0732-3123(96)90023-1

Ericsson, K. A., & Simon, H. A. (1980). Verbal reports as data. *Psychological Review, 87*(3). https://doi.org/10.1037/0033-295X.87.3.215

Ginsburg, H. P. (1981). The clinical interview in psychology research on mathematical thinking: Aims, rationales, techniques. *For the Learning of Mathematics, 1*(3), 4–11.

Goldin, G. A. (1997). Chapter 4: Observing mathematical problem solving through task-based interviews. *Journal for Research in Mathematics Education. Monograph, 9*(1997), 40–62. https://doi.org/10.2307/749946

Goulding, M., Hatch, G., & Rodd, M. (2003). Undergraduate mathematics experience: its significance in secondary mathematics teacher preparation. *Journal of Mathematics Teacher Education, 6*, 361–393. https://doi.org/10.1023/A:1026362813351

Graham, K. J., Li, Y., & Buck, J. C. (2000). Characteristics of mathematics teacher preparation programs in the United States: An exploratory study. *Mathematics Educator, 5*(1/2), 5–31.

Hodge, A. M., Gerberry, C. V., Moss, E. R., & Staples, M. E. (2010). Purposes and perceptions: What do university mathematics professors see as their role in the education of secondary mathematics teachers? *PRIMUS: Problems, Resources, and Issues in Mathematics Undergraduate Studies, 20*(8), 646–663. https://doi.org/10.1080/10511970902810733

Katz, V. J., & Barton, B. (2007). Stages in the history of algebra with implications for teaching. *Educational Studies in Mathematics, 66*, 185–201. https://doi.org/10.1007/s10649-006-9023-7

Knuth, E. J. (2002a). Secondary school mathematics teachers' conceptions of proof. *Journal for Research in Mathematics Education, 33*(5), 379–405. https://doi.org/10.2307/4149959

Knuth, E. J. (2002b). Teachers' conceptions of proof in the context of secondary school mathematics. *Journal of Mathematics Teacher Education, 5*, 61–88.

Lakatos, I. (1976). In J. Worrall & E. Zahar (Eds.), *Proofs and refutations*. UK: Cambridge University Press.

Larsen, S. P., & Zandieh, M. (2007). Proofs and refutations in the undergraduate mathematics classroom. *Educational Studies in Mathematics, 67*(3), 205–216. https://doi.org/10.1007/s10649-007-9106-0

Lave, J., & Wenger, E. (1991). *Situated learning: Legitimate peripheral participation*. New York: Cambridge University Press.

Miles, M. B., Huberman, A. M., & Saldaña, J. (2013). *Qualitative data analysis: A methods sourcebook* (3rd ed.). Thousand Oaks, CA: Sage Publications.

Moschkovich, J. N. (2004). Appropriating mathematical practices: A case study of learning to use and explore functions through interaction with a tutor. *Educational Studies in Mathematics, 55*(1–3), 49–80. https://doi.org/10.1023/B:EDUC.0000017691.13428.b9

Moschkovich, J. N. (2013). Issues regarding the concept of mathematical practices. In Y. Li & J. N. Moschkovich (Eds.), *Proficiency and Beliefs in Learning and Teaching* (pp. 257–275). Rotterdam: Sense Publishers.

Murray, E., Baldinger, E. E., Wasserman, N. H., Broderick, S., & White, D. (2017). Connecting advanced and secondary mathematics. *Issues in the Undergraduate Mathematics Preparation of School Teachers, 1*(August), 1–10.

National Council of Teachers of Mathematics. (1989). *Curriculum and evaluation standards for school mathematics*. Reston, VA: National Council of Teachers of Mathematics.

National Council of Teachers of Mathematics. (2000). *Principles and standards for school mathematics*. Reston, VA: National Council of Teachers of Mathematics Author.

National Governors Association Center for Best Practices, & Council of Chief State School Officers. (2010). *Common Core State Standards for Mathematics*. Washington, DC: National Governors Association Center for Best Practices, & Council of Chief State School Officers.

Novotná, J., & Hoch, M. (2008). How structure sense for algebraic expressions or equations is related to structure sense for abstract algebra. *Mathematics Education Research Journal, 20*(2), 93–104.

Pólya, G. (1957). *How to solve it* (2nd ed.). Princeton, NJ: Princeton University Press.

Rogoff, B. (1990). *Apprenticeship in thinking: Cognitive development in social context*. New York: Oxford University Press.

Silver, E. A., Clark, L. M., Ghousseini, H. N., Charalambous, C. Y., & Sealy, J. T. (2007). Where is the mathematics? Examining teachers' mathematical learning opportunities in practice-based professional learning tasks. *Journal of Mathematics Teacher Education, 10*(4–6), 261–277. https://doi.org/10.1007/s10857-007-9039-7

Simon, M. A. (1994). Learning mathematics and learning to teach: Learning cycles in mathematics teacher education. *Educational Studies in Mathematics, 26*(1), 71–94. https://doi.org/10.1007/BF01273301

Speer, N. M., Smith III, J. P., & Horvath, A. (2010). Collegiate mathematics teaching: An unexamined practice. *Journal of Mathematical Behavior, 29*(2), 99–114. https://doi.org/10.1016/j.jmathb.2010.02.001

Staples, M. E., Bartlo, J., & Thanheiser, E. (2012). Justification as a teaching and learning practice: Its (potential) multifaceted role in middle grades mathematics classrooms. *Journal of Mathematical Behavior, 31*(4), 447–462. https://doi.org/10.1016/j.jmathb.2012.07.001

Stylianides, A. J., & Ball, D. L. (2008). Understanding and describing mathematical knowledge for teaching: knowledge about proof for engaging students in the activity of proving. *Journal of Mathematics Teacher Education, 11*(4), 307–332. https://doi.org/10.1007/s10857-008-9077-9

Stylianides, G. J., Stylianides, A. J., & Philippou, G. N. (2007). Preservice teachers' knowledge of proof by mathematical induction. *Journal of Mathematics Teacher Education, 10*(3), 145–166. https://doi.org/10.1007/s10857-007-9034-z

Tabach, M., Levenson, E., Barkai, R., Tsamir, P., Tirosh, D., & Dreyfus, T. (2011). Secondary teachers' knowledge of elementary number theory proofs: the case of general-cover proofs. *Journal of Mathematics Teacher Education, 14*(6), 465–481. https://doi.org/10.1007/s10857-011-9185-9

Taylor, K. L., & Dionne, J.-P. (2000). Accessing problem-solving strategy knowledge: The complementary use of concurrent verbal protocols and retrospective debriefing. *Journal of Educational Psychology, 92*(3), 413–425. https://doi.org/10.1037//0022-0663.92.3.413

TEDS-M International Study Center. (2010). *Released Items: Future teacher mathematics content knowledge and mathematics pedagogical content knowledge - secondary*. MI: East Lansing.

Usiskin, Z. (1988). Conceptions of school algebra and uses of variables. In A. F. Coxford & A. P. Shulte (Eds.), *The Ideas of Algebra, K-12* (pp. 8–19). Reston, VA: National Council of Teachers of Mathematics.

Wasserman, N. H. (2016). Abstract algebra for algebra teaching: Influencing school mathematics instruction. *Canadian Journal of Science, Mathematics and Technology Education, 16*(1), 28–47. https://doi.org/10.1080/14926156.2015.1093200

Chapter 12
From Equations to Structures: Modes of Relevance of Abstract Algebra to School Mathematics as Viewed by Teacher Educators and Teachers

Josephine Shamash, Marita Barabash, and Ruhama Even

The Israeli high school curriculum in algebra deals mainly with the solution of different types of equations. However, modern algebra has a completely different viewpoint, and is concerned with algebraic structures and operations. The question then arises: What might be the relevance and contribution of an abstract algebra course for developing expertise in secondary school mathematics instruction? That is the focus of this chapter.

This chapter consists of five parts. The first part reviews relevant literature, which provides a basis for the study. In the second and third parts, the unique master's program and the specific abstract algebra course (which is one course in the master's program) are described. The fourth part of the chapter examines the relevance of the course to teaching high school mathematics, by analyzing didactic materials designed by six participating teachers that connect the mathematics learned in the abstract algebra course to the school curriculum and teaching. Finally, the fifth part discusses the findings and suggests directions for future research.

Background

The scholarly literature on the relevance and contribution of advanced mathematics studies to teaching secondary school mathematics is rather limited. It mainly

J. Shamash (✉) · M. Barabash
Department of Mathematics, Achva Academic College, Arugot, Israel

Department of Science Teaching, Weizmann Institute of Science, Rehovot, Israel
e-mail: josie.shamash@weizmann.ac.il

R. Even
Department of Science Teaching, Weizmann Institute of Science, Rehovot, Israel

N. H. Wasserman (ed.), *Connecting Abstract Algebra to Secondary Mathematics, for Secondary Mathematics Teachers*, Research in Mathematics Education,
https://doi.org/10.1007/978-3-319-99214-3_12

suggests two types of potential contributions in accordance with two dimensions of subject-matter knowledge: one type of potential contribution is at the level of specific content (e.g., algebra), and the other is the more general epistemological knowledge about what mathematics is and what doing mathematics entails. These two types are elaborated in the following paragraphs.

Assuming that knowledge of both school mathematics and academic mathematics is important for secondary school mathematics teachers, yet acknowledging the gap between academic and school mathematics, some researchers attempt to make explicit connections between school mathematics and academic mathematics at the level of specific content (e.g., Baldinger 2013; Christy & Sparks 2015; Dreher, Lindmeier, & Heinze 2016: Klein 1932; Murray et al. 2015; Wasserman 2016, 2017). For instance, Christy and Sparks (2015), and Wasserman (2016) explore the potential for aspects of abstract algebra to be influential in teaching the subject of solving equations in school mathematics.

Potential relevance at the level of specific content is associated also with the work of scholars who argue that teachers need to be provided with a body of mathematical knowledge that does not deviate from the material they teach in school (e.g., Wu 2011). Still, like those who aim at connecting academic and school mathematics, these scholars demonstrate that specific advanced mathematical studies (e.g., abstract algebra) are needed to achieve adequate teacher knowledge of the mathematical content that they need to teach at school. For example (emphasis in the original),

> This necessity that math teachers actually *know* the mathematics they teach sheds light, in particular, on why we want all high school teachers to know some abstract algebra: this knowledge allows them to really understand why there are only two arithmetic operations (+ and •) instead of four . . . (Wu 2011, pp. 380-381).

Another type of potential contribution of advanced mathematics studies for teaching secondary school mathematics—advocated in the literature mainly by mathematicians (e.g., Wu 2011; Ziegler & Loos 2014)—is at the more general epistemological level of knowledge about what mathematics is and what doing mathematics entails. Attempting to narrow the gap between "what (research) mathematicians take for granted as mathematics and what teachers and educators perceive to be mathematics" (Wu 2011, p. 382), these mathematicians suggest different approaches. For example, Wu emphasizes the importance of the fundamental principles of mathematics (e.g., definitions provide the basis for logical deductions, mathematics is coherent), while Ziegler and Loos (2014) attempt to broaden teachers' views of mathematics in multiple directions:

> . . . give them [teachers] a panoramic view on mathematics: . . . an overview of the subject, how mathematics is done, who has been and is doing it, including a sketch of main developments over the last few centuries up to the present (p. 9).

The existing empirical research literature on the relevance and contribution of advanced mathematics studies to teaching secondary school mathematics is scarce; it is mainly based on interviews with teachers (Zuo & Leung 2016). Most of the studies report on contributions at the level of knowledge about what mathematics is and what doing mathematics entails. For example, when interviewed about the relevance of advanced mathematics studies to their teaching, teachers referred to different aspects, such as: doing mathematics as problem solving (Even 2011), the role of intuition in doing mathematics (e.g., Adler et al. 2014; Even 2011), and the use of mathematics in other disciplines (e.g., Adler et al. 2014; Even 2011; Zazkis & Leikin 2010). Several studies found small contributions to the teaching of specific topics. For example, when interviewed, some teachers reported that they used the knowledge of specific topics they acquired during mathematics studies at colleges or universities to respond to students' questions or to enrich topics they taught (e.g., Even 2011; Zazkis & Leikin 2010). Additionally, a few studies found affective contributions. For example, some teachers mentioned a contribution to their personal confidence, or that experiencing difficulties when learning advanced mathematics made them more sensitive to students' emotional difficulties (e.g., Barton & Paterson 2009; Even 2011; Zazkis & Leikin 2010).

As can be seen, advanced mathematics studies at the college or university level are commonly considered to be essential in the professional education and development of secondary school teachers. Yet, the literature does not provide a sound answer to the question regarding the contribution and relevance of advanced mathematics courses for developing expertise in secondary school mathematics instruction. This chapter aims to address this problem by examining what might be the relevance of a carefully designed abstract algebra course for Israeli secondary school mathematics teachers to teaching secondary school mathematics. This course is part of a unique master's program, the Rothschild-Weizmann Program, which is described below.

The Rothschild–Weizmann Program

The Rothschild–Weizmann (RW) Program is a unique master's program without thesis, offered by the Weizmann Institute of Science. It is especially designed for experienced Israeli secondary school teachers of science and mathematics. The program has four strands—Mathematics, Physics, Chemistry, and Biology—and grants an M.Sc. degree in science and mathematics education. Admission to the program is selective and takes into account the applicants' background. A bachelor's degree in a relevant scientific field and 3 years of experience in teaching secondary school mathematics or science are required. Also taken into account is a candidate's potential to influence mathematics or science teaching at their school or region (as indicated by the application forms, the interviews, and the recommendation letters accompanying the applications). Each year, between five to twelve teachers are accepted into the program. They receive study grants and are exempt from paying tuition. 61 mathematics teachers from nine cohorts have graduated from the program as of 2018.

The program extends over 2 years (four semesters). During each year, the participating teachers spend 2 days per week on campus attending courses, and they continue to teach in parallel to their studies. Apart from three courses that are taught to all students, each of the four RW strands designs and conducts specific courses for its respective teachers.

One component of the mathematics strand of the program comprises mathematical studies tailored to the teachers in the program. The goal is to demonstrate that mathematics is a field that is very much alive, and that topics that are taught in high school are related to topics of current mathematical research (Even, Artstein, & Elbaum-Cohen, 2018). Thus, the main purpose of the mathematics component is to broaden secondary school teachers' perspective of, and understanding about, the discipline of mathematics and its relevance. Each of the eight mathematics courses in the program is independent of the other courses, and thus they can be offered on alternate years to teachers of two cohorts together. One of these mathematics courses is an abstract algebra course, developed and taught by the first author of this chapter. This course, which is the focus of this chapter, is described in detail in the following section. As it is offered every other year, some teachers take it during their second semester and some during their fourth (and last) semester in the program.

Another major component of the mathematics strand of the program focuses on important issues and aspects of mathematics education that are relevant to the teaching of secondary school mathematics (Even, Artstein, & Elbaum-Cohen, 2018). One of the five mathematics education courses deals with educational research and design linked to the learning and teaching of school algebra, developed and taught by the third author of this chapter. The course focuses on research, theory and practice related to the following aspects: a historical view of algebra and its development, conceptualizations of school algebra and its central concepts, aspects of algebra learning and understanding, algebra teaching, and algebra curriculum. This course is offered every other year to teachers of two cohorts together, during the same semester as the abstract algebra course.

Another mathematics education course, an integrative seminar—developed and taught by the second author of this chapter—aims to connect the mathematics learned in the mathematics courses to the school curriculum and teaching. To this end, the participating teachers are required to design or adapt teaching units, lesson plans or other teaching materials (of the school curriculum), demonstrating the impact of their academic studies of mathematics on their knowledge, teaching practices, attitudes and beliefs. The integrative seminar was not originally part of the RW program. It was added after the fifth year of the program, as the need for such a course became evident. Starting from the fifth cohort, the two-semester integrative seminar has been given every year to a single cohort of teachers during the second and third semesters of the four-semester program. So far, 36 teachers have taken this seminar. Since the abstract algebra course is offered to one cohort of teachers during their second semester in the program and to another cohort during the fourth semester (as the algebra course is offered every other year), every other cohort of teachers takes the integrative seminar before they take the algebra course. Of those who took the integrative seminar simultaneously with, or after the algebra course, four teachers chose to base the didactic materials they developed for the seminar on the abstract algebra course. These seminar works are analyzed in this chapter.

Additionally, the teachers are required to plan and carry out a final project, under the supervision of a mathematician, which includes a significant advanced mathematical module. The work on the project is accompanied by a seminar, also led by the second author, that involves reviewing mathematics and mathematics education papers, and presenting and discussing intermediate stages of the final projects. The final project seminar is held during the third and fourth semesters. The final project is presented both orally and in a written report. It requires the teachers to show some originality, either by solving an adequate mathematical problem, or organizing and presenting a topic otherwise scattered in the literature. Occasionally, suggestions for teaching the mathematical material to other teachers or to advanced high school students are included. Two final projects in the last 5 years were related to the abstract algebra course. They too are analyzed in this chapter.

The Abstract Algebra Course

Motivation for the Course

As mentioned above, the Israeli high school curriculum in algebra concentrates mainly on the solution of different types of equations. Many high school teachers have the impression that this is the main content of algebra, as evidenced by the students in our program during the years it has been running. During their mathematics degrees at university, most complete a basic (compulsory) course in linear algebra, but no courses in Galois theory or group theory. It is certainly true that the desire to solve certain types of equations gave rise to modern algebra. Since modern algebra today is concerned with algebraic structures and operations, the aim was to bridge this gap.

The motivation was to design a course for secondary school teachers that would provide an introduction to algebraic structures and modern abstract algebra, and link abstract algebra both to the high school curriculum in algebra and also to the mathematics curriculum in general. The underlying idea was the feeling that it is important for mathematics teachers to be aware of developments and current research in the discipline they teach, and that this kind of understanding and perspective are essential for teachers in planning high school syllabi and activities. The hope was that, after the course, teachers would be able to focus more on the "big ideas" in algebra as they design lessons and curricula, rather than concentrating primarily on developing algebraic technique and skills, though, of course, these are a pre-requisite for any study of abstract algebra.

Specific Aims for the Course

The basic aim of the algebra course in the program was to broaden the perspective and understanding of high school teachers by linking topics in high school algebra with modern algebra. The title given to the course, "Algebra: From Equations to Structures," reflects this aim. Up until almost 200 years ago, "algebra" was concerned with the solution of polynomial equations. After Abel and Galois, this was no longer the case. As a field of mathematical research, algebra became the study of algebraic structures and operations on those structures. This has caused a disconnect between what is regarded in high school mathematics as algebra, and what mathematicians regard as algebra. One of the aims of the course was to bridge this gap, and to show how the metamorphosis of algebra occurred, and why.

It was also necessary to develop enough background knowledge to enable a discussion of some of the most important theorems in algebra, including theorems proved in the last few years as examples of current research. Clearly, it was not possible—or desirable—in the framework of this course to prove the deep results in the main theorems, but the aim was to develop enough tools to discuss and understand them, and to understand why these theorems are important.

It was also regarded as central to the course that the teachers develop a clearer idea, by the end, of what algebra is, and also what algebra is not. Not all symbolic manipulation is algebra. On the other hand, some of what teachers regarded before the course as geometry and analysis connects directly to modern algebra.

Another aim was to make the course relevant to the teachers' work. In this, we viewed relevance in a variety of modes, from the narrow and specific, to the extremely broad:

1. Direct relevance to the high school algebra curriculum.
2. Relevance to other topics in the curriculum, not usually regarded by teachers and high school students as "algebra," but which would be regarded by mathematicians as algebra.
3. Relevance to non-algebraic school topics, such as compass-and-straightedge constructions in geometry, rotations and symmetry.
4. Providing a meta-mathematical perspective. The leap from equations to structures embodies the leap from the concrete to the abstract. Instead of finding numbers that solve equations, or even using formulas in which we can substitute numbers, structures involve operations on the roots of equations, on sets of numbers, and, indeed, on other operations. This is a step of abstraction. The example of algebra is but one area in which this occurred in the development of mathematics, and to understand what mathematics is, it is important to see such examples.
5. Relevance to their mathematical thinking and that of their pupils by providing examples to teachers of how a mathematician thinks.
6. Relevance to the mathematics teacher's role as the main source of knowledge to their pupils as to recent developments in the field, by allowing teachers to acquire an acquaintance with recent important theorems and developments in algebra. For interested pupils wanting light shed on mathematical topics that appear in popular books or in the media, a high school mathematics teacher is a natural person to whom they can turn. It was felt, in that respect, that the course could provide answers to those questions as far as algebraic developments are concerned.
7. Contributing to the self-confidence of the mathematics teachers by enhancing their mathematical knowledge in general.

The first three modes of relevance (1), (2), and (3) are mainly associated with contributions at the level of specific content. However, modes (2) and (3) are related also to contributions at the epistemological level of what mathematics is and what doing mathematics entails, as these modes are related also to the nature of mathematics as a field rich in connections. Modes (4), (5), and (6) are mainly associated with contributions to teachers at the epistemological level of what mathematics is: (4) focusing on striving for abstraction, (5), on ways of thinking about and doing mathematics, and (6), on the nature of the field as it develops continuously. Finally, mode (7) is associated with affective contributions.

Designing the Course

The 14-week long (semester) course consists of 4 weekly hours of lectures and 2 h a week of tutorials. The lectures include not only all the theoretical material and the proofs, but a large number of class activities where examples are worked through. The tutorials, run by a doctoral student in mathematics, provide a chance for teachers to review material and investigate additional examples to aid in assimilating the theoretical material. The teachers hand in weekly assignments, which are solved in the lecture sessions, and sometimes used as a point of departure for further activities. The evaluation for the course is based on the weekly assignments and a take-home exam at the end of the course.

One of the main problems in preparing the course, and during the teaching of the course the first time, was the lack of appropriate texts that would be accessible to the teachers. Most of the books containing the theoretical material, including those listed in the bibliography (included below), for instance, Gaal (1973), Isaacs (1994), Jacobson (1985), and Lang (1985), are advanced specialist texts in which only the introductory chapters in each are at a mathematical level appropriate to the teachers in the course. The solution to this was to write a course file, which grew to a full set of lecture notes and assignments that is at a level commensurate with the mathematical background of the teachers in the RW program and suited to their needs. (See Shamash, 2010, 2016.)

The two most important topics in high school algebra are, arguably, the solution of polynomial equations and solving systems of linear equations. These connect directly to some of the most important and fundamental theorems in abstract algebra. In order to present these topics and achieve the aims mentioned above, the following guidelines were used in designing the course:

- The development of algebra as a historical narrative.
- Stating and discussing the big theorems and results.
- Showing applications of abstract algebra.
- Adapting material to the mathematical background of the teachers.
- The use of detailed examples.

We describe each of these as they were applied in developing the course.

The Development of Algebra as a Historical Narrative

The quadratic formula, which is one of the most familiar formulas in high school mathematics, was a natural starting point for the course. The idea was to use this as motivation to follow, as a narrative, the historical attempts of mathematicians to solve polynomial equations of higher degrees, attempts which resulted in the development of group theory and field theory by Galois and Abel. In other words, algebraic structures grew out of a need to solve certain problems, and proved to be a much more fruitful way of viewing them.

This approach to the course led naturally to examining topics and fundamental theorems in both group theory and field theory. A number of popular mathematical texts (e.g., Derbyshire, 2008; Livio, 2006; Singh, 1999) take a historical perspective and were utilized at the beginning of the course; gradually, the mathematical content was "filled in." Proofs were included whenever they furthered the understanding of the teachers, and elucidated the concepts, but, of course, this was possible only when they were accessible to the audience. For instance, the Sylow theorems and Galois' theorems were stated but not proved during the course, whereas the homomorphism theorem for rings was proved in detail as the proof indeed elucidates the basic concepts of quotient structures, allowing the proof of the basic extension theorem for fields to be given in full as well.

Stating and Discussing the Big Theorems and Results

These included important theorems in commutative ring theory:

- The fundamental theorem of arithmetic.
- The homomorphism theorems for rings.
- The Chinese remainder theorem.

The homomorphism theorem in group theory, and the Sylow theorems and their implications for finite groups, were stated and discussed. All of these were illustrated with examples. In field theory, the basic field extension theorem, stating that every irreducible polynomial over a field has a root in an extension field, was given, followed by Galois' theorems. Also, the basic structure theorems of finite fields were developed and illustrated.

The main mathematical target of the course was to be able to state the results of Galois and Abel, as to the impossibility of solving the general polynomial equation of degree higher than 4, in a comprehensible way. After developing the algebra behind this, it was also possible to discuss constructibility by straightedge and compass, and to prove the impossibility of solving the three classical problems of ancient mathematics with ruler and compass—doubling a cube, trisecting an angle, and squaring a circle—and to see how these were consequences of the material discussed. This had the added advantage of connecting algebra to geometry, which (together with algebra) is the other major topic in the secondary school mathematics curriculum. However, along the "journey," the idea was to discuss many other major results in algebra in the past 150 years, and to understand recent directions that current research in algebra is taking.

It was a particular challenge to be able to discuss the main achievement of group theory in the last few decades, namely the classification of finite simple groups. This was stated as a theorem in the 1980s by Gorenstein, but gaps in the proofs were only completely resolved a few years ago. The quest, which culminated in the classification, started with the Feit–Thompson theorem, that groups of odd order are solvable, which was proved in 1962. In order merely to state these theorems, it is necessary to understand and develop a large amount of theory.

Applications of Abstract Algebra

Apart from being an interesting and fruitful area of mathematical research, algebra continues to have applications in many areas—often, quite unexpectedly. This was illustrated with the important area of public-key cryptography, describing, for example, the Rivest–Shamir–Adelman (RSA) algorithm from 1978 as an application of the Chinese remainder theorem. Cryptography, in itself, is, of course, a very active area of current mathematical research. Easy numerical examples of encryption and decoding were given, such as encrypting and decoding with very small primes like 3 and 5.

The application of linear algebra over finite fields in the area of error-correcting codes was also described, giving the Hamming codes as a specific example. These are linear codes which were invented in the 1950s and can correct single errors. Other specific examples of linear codes were given and tested for error correction. The full list of topics was as follows:

- Definitions of algebraic structures: Rings, fields, groups, vector spaces. Substructures.
- Ring theory: Ideals, homomorphisms, and quotients. Commutative rings, examples, special types of commutative rings. The Chinese remainder theorem and its application to RSA coding.
- Group theory: Introduction, subgroups, quotient groups. Sylow's theorems. Detailed examples.
- Field theory: Introduction. Extensions of fields.
- Cyclotomic fields, Galois groups, detailed examples.
- Finite fields: Detailed examples and properties.
- The general polynomial equation of degree n and solvability by radicals.
- Constructibility by ruler and compass.
- Squaring the circle, doubling the cube.
- Vector spaces: Vector spaces over fields of characteristic 2, and their uses in error-correcting codes.

Adapting Material to the Mathematical Background of the Teachers

The solution of systems of linear equations led to the development of linear algebra and the theory of vector spaces, which is fundamental to all areas of modern algebra. One of the pre-requisites for the teachers in the program was a B.Sc. in Mathematics, so it was assumed that they had some knowledge of linear algebra, but a brief review of linear algebra was given. No further background was assumed, so all other concepts and algebraic structures were defined and developed.

In order to do this efficiently, the mathematical structures were defined at the outset, but then studied one by one, starting with rings, followed by groups, then fields and Galois theory, and, finally, linear algebra over a field of finite characteristic. Each mathematical topic used all the previous ones, so the order chosen took this into account.

Use of Detailed Examples

Modern algebra is very abstract. In order to make the subject more concrete, the concepts were developed around examples (both mathematical and non-mathematical). These were investigated in detail, thus adding to the teachers' knowledge a number of interesting and basic examples and applications of algebraic structures. Pains were taken to make the course very computational, with an aim to make the subject matter more concrete. For instance, calculating specific examples of quotient rings and their arithmetic properties, and calculating small examples of Galois groups of certain polynomials, were intended to help elucidate the abstract theorems and to understand the structures. Finding solutions to systems of simultaneous congruences modulo various integers, or showing that there is no solution in some cases, was regarded as a productive way to understand the Chinese remainder theorem, as detailed below.

Examples from commutative ring theory:

- Much of commutative ring theory arises from the investigation of the properties of the ring of integers, and of the ring of polynomials over a field. These properties were then generalized to theorems. In this spirit, finding the greatest common divisor of two non-zero integers was discussed and also calculated for numerical examples in three different ways: (1) utilizing the Euclidean norm in the Euclidean algorithm, (2) using the fact that every ideal is principal, and (3) using the fundamental theorem of arithmetic to factor each number into primes. Each method illustrates another special property of the ring of integers.
- In discussing the Chinese remainder theorem, an algorithm to find an integer satisfying simultaneous congruences was shown and examples given, as well as examples in which there was no solution (when the conditions in the Chinese remainder theorem were not met).

Examples from finite group theory that were investigated in detail:

- The group of symmetries on a pair of symmetric reversible pants (also in Livio, 2006).
- The group of transformations on Loyd's "15 puzzle" (also in Livio, 2006).
- The group of transformations on the Rubik's cube.
- Using the Sylow theorems, it was shown that any group of order 15 is cyclic.
- The group of non-singular 2 by 2 matrices over a field of order 2.

Examples from field theory:

- Investigation of the arithmetic properties of quadratic extensions of the rational numbers, for example, by the square root of 2, and by i.
- Construction of a field of order 4 directly (using the field axioms), and as a quotient of polynomial rings over a field of order 2 by an ideal generated by a quadratic polynomial.
- The field of order 16 was investigated in great detail, its elements viewed in 3 different ways, which facilitated arithmetic in the field.
- The factorization of polynomials over the rational numbers and over finite fields of prime order was investigated. In particular, the factorization of $x^n - 1$. Specific examples for different values of n were calculated. One can generalize and prove some beautiful theorems over both kinds of fields. It is an interesting fact that for exponents less than 105 ($n < 105$), the factors over the rational numbers involve only polynomials whose coefficients are 0, 1 or -1. Contrary to expectations, this rule breaks down for $n = 105$.

Examples from Galois theory:

- Finding Galois groups of some specific polynomials over the rational numbers starting with $x^2 + 1$, and then for instance, $x^3 + 2x + 1$, $x^4 - 5x^2 + 1$, each illustrating different types of groups, viewing them as permutation groups on the sets of roots of the polynomials.
- Galois groups of cyclotomic fields, which follow naturally from the discussion of the polynomials $x^n - 1$ over the rational numbers.
- Finally, to illustrate solvability by radicals and constructibility, the construction of a golden rectangle was given, followed by the construction of a regular pentagon, which was given, both algebraically (constructing elements in quadratic field extensions, in fact building a root tower) and then geometrically.

The take-home exam that sums up the course also involves detailed investigation of specific examples, and provides an opportunity for the teachers to apply techniques and theorems learned in the course to some new examples not yet mentioned in class.

Bridging the Gap

Applying advanced mathematical studies learned in the framework of the Rothschild-Weizmann (RW) program to school mathematics teaching is by no means trivial. To support the participating teachers in this attempt, an integrative seminar was specifically designed to help the teachers bridge the gap between the advanced mathematical topics they are exposed to in the program, and the high school curriculum they teach in their classrooms.

To achieve this, the participating teachers are challenged to design or adapt teaching units, lesson plans or other teaching materials for school mathematics topics. Each teacher has to demonstrate how the advanced knowledge he or she has acquired expresses itself in school teaching and may potentially improve his or her quality of teaching (whatever the improvement is in the teacher's opinion, as long as it is justified in the work). The teacher is free to choose the scope of the unit, the grade level, the mathematical topic, and the level of teaching. The choice is subject to the seminar lecturer's approval, but is never forced upon the teacher. Therefore, the teachers' choices of topics for this task reflect their views and preferences of mathematical content.

In the course of the seminar, each student presents at least two "iterations" of the teaching unit to be discussed in the class, before the final version is ready and presented before the class. The aim of these iterative presentations is, first and foremost, to help every participant make the most of his or her work, both didactically and mathematically. The participants are experienced teachers and contribute to each other's work, mainly concerning the didactic aspects of each other's units. The instructor's main role is in helping to reveal or to strengthen the higher-mathematical background of the unit. The format in which everyone's work is discussed in the class enables exposure of each participant to a rich variety of examples. In each of them, higher mathematical knowledge, ways of thinking, and interconnections between various fields of mathematics are systematically revealed and the implications for teaching in school are discussed. In some cases, even if the topics themselves are not suitable for high school students, the higher mathematical knowledge might serve as a background for teaching other topics.

The integrative seminar takes place during the second and third semesters of the four-semester RW program. The seminar was not originally included in the RW program. It was added after 5 years' functioning of the program, once the leaders of the program recognized the need for such a course. So far, 36 student teachers have participated in the seminar since it started functioning.

Because the abstract algebra course is offered to one cohort of teachers during the second semester of their studies in the program and to another cohort during the fourth semester (as the algebra course is offered every other year), every other cohort of teachers takes the integrative seminar before they take the algebra course. Naturally, these teachers choose other mathematical courses as the basis for their seminar work. Of those who have taken the integrative seminar simultaneously with, or after, the algebra course, only a few teachers chose to connect their seminar assignments to the abstract algebra course. So far, 4 out of 36 students have chosen the algebra course as the basis for their seminar project.

Another medium, through which some teachers have shown connections between the advanced mathematical knowledge they acquired and their school teaching, is the final graduate project. For the final project, the teachers are asked to choose a field in mathematics in which they would perform supervised self-study of an advanced mathematical topic, paper or idea. Unlike the integrative seminar work, this choice need not necessarily be related to school mathematics. Nevertheless, some of them are partly motivated by school-related purposes—of these two such projects were related to abstract algebra. Thus, this chapter will refer to the work of six students: four students who composed teaching units and two students who wrote final projects.

In what follows, we present the four teaching units designed as part of the integrative seminar assignments (all intended for the most advanced 12th grade students), and the two final graduate projects that were related to the abstract algebra course. Most teachers who tackled algebraic topics, either as part of the integrative seminar tasks or the final projects, had high achievement in the algebra course: all but one had a grade of above 90 (out of 100). As can be seen below, the topics varied in content, as well as in the motivations for the teachers' choice.

Algebra as a Historical Narrative

Two teachers chose to design teaching units on historical and mathematical aspects of algebraic equations. These teachers did so because they were fascinated by the fact that, abstract as it is, modern algebra grew out of the search for solutions of polynomial equations expressible with radicals—a topic which was tangible enough for their students. Though the units differed from each other, they referred to similar historical facts, events, processes, and personalities. In particular, the teachers attempted to follow the ways of thinking of the personalities involved. They focused on the complex numbers which were already used by Cardano and Tartaglia, though not constructed rigorously until Euler and Gauss. They also focused on the evolution of the notion of "solution by radicals," and on the appearance of the abstract theories developed by Abel and Galois.

Though both teaching units focused on the historical perspective, which was one of the emphases of the algebra course, they actually seemed to address different relevance issues appearing in the course description. Their lesson materials began with a description for their students about a meta-mathematical perspective, which was one of the modes of relevance listed in the aims of the algebra course. They also described, to an extent that would be accessible for their students, a mathematicians' way of thinking, which was another mode of relevance listed for the course. The teachers did so by following the variety of paths in which theories related to polynomials and polynomial equations evolved, arriving at different mathematical discoveries in the course of this evolution.

One of the two teachers focused on cubic polynomial equations. He regarded his work as an enrichment unit focusing on the appearance of complex numbers in the search for a solution of cubic equations by radicals. The teacher wrote that his intention was "to follow with his students in the steps of Cardano and Tartaglia" to the first appearance of complex numbers, historically, in order for them to experience facing "something new and unfamiliar" in the course of their work and coping with it. The unit was intended for presentation after the students had some exposure to complex numbers. It comprised an overview of the appearance of complex numbers in solutions to cubic equations, from both historical and algebraic perspectives, and included examples and exercises. In particular, the unit included an example of a cubic equation with integer coefficients having real roots (including one integer root), though the intermediate steps of the procedure following the Cardano method yield complex-valued expressions:

$$x^3 - 15x - 4 = 0$$

$$x = u + v:$$

$$\text{i.} \quad u^3 + v^3 = 4$$

$$\text{ii.} \quad u^3 v^3 = 125$$

Therefore, u^3, v^3 must satisfy the equation

$$y^6 - \left(u^3 + v^3\right) y^3 + u^3 v^3 = 0$$

$$y^6 - 4y^3 + 125 = 0$$

$$y_{1,2}^3 = 2 \pm \sqrt{-121}, \text{ or } y_{1,2}^3 = 2 \pm 11\sqrt{-1}.$$

Substituting back to get x leads to $x_{1,2} = -2 \pm \sqrt{3}$; $x_3 = 4$.

The concluding remark of the unit described Hamilton's introduction of the quaternions in 1843 (without going into details). Thus, the work by this teacher spanned the historic evolution of number systems as the result of solving algebraic equations, and emphasized the interplay between algebraic equations—mathematical objects relatively familiar to the school students—and the evolution of numerical systems into algebraic structures.

The teacher reported having successfully taught the unit to a group of his advanced high school students. The discussion in the seminar group focused on the technicalities of teaching the unit, including the format of the lesson(s); whether the emphasis should be on exercises or on the historical background; and how many lessons should the unit include. Everybody in the group was more or less convinced that the material could and should be taught to an appropriate high school audience.

The other teacher chose to present an overview of algebraic equations and the history of the solution by radicals of quadratic, cubic, and quartic equations as a prelude to the emergence of Galois theory in the search for the solution by radicals of the quintic equation. As we shall see further on, this teacher implemented the mathematics teacher's role as a source of knowledge for his students in the evolution of modern mathematical thought—again, at a level accessible to his students. He also succeeded in presenting mathematical issues that arose in the course that were relevant for school mathematics. Unlike the previous teacher, his approach was wider, both mathematically and didactically, longitudinally encompassing the appearance of algebraic equations across the 6 years of the secondary and high school curriculum. This included manipulative, symbolic and functional-analytic aspects as a background to the presentation of the algebraic nature of polynomial equations.

In the algebra course, this teacher had been exposed to some "Big Theorems" and "Big Ideas" in mathematics; in particular, to the impossibility of the solution by radicals of the general polynomial equation of degree higher than 4, and to the Fundamental Theorem of Algebra. Therefore, he sought a way of enabling his students to encompass, to the best of their abilities, the vast mathematical perspective of what they had previously studied and of the new topics he presented to them. For the algebraic objectives of the unit, he listed Bézout's theorem for polynomials, division of polynomials with remainder, the Fundamental Theorem of Algebra, Viète's formulas, rational roots of polynomials with rational coefficients, Cardano's solution for the cubic equation, Ferrari's solution for the quartic equation, and, finally, the Abel–Ruffini impossibility theorem for quintics.

It is worth noting that, generally speaking, this teacher did not refer to the algebraic structures determined by taking quotients of the polynomial rings with ideals; instead, he used only division of polynomials by the polynomial that generates the ideal, so as to avoid proofs that would be inaccessible to his students. In his extensive overview, he also mentioned numerical and analytic approaches for finding roots of algebraic equations as alternatives to algebraic methods. This teacher motivated his choice, both of the topic and of the longitudinal nature of his work, by his intention to place the algebraic nature of polynomial equations in the context of school mathematics, reflecting in this way their place in mathematics. He asserted (and this was a major topic of the classroom discussion following his presentation) that a mathematics teacher must have an encompassing overview of a mathematical subject, even when he or she is teaching secondary school students to solve linear or quadratic equations. In this way, his or her teaching better reflects the mathematical nature of the topics being taught. Thus, for example, regarding the notion of a root of a polynomial, the teacher can treat it functionally, numerically, and algebraically, starting from first-order polynomials in order to give the students a richer and fuller picture.

These last two cases reflect "overview" suggestions for the introduction of algebraic topics in school, linking the historic evolution and the algebraic development of equations and their solutions to the school-students' knowledge and experiences.

Bringing Algebraic Structures to the Classrooms

Another teacher chose to expose her students to algebraic structures as a part of their regular curriculum-linked algebra learning. This choice was an implementation of one of the leading motivations of the algebra course—direct relevance to the high school algebra curriculum.

As part of a unit on complex numbers, this teacher was bold enough to reveal the underlying group structure of the unit circle, to define finite groups, group generators and other abstract concepts. In her opinion, this was both relevant and accessible to her students, provided she had adequately prepared them to grasp these ideas. She designed a unit focusing on the roots of unity as points on the unit circle in the complex plane, intended for 12th grade students. She claimed that the appearance of roots of unity is a "formative mathematical event" for her school students, encompassing polynomials, complex numbers, geometric sequences, geometry and algebraic structures. One can hardly argue with this assertion.

Having established the necessary techniques related to the trigonometric functions on the unit circle and the exponential form of complex numbers, she continued with the following steps:

- Definition of the circle group T—a multiplicative group on the unit circle (multiplication defined by the usual multiplication of complex numbers)—followed by the proofs of closure under operations, and the existence of inverses.
- Proofs that: T is isomorphic to $U(1)$—the group of unitary 1×1 matrices; T is isomorphic to ${}^{\mathbb{R}}/_{\mathbb{Z}} = \{\alpha + \mathbb{Z} | \, \alpha \in \mathbb{R}\}$—the quotient group defined by the equivalence of real numbers mod(1); T is isomorphic to $SO_{(2)}$—the group of rotations of the plane around the origin.
- Definition of the roots of unity, and the proof that they form a group.
- Proof that the roots of unity form an abelian group.
- Proof of a number of properties of the group of roots of unity of order $\boldsymbol{n}$, namely, that any such group is of order n; it is a cyclic group; any element of order k, where k is prime to n, is a group generator; the number of group generators is $\varphi(n)$—Euler's function.

What motivated the whole unit was the idea that behind the definitions of algebraic structures, and the verification that their properties hold for a specified set of objects, lay rather straightforward manipulations with complex numbers in exponential form, or with 2×2 rotation matrices, which require high school students only to be sufficiently proficient with the basic periodic properties of trigonometric functions.

The teacher took the trouble to emphasize the simplicity of "translating" the indicated properties of group structure into what she described as "school mathematical language." For example, "The fact that the number of generators for the group of nth roots of unity is $\varphi(n)$ (Euler's function), means simply that if k/n is a reduced fraction, i.e., if k and n are mutually prime, then $e^{2\pi i \frac{k}{n}}$ is a group generator."

As an example, she found all the generators of the 8th order roots of unity, which, technically speaking, is indeed a high school level exercise.

The class discussion that evolved after the presentation of her unit focused on whether the teachers should explicitly state the group structure and group properties standing behind a certain exercise, identity proof, etc., or leave each exercise as is, without relating it to an algebraic structure or to advanced knowledge in general. Some teachers actively supported the view that there was no need to mention, for example, the concept of a group generator to prove that the powers of $e^{2\pi i \frac{3}{8}}$ render all the roots of unity of order 8 exactly as the powers of $e^{2\pi i \frac{1}{8}}$ do. The discussion was not resolved.

Backing Up the Teachers' Mathematical Thinking and Self-Confidence

One of the modes of relevance listed in the aims of the algebra course was the enhancement of the self-confidence and personal self-esteem of the participants as mathematics teachers. This was reflected in the choice of a topic chosen by one of the teachers. This high school teacher, who usually teaches high-level classes, chose to write materials in which she would analyze, first and foremost for herself, the difference in viewing the complex numbers geometrically, as points in the Gaussian plane, and algebraically, as a field. She claimed that she had never considered the fact that the Gaussian plane and the vector representation of complex numbers, which are both very efficient and useful ways of looking at complex numbers, still fail to reflect the algebraic structure of the complex numbers as a number field. When asked why she was so insistent on this topic, she answered that once she had thought of this, she would be unable to go on teaching complex numbers at school until she had clarified the issue for herself. When she presented her work to the seminar group, several teachers admitted to having never thought of it before, and were glad to have been exposed to her analysis.

Algebra Applied to Geometry

One of the final graduate projects was directly associated with non-algebraic school topics, which is one of the modes of relevance listed in the aims for the algebra course. The teacher who worked on this project initially did not intend to choose a topic in algebra for his final project, but he changed his mind after he learned in one of the integrative seminar meetings about numbers (as points in the complex plane) which are not constructible by straightedge and compass. The fact that the solution to a classic geometric problem lay in extensions of number fields fascinated him and he asked to change the topic of his final project. Such applications of abstract algebra were among the direct aims of the algebra course, and applications of such an abstract and advanced topic, as field extensions, to what is usually considered as "elementary geometry" proved to be a convincing instance of the relevance of modern algebra to high school mathematics. Eventually his project resulted in a unit on geometric straightedge-and-compass constructions for high school students, with an emphasis on the algebraic side of the problem. He wanted to pass on to his students the idea of the impossibility of certain constructions as a general concept.

Another geometric result whose elegant proof is of an algebraic nature, is that $\sum_{k=1}^{n} \overrightarrow{OA_k} = \overrightarrow{0}$ (where A_1,A_2,... .A_n are the vertices of a regular n-gon, and O the center of its circumscribed circle). The proof follows directly from rather basic properties of the roots of unity and of their group structure, which was the focus of one of the teaching units which was not only written, but also taught by one of the teachers to her students. Any geometric proof of this fact is in one way or another related to the properties of the roots of unity.

General Algebraic Competence of Mathematics Teachers

Another final graduate project exemplifies the mode of relevance related to the mathematics teacher's role as the main source of knowledge to their pupils as to recent developments in the field. The teacher in question chose to study Rubik's cube as an example of how to make abstract algebra accessible to his students. While understanding that he would never teach the topic as part of the high school curriculum in his class, he wanted to demonstrate algebraic structures to his students by using the cube, as mathematical enrichment. He felt that since Rubik's cube is a popular game accessible to adolescents, it could be utilized to teach some group theory to high school students. For similar reasons, Rubik's cube was also utilized in the algebra course as an example to make group theory more concrete. The teacher's project focused on the group of transformations of the cube. He described a set of generators for the whole group, viewing it as a permutation group, and showed how certain sequences of moves stabilize certain tiles on the faces of the cube. He also presented sequences of moves that can be used for solving the cube, and gave an algorithm for the solution. He calculated the size of the group, which is very large. In fact, it has 43,252,003,274,489,856,000 elements! He also stated and gave proofs for some theorems that have been proved about Rubik's cube, and discussed the structure of the group of transformations.

In summing up the project, the teacher said that he thought one could expose high school students to group theory by using examples, such as Rubik's cube and other puzzles, such as Loyd's "15 puzzle," that was also described in the algebra course, and that these can help make a very abstract topic much more concrete.

Concluding Remarks

In this chapter we have examined the relevance of a course on abstract algebra to teaching high school mathematics, by analyzing didactic materials prepared by teachers in which they connected what they learned in the abstract algebra course, specifically designed for high school teachers, both to the school curriculum and to their teaching. The didactic materials the teachers designed reflected the contribution of their studies in the abstract algebra course both at the level of specific content (i.e., algebra), and at the level of the nature of mathematics.

At the level of specific content, the set of didactic materials demonstrated a view of algebra as a wide-ranging field concerned with structures and connected to other mathematical areas. For example, one of the seminar projects that presented algebra as a historical narrative demonstrates how the quest to find solutions to algebraic equations led to an extension of the number systems, exemplifying how new mathematical knowledge was created. Similarly, the project performed by the teacher who strived at understanding the difference between viewing the complex numbers as the Gaussian plane or as an algebraic number field dealt not with specific equations and their solutions, but, rather, with structures.

However, the materials also demonstrated various aspects related to the more general epistemological level of what mathematics is and what doing mathematics entails, situated in the context of algebra. Thus, for example, the second project related to the historical narrative gave a wide perspective on algebra and its connection to other fields of mathematics, in a form accessible to high school students: it encompassed different concepts related to the roots of polynomials, using not necessarily algebraic methods in order to compute the roots but stating the Fundamental Theorem which asserts their existence. The rich scope of results and approaches reflected in this project brings modern algebra to the high school classroom as an integral part of mathematics—flexible, vivid, posing questions, looking for answers, and posing new questions.

One of the important aspects of the algebra course for school mathematics, as viewed by teachers in the RW program, was related to the role of algebra in other school mathematics topics, mainly in geometry. The school mathematics curriculum is not rich in connections between various fields of mathematics, thus constructing an image of mathematics as a set of disjoint pieces of knowledge. The fact that the solution of classical geometric problems was eventually the result of the development of abstract algebra had its effect each time the teachers were exposed to questions about geometric constructions. The surprise was enhanced by the fact that constructible numbers are a "minority," contrary to the expectations of most teachers prior to the discussion of this issue. Having learned that π and e, being "special" mathematical constants, are non-constructible, the teachers sometimes received the impression that this was related to their "special" status and that most real numbers were constructible. The interconnectedness of the different areas of mathematics is illustrated by the origins of these constants: π originates from geometry, while *e* originates from analysis; it was geometry that posed the question of their constructibility, but it was algebra which eventually supplied the answer. In the algebra course, the teachers discovered, usually contrary to their previous intuitions and conceptions, that the seemingly "singular" non-constructible numbers are in a sense the "majority" of real numbers, and, therefore, studying them appears much more relevant than they might have imagined before.

Whereas the didactic materials the teachers designed revealed different modes of reference to their high school teaching associated with the abstract algebra course, it is worth noting that only a few teachers chose to use this course as the point of departure for their integrative seminar work. One reason for this might be that for teachers in the RW program, the course in abstract algebra is, allegedly, the quintessence of abstraction, much more so than the courses in the program in analysis and geometry, though these are also far from elementary. School algebra, concentrating mainly on the solution of different types of equations, and abstract algebra, which is concerned with algebraic structures and operations, appear to be very different and remote from each other. Thus, the course in abstract algebra might appear to be much less relevant than other mathematical courses whose connection to school mathematics follows directly from their names: geometry and analysis.

The fact that almost all the teachers who tackled algebraic topics in the didactic materials they designed received high grades in the algebra course might imply that using what was learned in the algebra course for teaching mathematics in high school requires a "mathematical maturity," in the sense of Steen (1983), which most teachers might need more time to develop. Thus, future research needs to re-address the question of relevance a year or more after the teachers have finished the program. Moreover, new modes of relevance for teaching school mathematics might emerge with time and practice; hence, our characteristics of relevance might be enriched accordingly.

References

Adler, J., Hossain, S., Stevenson, M., Clarke, J., Archer, R., & Grantham, B. (2014). Mathematics for teaching and deep subject knowledge: Voices of mathematics enhancement course students in England. *Journal of Mathematics Teacher Education, 17*(2), 129–148.

Baldinger, E. (2013). Connecting abstract algebra to high school algebra. In M. Martinez & A. Castro Superfine (Eds.), *Proceedings of the 35th annual meeting of the North American chapter of the International Group for the Psychology of mathematics education* (pp. 733–736). Chicago, IL: University of Illinois at Chicago.

Barton, B., & Paterson, J. (2009). *Teachers learning mathematics: Professional development research. Report of a TLRI project*. Wellington: NZ Council for Educational Research Retrieved from: http://tlri.org.nz/sites/default/files/projects/9256-FinalReport.pdf

Christy, D., & Sparks, R. (2015). Abstract algebra to secondary school algebra: Building bridges. *Journal of Mathematics Education at Teachers College, 6*(2), 37–42.

Dreher, A., Lindmeier, A., & Heinze, A. (2016). Conceptualizing professional content knowledge of secondary teachers taking into account the gap between academic and school mathematics. In C. Csíkos, A. Rausch, & J. Szitányi (Eds.), *Proceedings of the 40th Conference of the International Group for the Psychology of Mathematics Education*. PME: Szeged, Hungary.

Even, R. (2011). The relevance of advanced mathematics to teaching secondary school mathematics: Practitioners' views. *ZDM, 43*(6-7), 941–950.

Even, R., Artstein, Z., & Elbaum-Cohen, A. (2018). The Rothschild-Weizmann master's program for practicing mathematics teachers. In N. Movshovitz-Hadar (Ed.), *K-12 mathematics education in Israel: Issues and challenges* (Vol. 13, pp. 235–242). Singapore: World Scientific Publication. isbn:978-981-3231-18-4.

Klein, F. (1932). Elementary mathematics from an advanced standpoint: Arithmetic, algebra, analysis (Vol. 1, E. R. Hedrick & C. A. Noble, trans.). New York: Macmillan. (Original work published 1908).

Murray, E., Baldinger, E., Wasserman, N., Broderick, S., White, D., Cofer, T., et al. (2015). Exploring connections between advanced and secondary mathematics. In T. G. Bartell, K. N. Bieda, R. T. Putnam, K. Bradfield, & H. Dominguez (Eds.), *Proceedings of the 37th Annual Meeting of the North American Chapter of the International Group for the Psychology of Mathematics Education* (pp. 1368–1376). East Lansing, MI: Michigan State University.

Steen, L. A. (1983). Developing mathematical maturity. In A. Ralston & G. S. Young (Eds.), *The future of college mathematics* (pp. 99–110). New York, NY: Springer.

Wasserman, N. (2016). Abstract algebra for algebra teaching: Influencing school mathematics instruction. *Canadian Journal of Science Mathematics and Technology Education, 16*(1), 28–47.

Wasserman, N. (2017). Making sense of abstract algebra: Exploring secondary teachers' understanding of inverse functions in relation to its group structure. *Mathematical Thinking and Learning, 19*(3), 181–201.

Wu, H. (2011). The mis-education of mathematics teachers. *Notices of the AMS, 58*(3), 372–384.

Zazkis, R., & Leikin, R. (2010). Advanced mathematical knowledge in teaching practice: Perceptions of secondary mathematics teachers. *Mathematical Thinking and Learning, 12*(4), 263–281.

Ziegler, G. M., & Loos, A. (2014). *Teaching and Learning "What is Mathematics."* Retrieved from: https://www.discretization.de/media/filer_public/2014/09/05/20140413icm-proceedings-ziegler.pdf

Zuo, H. & Leung, F. K. S. (2016). Senior secondary school teachers' advanced mathematics knowledge and their teaching in china. Paper presented at the 13th International Congress on Mathematics Education.

Course Bibliography

Bergen, J. (2010). *A concrete approach to abstract algebra: From the integers to the insolvability of the quintic*. Burlington, MA: Elsevier, Academic Press.

Derbyshire, J. (2008). *Unknown quantity: A real and imaginary history of algebra*. London: Atlantic Books.

Gaal, L. (1973). *Classical Galois theory*. New York, NY: Chelsea Publishing.

Isaacs, I. M. (1994). *Algebra: A graduate course (Graduate studies in mathematics, volume 100)*. Providence, Rhode Island: American Mathematical Society.

Jacobson, N. (1985). *Basic Algebra I*. Mineola, New York: Dover Publications.

Lang, S. (1985). *Algebra (Graduate texts in mathematics)*. Boston MA: Addison Wesley Publishing Company.

Livio, M. (2006). *The equation that couldn't be solved: How mathematical genius discovered the language of symmetry*. New York, NY: Simon & Schuster Paperbacks.

Shamash, J. (2010). *Algebra: From equations to structures (course file)*. Rehovot, Israel: Rothschild-Weizmann Programme for M.Sc. in Science Education, Department of Science Teaching, Weizmann Institute of Science.

Shamash, J. (2016). *Algebra: From equations to structures (lecture notes)*. Rehovot, Israel: Rothschild-Weizmann Programme for M.Sc. in Science Education, Department of Science Teaching, Weizmann Institute of Science.

Singh, S. (1999). *The code book: The science of secrecy from ancient Egypt to quantum cryptography*. New York, NY: Anchor Books, Random House Inc..

Chapter 13
Using Geometric Habits of Mind to Connect Geometry from a Transformation Perspective to Graph Transformations and Abstract Algebra

Yvonne Lai and Allan Donsig

Mathematicians and mathematics educators see potential in using ideas from abstract algebra to deepen teachers' knowledge. Many programs of teacher preparation provide opportunities for learning abstract algebra (e.g., Tatto, Schwille, Senk, Ingvarson, Peck, & Rowley 2008); policy guides advocate for abstract algebraic ideas as a way to cohere concepts across K-16 and beyond (e.g., Conference Board of the Mathematical Sciences 2012); and there are potential links not just among abstract algebra and the content of secondary mathematics but also to is teaching (e.g., Wasserman 2016). Nonetheless, these links do not serve teachers well if they appear irrelevant to secondary mathematics teaching or its concerns (e.g., Ticknor 2012).

In this chapter, we propose that teaching geometry from a transformation approach provides an opportunity to showcase abstract algebraic ideas in ways that are accessible and relevant to secondary mathematics. To make this argument, we show how Geometric Habits of Mind (GHOM), first proposed for teaching and learning geometry in grades 5–10 (Driscoll et al. 2007), can be applied to abstract algebra so as to promote mathematical coherence spanning grades 5–16. We use mathematical coherence to refer to ways that the mathematics of secondary geometry and algebra and of university abstract algebra can inform each other, especially in two key modes of mathematical work: problem solving and theory building (Gowers 2000).

We focus on the content of secondary level geometry from a transformation approach and graph transformations as they appear in secondary level algebra. We take abstract algebra at the university level to include the following: linear algebra and group theory; constructions of groups and rings using matrices; and

Y. Lai (✉) · A. Donsig
Department of Mathematics, University of Nebraska-Lincoln, Lincoln, NE, USA
e-mail: yvonnexlai@unl.edu

N. H. Wasserman (ed.), *Connecting Abstract Algebra to Secondary Mathematics, for Secondary Mathematics Teachers*, Research in Mathematics Education,
https://doi.org/10.1007/978-3-319-99214-3_13

representation theory, in which algebraic objects and their interactions are mapped to matrix groups so as to leverage their linear algebraic properties. Our stance aligns with well-known treatments of abstract algebra (e.g., Artin 1991; van der Waerden 1931/1949).

This chapter is organized as follows. We discuss:

- How the term "transformation approach" has been used, and how attending to GHOM can integrate geometry, graph transformations, and abstract algebra.
- Three situations where GHOM connect abstract algebraic ideas, geometry from a transformation approach, and graph transformations:
  - Reconciling three common, yet different, and seemingly contradictory definitions for congruence from a transformation approach,
  - Using transformations to understand conventions for defining families of functions and shapes, and
  - Recognizing, defining, and naming the geometric properties of transformations of the plane in terms of linear algebra.

We conclude with implications for preparation for secondary teaching.

Terminology. We use "transformation approach," "transformation view," and "transformation perspective" interchangeably. We use "teachers" to refer to prospective and practicing teachers. We use the term "function" to refer to functions where the domain and codomain are subsets of $\mathbb{R}$. We assume that all equations are real and in two variables. "Transformations" refer to bijective maps from $\mathbb{R}^2$ to $\mathbb{R}^2$.

Geometry from a Transformation Approach

One of the most significant shifts to mathematics standards recommended by the Common Core State Standards for Mathematics (CCSSM; National Governors Association Center for Best Practices & Council of Chief State School 2010) is its "transformation" approach to geometry. As the standards state:

> The concepts of congruence, similarity, and symmetry can be understood from the perspective of geometric transformation. Fundamental are the rigid motions: translations, rotations, reflections, and combinations of these, all of which are here assumed to preserve distance and angles (and therefore shapes generally). …In the approach taken here, two geometric figures are defined to be congruent if there is a sequence of rigid motions that carries one onto the other. …Similarity transformations (rigid motions followed by dilations) define similarity in the same way that rigid motions define congruence,… (p. 74).

This approach contrasts to those based more closely on Euclid's *Elements* and its postulates for congruence of segments and angles. Developing geometry from a transformation approach, as opposed to an *Elements* approach, has taken on a renewed relevance with the release of the Common Core; however, it is worth noting that mathematics educators have called for a transformation approach to geometry since the 1900s (Betz 1909; Usiskin & Coxford 1972), and multiple studies and debates about this approach date back to at least the 1960s (for a review, see Suydam 1985).

Why Refocus from Content to Habits of Mind

The Common Core's conceptualization of transformation approach to geometry resembles that of previous texts; for instance, Usiskin and Coxford (1972) wrote:

> The transformation approach described here [in a textbook written by the authors] has six major features: (1) an informal introduction to postulates and theorems, (2) the postulation of preservation properties of reflections, (3) the use of reflection proofs to introduce the student to writing his own proofs, (4) a study of symmetry as part of the logical development, (5) a general treatment of congruence in terms of transformations, and (6) a corresponding treatment of similarity via transformations. (p. 21)

We contend that such conceptualizations may obscure the affordances of geometry from a transformation approach. The benefits of a transformation approach have more to do with its pedagogical and mathematical applications beyond geometry than it does with specific geometric content. It is worth considering the features of the transformation approach that transfer to these other areas.

Pedagogically, "transformations provide unity" to the study of congruence, similarity, and symmetry, and the intuitions on which they rest (Usiskin & Coxford 1972). Mathematically, a transformation approach offers deeper connections between algebra and geometry due to its emphasis on transformations as maps; it allows one to consider congruence and similarity of graphs and other curves that is otherwise impossible using axiomatic Euclidean geometry. A transformation approach to geometry supports reasoning such as seen in algebra at the upper secondary level (Douglas & Picciotto 2017). In geometry, students study rigid motions to prove congruence and similarity. In algebra, students do not construct formal proofs of congruence or similarity but they do examine graph transformations. A transformation approach is more closely aligned with advanced mathematics, including matrix theory (Usiskin & Coxford 1972). In university level linear algebra, students do not focus on congruence or similarity, however, they do examine properties of different transformations of the plane and higher dimensions.

We argue that refocusing a transformation approach on habits of mind makes its mathematical benefits more visible. In particular, we contribute a potential extension of Driscoll et al.'s Geometric Habits of Mind (GHOM) framework that supports the teaching and learning of geometry from a transformation approach, while also promoting connections to graph transformation and abstract algebra.

Proposed Extension of the Geometric Habits of Mind Framework

As Cuoco et al. (1996) argued, "Although it is necessary to infuse courses and curricula with modern content, what is even more important is to give students the tools they will need in order to use, understand, and even make mathematics that does not yet exist" (p. 376). To do so, they propose that curricula be organized around *mathematical habits of mind*—productive ways of thinking that promote the learning and use of mathematical ideas.

Building on these ideas, Driscoll and colleagues developed a framework for habits of mind geared specifically for geometry. This framework, *Geometric Habits of Mind* (GHOM) is premised on the following principles: "Each GHOM should represent mathematically important thinking"; "Each GHOM should connect to helpful findings in the research literature on the learning of geometry and the development of geometric thinking"; "Evidence of each GHOM should appear often in work with teachers and students"; and "GHOMs should lend themselves to instructional use" (Driscoll et al. 2007, pp. 9–10). From these principles coupled with analyses of student artifacts, examinations of research literature, and discussions with teachers and mathematicians, Driscoll and colleagues selected four GHOMs that constitute their framework:

(1) *Reasoning with relationships*;
(2) *Generalizing geometric ideas*;
(3) *Investigating invariants*; and
(4) *Balancing exploration and reflection.*

In Table 13.1, we describe how Driscoll et al. (2007) conceptualized each GHOM, and we suggest how these GHOMs might extend from geometry at the secondary level to graph transformations and abstract algebra. In suggesting extensions for (1)–(3), we take on Driscoll et al.'s original conceptions as part of the GHOM framework while also adding to it in ways that are consistent with the use of the original and that align with mathematics education literature. We do not suggest an extension for (4), a metacognitive skill which extends across mathematics.

Table 13.1 Proposed extension of Geometric Habits of Mind, with example connections to algebra

Reasoning with relationships
Original conception. Looking for, making, and using relationships—such as congruence, similarity, or parallelism—among geometric figures
Extension. Looking for and using relationships among relationships—such as finding different sequences of transformations to establish congruence or similarity, understanding why these different sequences accomplish the same mathematical result
Connection to graph transformations. Finding different sequences of transformations between graphs of $y = f(x)$ and $y = cf(ax + b) + d$ for a function f; understanding the notion of "family" of functions or of equations as being defined by linear transformations. For example, all ellipses in standard form—with axes parallel to the x- and y-axes—can be generated through horizontal and vertical translations and scalings of any given ellipse in standard form
Connection to Abstract Algebra. Finding different generating sets for plane congruence transformations or plane similarity transformations; decomposing a given linear transformation into a product of constituent transformations; understanding why the same group can be constructed using different generating sets

(continued)

Table 13.1 (continued)

Generalizing geometric ideas
Original conception. Wanting to understand the "always" and "every" related to geometric phenomena. Includes asking questions such as "Can I think of examples where this would not be true? If so, should I change my generalization?" "Would this apply to other dimensions?" Shifts attention from a given set of objects to a larger set containing the given one
Extension. Generalizing from behavior on points to behavior on shapes to behavior on the entire plane, for instance generalizing from instances of particular types of transformations—such as translations, rotations, reflections, glide reflections, dilations, and vertical or horizontal scalings—to constructing definitions of these transformations. Specializing from behavior on plane to shapes to points. Drawing analogies between different families of transformations and their implications, such as congruence transformations, similarity transformations, and graph transformations
Connection to graph transformations. Using the description of a transformation as defined on one point (e.g., (x, y) is mapped to $(x + 1, y + 2)$ to construct how this transformation would map a graph of an equation or function (such as an ellipse or parabola) or using knowledge of how a transformation would map a graph to determining how a particular point would behave (e.g., constructing points on the graph of $y = f(ax + b)$ given points on the graph of $y = f(x)$ or vice versa)
Connection to Abstract Algebra. Generalizing how a linear transformation maps basis vectors to how it would map particular figures on the plane or the entire plane; Recognizing, articulating, and naming which plane transformations are, or are not, linear transformations
Investigating invariants
Original conception. Analyzing which attributes of a figure stay the same and which change when the figure is transformed in some way. The idea of invariants is fundamental in distinguishing one kind of geometry from another. Includes asking questions such as "What would happen if this transformation were applied over and over and over again?"
Extension. Analyzing what a particular transformation or type of transformation keeps invariant. Classifying congruence transformations into types based on how they change attributes of a figure or of the entire plane (e.g., translations have no fixed points; rotations have exactly one fixed point; reflections fix a line and also reverse orientation)
Connection to graph transformations. Understanding vertical and horizontal shifts do not change the dimensions of graphs, but that vertical and horizontal scalings do. Constructing images of graphs under compositions of vertical or horizontal transformations
Connection to Abstract Algebra. Classifying linear transformations into types based on how they change attributes of a figure or of the entire plane (e.g., distinguishing isometries from non isometries, or between conformal and anti-conformal isometries)
Balancing Exploration and Reflection
Trying various ways of solving a problem and stepping back and taking stock. Asking questions such as "What did I learn from that?" "If I already had the solution, what would it look like?" Includes testing conjectures and working backwards

Our extension for (1) aligns with a developmental view of mathematical thinking, where we extend the GHOM from looking for and using relationships to looking for and using relationships among relationships [cf. Van Hiele Levels (Van Hiele-Geldof 1957) and APOS (Dubinsky & Mcdonald 2001)]. Furthermore, our extension aligns with the mathematical view of seeing geometry as the study of objects through studying actions (such as transformations) on these objects (Klein 1939/2004).

Our extension for (2) draws on the original conception's focus on shifting attention from a given set of objects to a larger set containing the given one, while also keeping in mind Polya's (1954) argument that reasoning in mathematics not only involves generalizing, but also specializing and analogy—and that in some cases the latter two are more effective than generalizing. For instance, Mason (2018) described a geometric transformation problem, involving finding a center of dilation, where specializing to thinking about the image of a single point is more effective than thinking about the image of larger sets such as line segments.

Our extension for (3), similarly to our extension for (1), is based on a developmental view that mathematical thinking progresses when learners can process relationships as objects and ultimately families of relationships. Driscoll et al. (2007) conception for (3) concerns attributes for figures; we extend this to attributes of families of transformations such as all translations, all rotations, or all reflections.

Using GHOM to See Connections Among Transformations at the Secondary Level and Abstract Algebra

To illustrate the potential for focusing on GHOM as a way to make connections visible, we will illustrate three situations where transformations at the secondary level, such as geometric and graph transformations, connect via GHOM to abstract algebra. In each, we represent connections as in Fig. 13.1. We use the idea that mathematical activities can be modeled as either problem solving or theory building, a dichotomy that while an oversimplification has also been productive for reflecting on mathematics and its teaching (Bass 2017; Gowers 2000).

Consistent with the literature on problem solving (e.g., Schoenfeld 1992), we take problem solving here to include the ability to solve the sorts of tasks that often come up in mathematics courses, such as finding examples and constructing proofs, particularly when these tasks require non-routine use of knowledge.

We use Bass' (2017) definition of "theory building practices" as "creative acts of recognizing, articulating, and naming a mathematical concept or construct that is demonstrably common to a variety of apparently different mathematical situations,

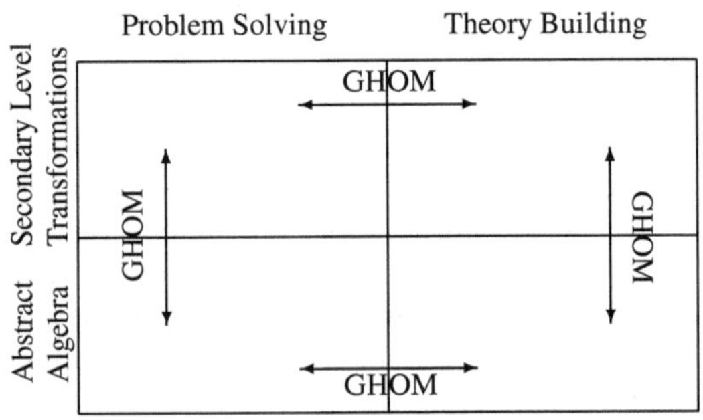

Fig. 13.1 GHOM as connectors

a concept or construct that, at least for those engaged in the work, might have had no prior conceptual existence" (p. 230). Bass's conceptualization has been used to inform instructional tasks for pre- and in-service secondary teachers.

As Bass (2017) observed, there has been much attention in education on problem solving and substantially less on theory building. Moreover, as Bass demonstrated, without some analysis of commonalities, it would be hard to connect problem solving in one area to another. We see this chapter as highlighting where both theory building and problem solving contribute to articulating connections among transformations at the secondary level and abstract algebra; and demonstrate that GHOMs can be used to make those connections.

We now turn to our three situations.

Reconciling Definitions for Congruence

Even among materials presenting geometry from a transformational approach, definitions for congruence appear to differ:

1. "Two figures in the plane are said to be *congruent* if there exists a finite composition of basic rigid motions [translations, rotations, and reflections] that maps one figure onto the other" (Eureka Math 2015, Geometry Module 1, Lesson 19, p. 161);
2. "Two figures F and G are *congruent figures*, written $F \cong G$, if and only if G is the image of F under a reflection or composite of reflections" (Coxford, Usiskin, & Hirschhorn 1991, Lesson 6–5, p. 279);
3. "A set A is *congruent* to a set B (and write $A \cong B$) if B is the image of A by a rigid motion" (Park City Mathematics Institute 2019, Geometry Transformed Class 1 Facilitator's Guide, p. 3).

The first definition implies that showing congruence requires exhibiting a sequence of translations, rotations, and reflections that map one figure to the other. The second definition appears on the surface to contradict the first definition: are no translations or rotations allowed? Finally, the last definition suggests that a sequence may not be sufficient one needs to find exactly one rigid motion to map one figure to the other, and no more than one.

In fact, these definitions are mathematically equivalent. There is no mathematical contradiction between defining congruence in any of these ways, even if the definitions result in different pedagogical implications. For instance, it is often easier for students to find a sequence of rigid motions to map a given figure onto another, making the first definition more accessible than the third. The second definition (which in its textbook is presented directly following a definition very close to the first) has the implication that students may only look for reflections when attempting proofs on congruence. The third underscores the fact that the composition of any

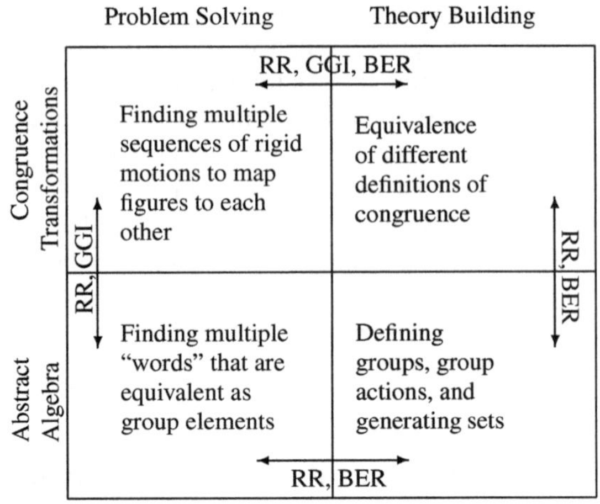

Fig. 13.2 Connecting the activities of establishing congruence with the activities of establishing isomorphisms and equivalence of group elements. RR *reasoning with relationships*, GGI *generalizing geometric ideas*, BER *balancing exploration and reflection*

sequence of rigid motions is itself a rigid motion. In practice, it may well be worth using congruence as a context for discussing the affordances of mathematically equivalent definitions that do not appear equivalent.

We highlight the role of three GHOMs to understand the mathematical equivalence of these definitions: *reasoning with relationships*, *generalizing geometric ideas*, and *balancing exploration and reflection*. The theoretical underpinnings of the argument are twofold, that of generators and relations for groups (e.g., Gallian 2013, Chapter 26), and that of group actions (e.g., Artin 1991, Chapter 6). Problem solving activities that often arise when studying group presentations include finding different "words" that are equivalent as group elements (e.g., in the rotational symmetry group of a hexagon, if ρ is rotation by $\pi/3$ radians, the "word" $\rho\rho\rho\rho\rho\rho\rho\rho$ is equivalent to the "word" $\rho\rho$). When the group studied is the group of congruences of the Euclidean plane, finding "words" that are equivalent as group elements is mathematically equivalent to finding different sequences of rigid motions that map the same two (sufficiently asymmetric) figures to each other.

To demonstrate the use of these GHOMs, we sketch mathematical reasoning for showing the equivalence of these definitions, first by showing the equivalence of Definitions 1 and 3, then by showing the equivalence of Definitions 1 and 2. In each section, we describe the primary GHOMs used (Fig. 13.2).

Equivalence of 1 and 3: Connections to Groups and Their Actions

Showing the equivalence of Definitions 1 and 3 requires two steps: first, that the finite composition of a sequence of rigid motions is itself a rigid motion; second, that any rigid motion of the plane can be written as a translation, rotation, reflection, or its composition.

A rigid motion is defined as a transformation of the plane that preserves distance and angle measure. Therefore, to show that the finite composition of a sequence of rigid motions is itself a rigid motion, we must show that the result of the composition also preserves distance and angle measure.

With the knowledge that any finite composition of rigid motions is itself a rigid motion, we must then show that it is sufficient to compose translations, rotations, and reflections. This is usually taken as an assumption in geometry curricula and standards (e.g., CCSSM), but it is possible to understand this classification of rigid motions by analyzing how a coordinate plane must be transformed if distance and angle measure are invariant. The defining features of a standard coordinate system on a plane are the location of the origin; perpendicular x- and y-axes with positive and negative directions; and a unit distance. With these features, we can construct any coordinate.

Rigid motions are often thought of as maps from points to points; but they are in fact defined as maps from the plane to the plane, because they are instances of linear or affine transformations such as studied in linear algebra. We now think of a plane that is "decorated" with a coordinate system, and how a rigid motion would move this system. Suppose that we know the image of the origin. The image of the x- and y-axes must extend from the image of the origin, because rigid motions preserve lines. The image axes must be perpendicular because rigid motions preserve angles. We can break the image possibilities into several cases, including:

- The image axes are parallel to the original axes, and the positive and negative directions of both x- and y-coordinates face the same way.
- The image axes are parallel to the original axes, but the positive and negative directions only face the same way for one coordinate.
- The image axes are not parallel to the original axes, and if they were rotated to be parallel, the positive and negative directions of both x- and y-coordinates would face the same way.

These are not all the cases, but representative ones used to illustrate the kind of reasoning used. In the first example case, the rigid motion is a translation because there is a vector such that translation by that vector will map the original to the image. In the second example case, if the image origin is the same point as the original origin, the rigid motion is a reflection about the axis where the positive and negative directions face the same way. Otherwise, it is a glide reflection, which is a composition of a translation and reflection. In the third case, if the image origin is the same point as the original origin, the rigid motion is a rotation about the origin. Otherwise, the rigid motion is a rotation about the origin followed by a translation carrying the original origin to the image origin. In each of these cases, the rigid motion can be described as a composition of translation, reflection, or rotation; and the same can be done for the remaining cases. Hence every rigid motion is the composition of a translation, reflection, or rotation.

We additionally note that the finite composition of any translation, reflection, or rotation is in fact either a single translation, reflection, rotation, or glide reflection. One way to show this is to construct the translation, reflection, rotation, or glide reflection needed in each of the cases of the coordinate system image.

Generalizing Geometric Ideas

This habit of mind is used to finding different sequences that result in the same mapping. The mathematical reasoning used to show that any rigid motion is the composition of a translation, reflection, or rotation goes back and forth between generalizing and specializing from points to larger sets such as image axes or the entire plane. This way of generalizing geometric ideas is also useful in working with group actions in abstract algebra, where looking at attributes or envisioned decorations of the objects can be helpful for visualizing and computing the images of actions. These activities lead to finding "words" that are equivalent as group elements.

Balancing Exploration with Reflection

Stepping back and taking stock of the fact that rigid motions are closed under composition opens the door to theory building in at least two ways. First, one could compare the set of rigid motions to other sets such as those of permutation groups or integers. Each of these sets has associated with it an operation; each element of the set has an inverse with respect to that operation; and there is an identity element. So rigid motions can be used as a way of building toward the theory of defining groups. Second, to realize that the result of the computations done on the image axes means the equivalence of Definitions 1 and 3 requires stepping back from the computations and taking stock of its implication. With the knowledge that rigid motions form a group, we can step back more and ask, what does it mean that we can generate all the rigid motions using a limited set of elements? Would there be a way of generating this same group using even fewer elements? What would it mean to generate? It is this line of questioning that leads to the notion of isomorphism of different group presentations as well as to the proof that Definitions 1 and 2 are equivalent.

Equivalence of 1 and 2: Connections to Generating Sets for Groups

Consider the activity: Using Geogebra, draw two parallel lines. Compose reflections about these lines. What does the resulting transformation look like? What happens when you move the parallel lines by changing their direction or moving them closer or further apart? Now construct two lines that intersect, and compose the reflections about these lines. What do you notice about the resulting transformation? What do you wonder?

A beautiful fact of rigid motions is that all rigid motions are generated by reflections. In other words, if we were only allowed to transform shapes using reflections, we would still be able to construct all possible rigid motions, because any translation, rotation, or glide reflection can be obtained by a sequence of reflections. The proofs underlying the above activity rely on analyzing the vector between a point and its image after the reflections about two parallel lines, and the angle made by a point and its image after the reflections about two intersecting lines. (More precisely, if translation by vector v maps the first line of reflection to the second line of reflection, then the resulting translation is by the vector $2v$; and if the counterclockwise angle measure from the first line of reflection to the second is θ, then the resulting rotation is by angle 2θ and its center is the intersection of the lines of reflection.) Finally, because a glide reflection is defined as a composition of a translation and a reflection about a line parallel to the translation direction, we can conclude that glide reflections are also compositions of reflections.

The reasoning here depends on first classifying all rigid motions to be translations, rotations, or glide reflections. Then, we decomposed glide reflections into a composition of a translation and a rotation, and decomposed translations and rotations into reflections. Analyzing these relationships yielded the equivalence of Definitions 1 and 2. By generalizing from points to the plane, we showed the equivalence of Definitions 1 and 3, which implies the equivalence of all three definitions.

Reasoning with Relationships

This habit of mind is used to understand how translations, rotations, and glide reflections can be formed by composing reflections, and that, moreover, any composition of reflection must be a reflection, translation, rotation, or glide reflection. It is also reasoning with relationships that allows one to show that different composition of elements of a group ("words") constitute equivalent group elements. Although the reasoning to show that reflections generate rigid motions can be done geometrically, once it is done, it can be summarized and represented algebraically: e.g., one can write that $R_M \circ R_L = \rho_{O,2\theta}$, where O is the intersection of L and M and θ is the counterclockwise angle from L to M.

Generalizing Geometric Ideas

The algebraic representation described above is analogous to that used when working with group presentations to establish relations that are helpful for showing the equivalence of words in a particular group presentation. Seeing the analogy between these representations is a form of generalizing geometric ideas.

Balancing Exploration with Reflection

Stepping back from the computations there is again an opportunity to build the theory of equivalence of groups with different presentations. Rigid motions present one instance of showing that there can be multiple ways of generating the same group. One can also find multiple generating sets for permutation groups, matrix groups, dihedral groups, or other groups that arise as examples in an abstract algebra course. From here one could construct a theory of generators and relations, and discuss the affordances of different generators for working with different problems. Finally, it is stepping back from the theory of group presentations, and analyzing the types of relationships used that allows one to see that the equivalence of these definitions is just one instance of the fact that there can be multiple sets of generators and relations for the same group, and that they each may come with different affordances.

Understanding Conventions for Defining Families of Functions and Shapes

In secondary mathematics, graph transformations arise in working with functions and equations. Related to the notion of graph transformations are definitions of families of functions and equations, which are closed under graph transformations. For instance, exponential and sinusoidal functions remain exponential and sinusoidal, and ellipses and parabolas remain ellipses and parabolas under any horizontal and vertical scale or translation. Yet this invariance does entirely not carry over from high school algebra to high school geometry. For instance, circles are treated differently:

- In high school algebra, circles are a special member of the family of ellipses;
- In high school geometry, circles are treated as a family of shapes in and of themselves, whose theorems and properties largely do not extend to ellipses; transforming circles by a horizontal or vertical scale would result in a non-circle.

To bring coherence to these ideas, we describe how graph transformations sit in a larger theory of affine transformations, and how naming families of shapes is determined by a choice of transformation group. We highlight how GHOMs promote these connections (Fig. 13.3).

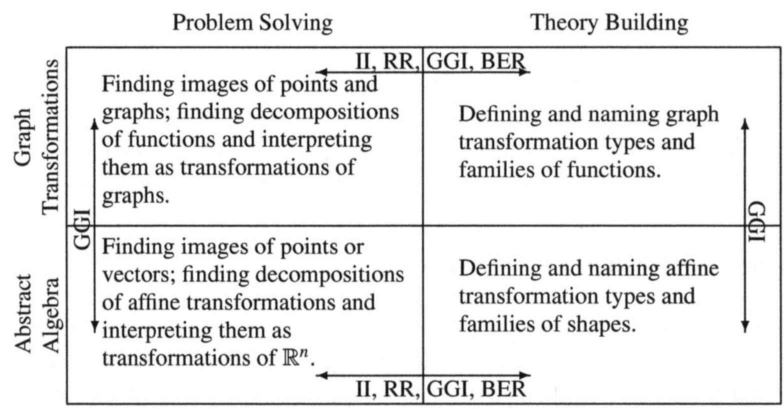

Fig. 13.3 Connecting the activities of transformations of graphs of functions with those of affine transformations. Key: II = investigating invariants; GGI = generalizing geometric ideas; RR = reasoning with relationships; BER = balancing exploration and reflection

Graph Transformations: Connections to Affine Transformations

The most common transformation studied in high school is that taking $x \mapsto ax + b$ and $y \mapsto cy + d$, where $a, b, c, d \in \mathbb{R}$ are constants such that $a, c \neq 0$. These can show up in transformations of graphs of functions, such as mapping the graph of $y = f(x)$ to the graph of $y = c \cdot f(ax + b) + d$, and also as transformations of equations, such as from $x^2 + y^2 = 1$ to $(ax + b)^2 + (cy + d)^2 = 1$.

The study of these transformations connects to the study of affine transformations in at least two ways: interpreting decompositions of transformations as actions on a space (such as a graph or of an entire vector space); and recognizing, defining, and naming transformations of the space by first understanding what transformations do to a single point of the space. We consider affine transformations to be a part of linear algebra, as they consist of linear transformations, combined with translations, and they act on vectors. The GHOMs of reasoning with relationships and generalizing geometric ideas are central to the connection, and balancing exploration with reflection comes into play as well.

To see these connections, we use a task[1] that the first author has used with pre-service secondary teachers.

1. Find as many ways as you can to decompose the function

$$g(x) = (2x + 3)^4 + 7$$

using linear functions and $f(x) = x^4$.

2. Find as many ways as you can to transform the graph of $f(x) = x^4$ to the graph of $g(x) = (2x + 3)^4 + 7$. (Figure 13.4 shows these graphs.)

The intentions of this question are to invite teachers to see decomposition as intrinsic to the study of algebra, to foster curiosity about issues of order of transformations, to relate algebraic and graphical representations of transformations, and to create intellectual need (Harel 1998) for coming up with notation for types of transformations.

Through this task, teachers have conjectured that:

- "Input" and "output" compositions and transformations can commute, meaning that compositions that change the input to f can commute with compositions that change the output of f.

[1]This task typically is the culmination of a sequence in which teachers consider similar questions for less mathematically complicated functions, such as $g(x) = (x + 3)^4$, $g(x) = (2x)^4$ or $g(x) = x^4 + 7$.)

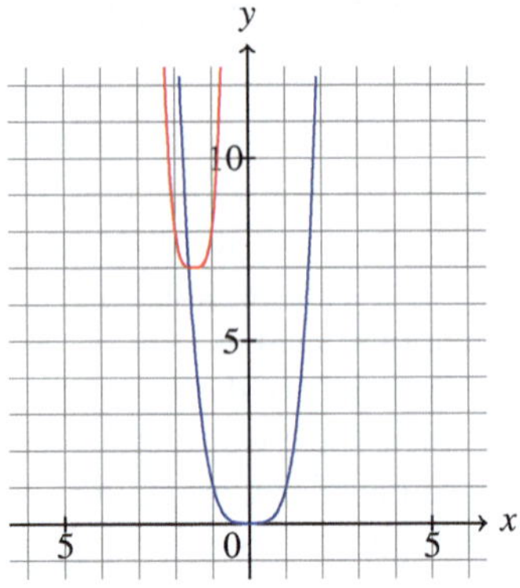

Fig. 13.4 Graphs of $f(x) = x^4$ and $g(x) = (2x + 3)^4 + 7$ on a unit square grid

- "Input" compositions and transformations do not necessarily commute with each other, nor do "output" compositions with each other.
- A horizontal scale by factor $\frac{1}{a}$ maps the graph of f to the graph of $y = f(ax)$.
- Translation by $(-b, 0)$ maps the graph of f to the graph of $y = f(x + b)$
- The graph of $y = f(ax + b)$ is a composition of translation by $(-b, 0)$ followed by a horizontal scale by factor $\frac{1}{a}$.
- Translation by $(0, d)$ maps the graph of f to the graph of $y = f(x) + d$.

And, when working with composing transformations, the teachers find a need to come up with notation for these transformations, such as:

- Horizontal scaling by a factor of $\frac{1}{2}$, mapping $(x, y) \mapsto (\frac{1}{2}x, y)$, and denoted $H_{\frac{1}{2}}$.
- Translation by vector $(-3, 0)$, mapping $(x, y) \mapsto (x - 3, y)$, and denoted $T_{(-3,0)}$.
- Translation by vector $(0, 7)$, mapping $(x, y) \mapsto (x, y + 7)$, and denoted $T_{(0,7)}$.

In doing so, they have then opened a door to generalizing to horizontal scalings of different factors, as well as translations by different vectors, and how these transformations may interact with each other.

Generalizing Geometric Ideas

There are two ways that generalizing geometric ideas arises. First, the pattern of reasoning for this task, going from a sequence of algebraic decompositions to actions on a graph, is analogous to going from a sequence of matrix decompositions to actions on a vector space. By exploiting this analogy, one can begin to ask questions such as how standard decompositions of linear transformations, e.g., LU decompositions or those related to eigenvectors, can be interpreted geometrically.

Second, to make definitions for the transformations requires generalizing from points to a larger space, such as a graph or a vector space; and to use these definitions sometimes requires specializing from the space to individual points. As has been written elsewhere, understanding how to go from images of points to images of an entire graph can benefit from both covariational and correspondence reasoning—for example, an explanation and suggested task for this can be found in Johnson (n.d.). Effective forms of both these kinds of reasoning require generalizing from points to a larger space and then specializing from a larger space to points.

Reasoning with Relationships

Another analogy that becomes accessible via reasoning with relationships is that of recognizing composition dependence. In linear algebra, a big deal is often made of the fact that matrices—unlike real numbers—do not commute. However, undergraduate mathematics students have in fact seen another example of objects that do not commute: transformations of graphs. The correspondence between graph transformations and affine transformations is through reasoning with relationships in the graphical representation.

Balancing Exploration with Reflection

Theorizing about composition dependence and naming graph transformation types is made possible by balancing exploration with reflection. In this case, teachers are stepping back and asking how to name the commonalities between different transformations, as well as make these commonalities precise.

Naming Families of Functions and Equations: Connections to Naming Families of Shapes by Choosing Group Actions

Thus far we have described how graph transformations can be applied to shapes commonly encountered in algebra at the secondary level—such as lines, parabolas, exponential curves, or trigonometric curves—but not what these transformations bring to the concepts behind these shapes.

Motivating Example: Invariance of x- and y-"Directions"

As a first step to unpacking this issue, consider the question:

> Draw some lines in the plane. Then transform these lines using a variety of horizontal and vertical scalings. What do you notice about the before and after directions of the lines? Are there lines that stay the same direction?

The only directions held invariant by all horizontal and vertical scalings are the horizontal and vertical directions, in other words, the directions given by the x-axis and y-axis. If we were to begin with a unit square grid on the plane aligned with the x-axis and y-axis, horizontal and vertical scalings would map the squares to rectangles, and the rectangles' sides would be parallel to the x-axis and y-axis.

Explaining Conventions for Defining Families of Functions

When we apply horizontal and vertical scalings to graphs of common families of functions or equation, their relationship to the x-axis and y-axis cannot change substantially. If the graph began as symmetric to one of the axes, the image graph is symmetric to that axis. If the graph were parallel to an axis, the image graph is parallel to that axis. If the graph was perpendicular to an axis, the image graph is perpendicular to that axis. If the graph moves away or toward an axis, the image graph also moves away or toward an axis. The graph may be stretched in some fashion, but fundamental features, such as the number of intercepts or the number of relative maxima or minima, cannot change.

Consistent with this reasoning, many definitions of functions and equations studied in algebra can be interpreted in terms of scaling and translation. A linear function is any function of the form $f(x) = ax + b$; the graphs of all such functions are scalings and translations of the graph of function $f(x) = x$. Any quadratic function can be written in the form $f(x) = a(x - h)^2 + k$; the graphs of these functions are scalings and translations of the graph of $f(x) = x^2$. Similarly, all exponential functions can be thought of as a scaling and translation of $f(x) = 10^x$, and all sinusoidal functions can be thought of as a scaling and translation of $f(x) = \sin(x)$. The fact that these functions can be decomposed into compositions of parent functions and linear scaling or translation means that any algebraic properties, such as the degree of a polynomial, or the property of being exponential, do not change under composition. The geometric interpretation of this algebraic fact is that their fundamental shape is preserved by the associated transformations of their graphs.

Investigating Invariance

The theme of the reasoning above is investigating invariance, both of the coordinate system, and of attributes of functions and their graphs, such as maxima, minima, degree, or periodicity. Because the standard graph transformations keep these attributes constant, the images of the transformations are said to represent members of the same family. Hence circles are in the same family as ellipses, all parabolas are in one family, all lines are in one family, and all sinusoidal functions are in one family. And, a way to generate all members of a family is to perform all possible transformations $(x, y) \mapsto (ax + b, cy + d)$, where $a, b, c, d \in \mathbb{R}$ and $a, c \neq 0$, on any one individual member of a family.

Generalizing Geometric Ideas

To see the connection of the above to abstract algebra—and to return to the circle example at the beginning of this section—we note that an explicit choice of transformation group was made. This group is $\{(x, y) \mapsto (ax + b, cy + d) \mid a, b, c, d \in \mathbb{R}, a, c \neq 0\}$ with the operation of composition. This group of transformations has the property that it keeps the x- and y-"directions" constant, which is what allows it to interact so nicely with functions typically studied in secondary mathematics.

However—to generalize the geometric idea of choosing a transformation group—we could have made a different choice of group. For instance, we might have instead chosen to see families would have been generated had we chosen the group of rigid motions, or the group generated by dilations and rigid motions. In the case of rigid motions, the "directions" of the coordinate plane would certainly not be invariant under transformation but the attributes of distance and angle would be. The group of rigid motions is also known as the group of "congruence transformations" because two figures are congruent if and only if there is a congruence transformation mapping one to the other. The "families" of shapes possible under rigid motions are all those shapes which are congruent to each other. And by considering the case of the group generated by dilations and rigid motions, the family of a circle are circles of all possible radii. This group is also known as the group of "similarity transformations," because two figures are similar if and only if there is a similarity transformation mapping one to the other. The notion that one can name families of shapes by choosing a transformation group traces back to Klein's Erlanger Program, and is possible due to the study of groups.

Linearity and Plane Transformations

Transformations of the plane are a basic topic in secondary mathematics and many of these transformations are linear. In our experience, some undergraduates and some teachers, including those who have completed a full course in linear algebra, may not find accessible the concept of linearity or how to visualize transformations. It can often be difficult to connect the concept image of a plane transformation to the abstract conditions defining linear transformations. This is one aspect of a known concern, that learners may find it challenging to integrate pre-existing knowledge into an abstract, axiomatic approach to linear algebra (Dorier & Sierpinska 2001), or in Tall and Vinner's (1981) terms, to connect concept image to concept definition.

Below we outline a discovery task relating plane transformations and linear transformations. We illustrate the potential of this task to engage teachers in GHOMs. This task has been used in a class for in-service teachers focusing on

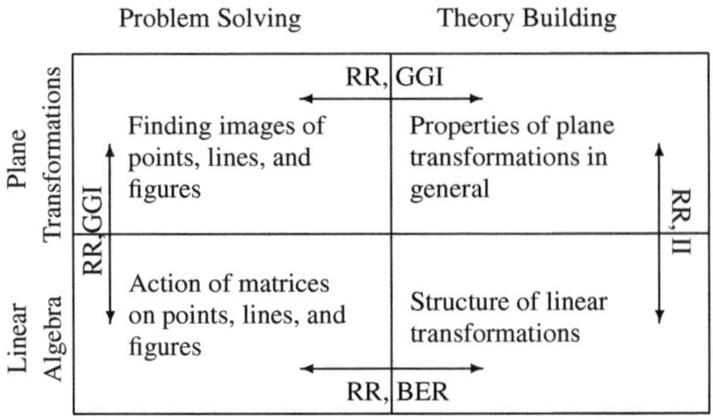

Fig. 13.5 Connecting plane transformations and linear transformations. RR *reasoning with relationships*, II *investigating invariants*, GGI *generalizing geometric ideas*, BER *balancing exploration and reflection*

material for precalculus, Math 811T (Functions for High School Teachers), taught at our institution roughly ten times in the last decade, typically with 8–20 in-service teachers working together for 4 h a day every day for 2 weeks.

The purpose of this task is to enhance teachers' understanding of both plane transformations and linearity by recognizing which plane transformations are linear and showing the common properties of linear transformations unify a diverse range of plane transformations, with a particular emphasis on the geometric interpretation of linearity. Recognizing linear transformations enables use of their common properties to simplify and unify a wide range of geometric phenomena, which indeed is the value of abstraction across mathematics: "diverse phenomena are made to all appear as particular manifestations of a single unifying theoretical construct" (Bass 2017, p. 230).

We operate from the premise that when a teacher's concept image and concept definitions are more tightly connected, that there is more fluid connection between properties of linear algebra and particular transformations—and, moreover, that this will promote appreciation for the special features of those transformations that arise most often in school curricula, such as congruence transformations, similarity transformations, and function transformations. Figure 13.5 summarizes how theory building, in terms of the general properties of each type of transformation, explains and unifies the specific calculations and problems carried out with concrete transformations of each type. GHOMs connect different approaches and the types of transformations.

Plane transformations also connect to group theory, in the sense that many families of transformations, such as, say, translations, rotations, dilations, horizontal shears, and vertical stretches, each forms a group under composition. From the perspective of abstraction, that is, recognizing the common structures that unify the many different examples of groups, each family is not seen *sui generis*, but rather as an instance of the common framework. Moreover, this framework can be used across the school curriculum. For example, a defining property of groups is the existence of inverses. This can be used in arithmetic problems, where a student can check her answer to a subtraction problem, say $5.07 - 0.3$, by adding 0.3 to her answer to see if that results in 5.07. Similarly, a student who has transformed the graph of a function can apply the inverse transformation to the new graph and check if the result matches the original graph. This common, almost universal, strategy works whenever there is an underlying group and, it seems to us, a teacher who thinks of it in terms of groups will not need to understand it afresh each time.

Discovery Task

To set the pedagogical context, after developing various families of functions, such as exponential and logarithmic, trigonometric, and polynomial, the course turned to the use of plane transformations to build a wide range of functions out of these basic families. Students had just studied a variety of transformations, including translations, rotations, reflections, horizontal and vertical stretches, horizontal and vertical shears, and inversion with respect to the unit circle. In previous work, the teachers had studied each of these transformations individually and could describe their action on points. Moreover, they had repeatedly looked at the images of equations, particularly lines and circles, under transformations. For example, in studying inversion, which maps (x, y) to $(x/(x^2 + y^2), y/(x^2 + y^2))$, the teachers have shown that this map is self-inverse (applying it twice sends a point back to itself), maps lines to circles passing through the origin and vice versa, and maps circles not passing through the origin to other such circles.

The task begins with each teacher (or group of teachers) picking six transformations, one of each type listed above (of course, they have only one example of inversion). Then the students look at the images of the standard basis vectors, then small integral multiples under each transformation, and then several sums and differences of these multiples. That is, they plot the images of $(k, 0)$, $(0, k)$ such that $k = 1, 2, 3$, and, say, $(\pm 2, \pm 1)$. Teachers quickly point out that most of these transformations have the common property that the axes are mapped to lines through the origin, although the image lines need not be perpendicular. For example, they might end up with a diagram such as in Fig. 13.6.

After being reminded of the definition of linearity, they are asked to explain what linearity requires of the various images they have computed. Quickly, they sort out which transformations are linear and which are not.

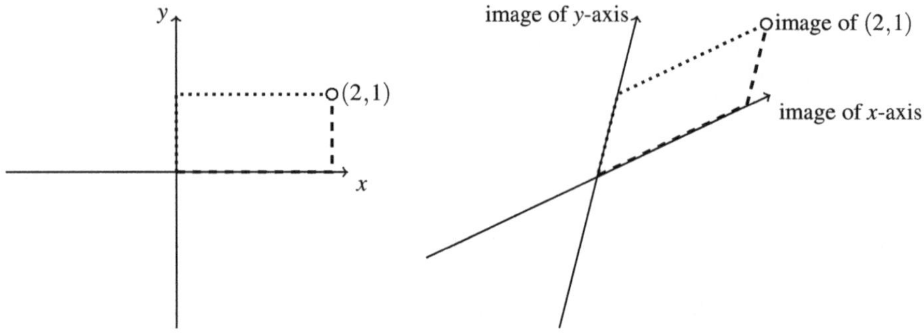

Fig. 13.6 Images of axes and a point under a linear transformation

Generalizing Geometric Ideas

This sorting and recognition of linear transformations requires working out the images, not only of points, but also of the axes and finally recognizing that the image of each point in the plane can be obtained from the images of the two basis vectors on the axes, as suggested in the diagram above. So they are generalizing their geometric idea of the action of transformations on points to its action on the coordinate system as a whole.

Experienced teachers are highly sensitive to the key features of secondary mathematics and very quick to attend to geometric and algebraic features. However, recognizing how linear algebra can be used to understand secondary mathematics requires theory building to recognize the abstract properties of plane transformations and connect them to the theory of linear transformations. In other words, the participants' problem solving work in plane transformations and in linear algebra needs to be connected by using theory-building in both areas to connect the underlying concepts in each area.

Next, participants are asked which linear transformations preserve distances and angles. Rigid motions, which preserve distances and angles, are fundamental, but identifying them (or at least, the linear rigid motions) from first principles focuses attention on the definition and on how rigid motions embed in the family of all linear transformations of the plane. Having a range of linear transformations, in particular, stretches and shears, participants can establish that distances and angles are not preserved, or even scaled by a common factor, by all linear transformations.

Teachers establish for themselves geometric properties of the transformations, such as that lines are mapped to lines but circles are not mapped to circles. There are multiple approaches possible; that is, finding the image of a locus can be done either by transforming the equation or graphing the images of points. The participants go on to conjecture that ellipses are mapped to ellipses. When there is more time, it may be helpful to ask whether conic sections are mapped to conic sections.

Investigating Invariants

This work is a direct application of this habit, as invariance properties, of distances or of signed angles of geometric loci, are central to this classification. For distances and signed angles, the focus is on identifying which families of transformations leave invariant all possible elements of the family while for others, such as the images of lines, circles, and ellipses, it is on identifying which families are left invariant by all possible transformations.

It is worth emphasizing the extent to which teachers spontaneously identify invariants, based on their experience. It often suffices to ask "what happens to distances?" and the teachers themselves identify which transformations preserve distances and, more subtly, which scale distances by a uniform constant. A significant challenge is overgeneralizing from limited data. For example, some may conjecture that a transformation with determinant one must be an isometry; a counterexample, for the form of shears, shows this is false. Thus, a large family of varied examples is essential to successfully identifying true invariants. This shows the interplay between this habit, Investigating Invariants, and another, Balancing Exploration and Reasoning.

Using the images of vectors computed above, participants can show that the image of a general vector (x, y) can be expressed as $x \cdot T(1, 0) + y \cdot T(0, 1)$. Using matrix notation, they then show that one transformation, call it T,

$$T \begin{bmatrix} x \\ y \end{bmatrix} = \begin{bmatrix} a & b \\ c & d \end{bmatrix} \begin{bmatrix} x \\ y \end{bmatrix},$$

where $T(0, 1) = (a, c)$ and $T(1, 0) = (b, d)$. This is a basic result in linear algebra, that the columns of the matrix of a linear transformation are given by the images of the basis vectors, and typically, many participants have seen it in a linear algebra course.

Generalizing Geometric Information

Understanding the above-mentioned result geometrically, and so grounding it in their prior (and highly informed) geometric knowledge, extends that geometric information to a broader context. Working out the details of the verification for several families of transformations justifies this generalization, and so, connects the theoretical linear algebra result with concrete, geometrical transformations.

Looking at each family of transformations, the teachers develop matrix forms along the following lines:

Transformation	Image of (x, y)	Matrix	Det.
Rotation by θ about origin	$(x\cos\theta - y\sin\theta, x\sin\theta + y\cos\theta)$	$\begin{bmatrix} \cos\theta & -\sin\theta \\ \sin\theta & \cos\theta \end{bmatrix}$	1
Rotation by π about origin	$(-x, -y)$	$\begin{bmatrix} -1 & 0 \\ 0 & -1 \end{bmatrix}$	1
Rotation by $\pi/2$ about origin	$(-y, x)$	$\begin{bmatrix} 0 & -1 \\ 1 & 0 \end{bmatrix}$	1
Reflection in x axis	$(x, -y)$	$\begin{bmatrix} 1 & 0 \\ 0 & -1 \end{bmatrix}$	-1
Reflection in line $y = x$	(y, x)	$\begin{bmatrix} 0 & 1 \\ 1 & 0 \end{bmatrix}$	-1
Dilation by a factor of k	(kx, ky)	$\begin{bmatrix} k & 0 \\ 0 & k \end{bmatrix}$	k^2
Stretch in x direction	(kx, y)	$\begin{bmatrix} k & 0 \\ 0 & 1 \end{bmatrix}$	k
Stretch in y direction	(x, ly)	$\begin{bmatrix} 1 & 0 \\ 0 & l \end{bmatrix}$	l
Shear in x direction	$(x + ky, y)$	$\begin{bmatrix} 1 & k \\ 0 & 1 \end{bmatrix}$	1
Shear in y direction	$(x, y + kx)$	$\begin{bmatrix} 1 & 0 \\ k & 1 \end{bmatrix}$	1

Reasoning About Relationships

The starting point of the above table is the basic fact mentioned above, that the columns are given by the images of the usual basis vectors under the particular transformation. The table then connects the ideas of linear algebra (such as multiplying a matrix by a vector (x, y)) and those of plane transformations (such as the action of a transformation on a point (x, y) in the plane). It might seem that 2-by-2 matrices are too simple to require the abstract methods of linear algebra. Nonetheless, these methods provide a common framework for these transformations and, although it can be extended much further, linearity is already an informative point of view in understanding plane transformations.

Moreover, the families of matrices associated with various types of linear plane transformations show the relationships within and between these families. For example, the self-inverse property of reflections is seen in the fact that the squares of the associated matrices are the identity. Algebraic results can be derived from these matrix forms; for example, knowing that compositions of rotations are rotations gives, under matrix multiplication, the addition formula for sine and cosine.

Building this table also balances reasoning with exploration, as the matricial forms are obtained by exploration and then organized by reasoning into coherent families, such as all reflections, or all horizontal and vertical and two-way stretches.

A particularly striking application of the accessibility of plane transformations is relating the area of a shape to its image shape under various transformations, already mentioned above. Applying a range of transformations to the rectangle with coordinates $(\pm 1, \pm 1)$, participants quickly conjecture that areas are multiplied by a constant factor. Comparing the results with the matrix forms given above suggests that the factor is the absolute value of the determinant. Of course, this fact is stated in every multivariable calculus text, so the participants are understanding something they once formally "knew." Finally, teachers spontaneously look for a meaning for the sign of determinant, which for isometries, corresponds to whether the transformation is conformal or not. (Recall that isometry is conformal or anti-conformal according to whether it preserves or reverses signed angles.)

Identifying Invariants

Recognizing that linear transformations scale areas fits with this geometric habit of mind. This recognition depends on having a plentiful supply of examples and a computationally accessible context, so conjectures can be tested quickly against a range of examples. Looking at the images of shapes is essential to recognizing how linear transformations change area. For example, however counter-intuitive it may be that shears preserve area while changing angles and distances, it does not take participants long to examine enough examples to be convinced.

Once the conjecture is made and tested, providing justification brings up different issues, and linear algebra becomes essential. The emphasis here is not finding formal proofs of conjectures but forming plausible conjectures and testing them against a range of examples. Plausibility arguments, rather than formal proofs, are used to support general arguments. For example, to justify that linear transformations change areas by a constant factor, the first step is for the participants to show that various squares with horizontal and vertical sides are all mapped to similar figures, showing the area scaling result for figures of this form. To extend to general figures, the instructor sketches the idea of approximating an arbitrary shape by small squares, explaining why a similar result might hold for any figure. To those who understand Riemann sums conceptually, the argument is plausible, although making it a formal argument is beyond the scope of the course.

Balancing Exploration and Reasoning Throughout this development, teachers are asked to form and test conjectures against examples. By tying geometric and algebraic formulations together, a transformation approach develops facility in moving between these two representations. For each representation, different aspects are intuitive or not and so the ability to move between them has a clear value.

Another commonly listed habit of mind, applying past knowledge to new situations, certainly arises in applying linear algebra, something many teachers already knew, to a context where they had not previously used it. So the transformational approach gives new justifications for familiar facts; for example, compositions of two reflections cannot themselves be reflections, since reflections are anti-conformal and the composition of two anti-conformal isometries is necessarily conformal.

Conclusion

We have described three mathematical contexts where the signature habits of mind of a transformational approach can serve to explain and unify ideas.

In the first context (Reconciling Definitions for Congruence), exploring the structure of rigid motions as an algebraic family allows for teachers to make sense of apparently contradictory definitions that they may encounter in secondary and teacher education curricula. Generalizing from points to the plane using the algebraic structure of the coordinate system supports a explanation of the classification of rigid motions of the plane.

In the second context (Transformations of Graphs of Functions and Equations), structural decomposition of functions and their transformations provide a systematic way to conjecture the form of a function composition from the form of its associated transformations. Then, one can explain why this conjecture must be true through analysis of the outputs and inputs of the graph as a whole as well as individual points on the graph. Considering mathematical equivalence in terms of common graph transformations introduced in algebra provides an explanation of conventions on defining families of functions and equations.

In the third context (Linearity and Plane Transformations), comparing images of different linear transformations allows teachers to apply their existing knowledge of linear algebra to give new insights about a familiar context. Identifying families of transformations and examining invariants gives teachers appreciation for why rigid motions form a special family of transformations, that there are a variety of ways of transforming shapes in the plane, and how mathematical equivalence of shapes can be defined differently depending on the family of transformations considered.

Each context provides opportunities for teachers to develop habits of mind, in the sense that they highlight ways of approaching mathematical problems as opposed to emphasizing particular theorems, procedures, or proofs. The habits of mind that we discuss here are well-suited to studying geometry from a transformational approach. As mathematicians and mathematics educators have argued, transformation approaches to geometry have the potential to help secondary students form stronger conceptual links between geometry and algebra. We contribute our conceptualization of a "transformation approach," which can potentially clarify how and why the approach connects algebra and geometry. By identifying signature traits of transformation approaches, we provide descriptions of how reasoning in geometry can apply to algebra and vice versa. We do not claim that our list of traits is exhaustive, or even optimal. We do claim that these are traits that are useful for geometry from a transformation perspective, and that the general approach of conceptualizing an approach through its habits of mind is profitable.

A transformation approach, whether applied to geometry or algebra or even advanced mathematics, would not be possible without significant developments in exposition and research of what is now known as abstract algebra. The text of van der Waerden revolutionized the way that generations of mathematicians conceived of abstract algebra. Without Klein's Erlangen program, geometries would still be most known by how well they adhere to the postulates of Euclid's *Elements* rather than by the families of motions that preserve fundamental geometric properties, and how these families of motions can be understood as families of matrices with beautiful group theoretic properties. Embracing a transformation approach means connecting geometry and algebra in ways that support students and teachers' potential access to advanced mathematics.

We have thus far only addressed the mathematical aspects of a transformation approach, characterizing them in terms of habits of mind and particular contexts in which these habits can be situated. We have also proposed that these situations support explanations of conventions and procedures in secondary mathematics. Yet as Harel and Sowder (2008) have observed, explanations in teaching are a process of *ascertaining* and *persuading*. When ascertaining, a teacher seeks an explanation that convinces themselves. When persuading, a teacher seeks an explanation that may convince another, such as a student. These situations, as described, support the process of ascertaining. The process of ascertaining is vital to persuading, in the same way that knowing how to solve a problem supports hearing students' thinking on a problem (e.g., Smith & Stein 2011). In other words, it is important but not enough for teacher education to provide teachers opportunities to learn useful concepts or habits of mind.

As scholars have previously noted, teachers must also have opportunities to situate these concepts or habits of mind in actual practice (Ball & Cohen 1999; Ball & Forzani 2009); and more recently, educators have taken on this stance to create activities where teachers apply advanced mathematical knowledge to secondary mathematics teaching situations (e.g., Heid, Wilson, & Blume 2015), with promising early results (e.g., Wasserman, Weber, & McGuffey 2017). In line with these contributions, our future work involves creating activities where teachers explicitly apply the ideas to teaching situations. In doing so, we pursue situations that have the potential to develop the habits of mind of a transformation approach in both algebra and geometry. We believe that this approach has the potential to fulfill the promise of using abstract algebra to deepen teachers' knowledge of secondary mathematics and its teaching.

Acknowledgements We are grateful to the reviewers and the editor for helpful comments, and to Erin Baldinger for pointing us to Driscoll et al.'s (2007) Geometric Habits of Mind framework.

References

Artin, M. (1991). *Algebra.* Upper Saddle River, NJ: Prentice Hall.

Ball, D. L., & Cohen, D. K. (1999). Developing practice, developing practitioners: Toward a practice-based theory of professional education. In G. Sykes & L. Darling-Hammond (Eds.), *Teaching as the learning profession: Handbook of policy and practice* (pp. 3–32). San Francisco, CA: Jossey-Bass.

Ball, D. L., & Forzani, F. (2009). The work of teaching and the challenge for teacher education. *Journal of Teacher Education, 60*(5), 497–511.

Bass, H. (2017). Designing opportunities to learn mathematics theory-building practices. *Educational Studies in Mathematics, 95*(3), 229–244.

Betz, W. (1909). Intuition and logic in geometry. *The Mathematics Teacher, 2*(1), 3–31.

Conference Board of the Mathematical Sciences. (2012). *The mathematical education of teachers II.* Providence RI and Washington DC: American Mathematical Society and Mathematical Association of America.

Coxford, A., Usiskin, Z., & Hirschhorn, D. (1991). *The University of Chicago School Mathematics Project: Geometry.* Glenview, IL: Scott & Foresman.

Cuoco, A., Goldenberg, E. P., & Mark, J. (1996). Habits of mind: An organizing principle for mathematics curricula. *The Journal of Mathematical Behavior, 15*(4), 375–402.

Dorier, J. L., & Sierpinska, A. (2001). Research into the teaching and learning of linear algebra. In D. Holton (Ed.), *The teaching and learning of mathematics at university level - An ICMI study.* New ICMI Study Series (Vol. 7, pp. 255–273). Dordrecht: Kluwer.

Douglas, L., & Picciotto, H. (August, 2017). *Transformational proof in high school geometry: A guide for teachers and curriculum developers.* Retrieved from https://www.mathedpage.org/transformations/proof/transformational-proof.pdf

Driscoll, M. J., DiMatteo, R. W., Nikula, J., & Egan, M. (2007). *Fostering geometric thinking: A guide for teachers, grades 5–10.* Portsmouth, NH: Heinemann.

Dubinsky, E., & Mcdonald, M. A. (2001). APOS: A constructivist theory of learning in undergraduate mathematics education research. In D. Holton (Ed.), *The teaching and learning of mathematics at university level - An ICMI Study.* New ICMI Study Series (Vol. 7, pp. 275–282). Dordrecht: Kluwer.

Eureka Math. (2015). Geometry Module 1: Teacher materials. Retrieved from https://www.engageny.org/resource/geometry-module-1

Gallian, J. (2013). *Contemporary abstract algebra.* Boston, MA: Brooks/Cole.

Gowers, W.T. (2000). The two cultures of mathematics. Retrieved from https://www.dpmms.cam.ac.uk/~wtg10/2cultures.pdf

Harel, G. (1998). Two dual assertions: The first on learning and the second on teaching (or vice versa). *American Mathematical Monthly, 105*(6), 497–507.

Harel, G., & Sowder, L. (2008). Toward a comprehensive perspective on proof. In F. Lester (Ed.), *Handbook of research on teaching and learning mathematics* (2nd ed.). Greenwich: Information Age Publishing.

Heid, M. K., Wilson, P. & Blume, G. W. (Eds.). (2015). *Mathematical understanding for secondary teaching: A framework and classroom-based situations.* Charlotte, NC: Information Age Publishing.

Johnson, H. L. (n.d.) Are transformations of functions giving your students trouble? Try a covariation approach. Retrieved from https://hthrlynnj.com/investigating-students-math-reasoning/are-transformations-of-functions-giving-your-students-trouble-try-a-covariation-approach/

Klein, F. (1939/2004). *Elementary mathematics from an advanced standpoint: Geometry* (Vol. 2). Mineola, NY: Dover Publications.

Mason J. (2018). Combining geometrical transformations: A meta-mathematical narrative. In R. Zazkis & P. Herbst (Eds.), *Scripting approaches in mathematics education*. Advances in Mathematics Education (pp. 21–51). Dordrecht: Springer.

National Governors Association Center for Best Practices & Council of Chief State School. (2010). *Common core state standards for mathematics*. Washington, DC: Authors.

Park City Mathematics Institute. (2016). Geometry transformed. Retrieved from http://projects.ias.edu/pcmi/hstp/sum2016/wg/geometry/abstract.geometry.html

Polya, G. (1954). *Mathematics and plausible reasoning* (Vol. 1). Princeton, NJ: Princeton University Press.

Schoenfeld, A. H. (1992). Learning to think mathematically: Problem solving, metacognition, and sense-making in mathematics. In D. Grouws (Ed.), *Handbook for research on mathematics teaching and learning* (pp. 334–370). New York, NY: MacMillan.

Smith, M. S., & Stein, M. K. (2011). *Five practices for orchestrating productive mathematics discussions*. Reston, VA: National Council of Teachers of Mathematics.

Suydam, M. N. (1985). The shape of instruction in geometry: Some highlights from research. *The Mathematics Teacher, 78*(6), 481–486.

Tall, D., & Vinner, S. (1981). Concept image and concept definition in mathematics with particular reference to limits and continuity. *Educational Studies in Mathematics, 12*(2), 151–169.

Tatto, M. T., Schwille, J., Senk, S., Ingvarson, L., Peck, R., & Rowley, G. (2008). *Teacher education and development study in mathematics (TEDS-M): Conceptual framework*. East Lansing, MI: Teacher Education and Development International Study Center, College of Education, Michigan State University.

Ticknor, C. S. (2012). Situated learning in an abstract algebra classroom. *Educational Studies in Mathematics, 81*(3), 307–323.

Usiskin, Z. P., & Coxford, A. F. (1972). A transformation approach to tenth-grade geometry. *The Mathematics Teacher, 65*(1), 21–30.

Van Hiele-Geldof, D. (1957). *De didaktiek van de meetkunde in de eerste klas van het V.H.M.O.* Unpublished doctoral dissertation, University of Utrecht.

van der Waerden, B. L. (1931/1949). *Modern algebra*. New York, NY: Frederick Ungar Publishing Co.

Wasserman, N. (2016). Abstract algebra for algebra teaching: Influencing school mathematics instruction. *Canadian Journal of Science, Mathematics and Technology Education, 16*(1), 28–47.

Wasserman, N., Weber, K., & McGuffey, W. (2017). Leveraging real analysis to foster pedagogical practices. In A. Weinberg, C. Rasmussen, J. Rabin, M. Wawro, & S. Brown (Eds.), *Proceedings of the 20th Annual Conference on Research in Undergraduate Mathematics Education (RUME)* (pp. 1–15). San Diego, CA: RUME.

Chapter 14
Developing a Structural Perspective and Its Role in Connecting School Algebra and Abstract Algebra: A Factorization Example

Younhee Lee and M. Kathleen Heid

Introduction

The study of algebra in the context of school mathematics is often limited to generalized arithmetic, the study of procedures for solving equations, or the study of quantitative relationships (Usiskin, 1988). Much of the work of school algebra involves performing a limited set of mathematical operations on specific types of mathematical elements. Each of these operations requires that students first recognize the elements on which the specific operations are permitted or defined. This ability to recognize the elements and the applicable operations underpins an important aspect of mathematical thinking, what Cuoco and his colleagues (i.e., Cuoco, Goldenberg, & Mark, 1996; Mark, Cuoco, Goldenberg, & Sword, 2010) might refer to as a mathematical habit of mind. The mathematical habit of mind on which this chapter focuses, *a structural perspective*, underpins the study of abstract algebra, and we propose that focusing instruction in abstract algebra on developing a structural perspective can positively and productively affect prospective teachers' understanding and future teaching of school algebra.

Y. Lee (✉) · M. K. Heid
Department of Curriculum and Instruction, The Pennsylvania State University, University Park, PA, USA
e-mail: yul182@psu.edu

N. H. Wasserman (ed.), *Connecting Abstract Algebra to Secondary Mathematics, for Secondary Mathematics Teachers*, Research in Mathematics Education,
https://doi.org/10.1007/978-3-319-99214-3_14

Literature Review: The Role of Mathematical Structures in Understanding Mathematics

Developing understanding of mathematical structures is one of the ultimate goals of learning school algebra (Usiskin, 1988). Yet developing such understanding is known to be challenging for many secondary students. Over the past few decades, researchers have analyzed ways of understanding mathematics that highlight the role of mathematical structure.

Over two decades ago, Sfard (1991) observed that understanding embodied in the study of mathematics took on different forms. She distinguished understanding a mathematical entity *structurally* from understanding it *operationally*. Whereas students with a structural understanding may understand polynomials as mathematical objects, students with an operational understanding may think of polynomials only in the context of procedures performed on them, such as factorization or finding zeros. Students with a structural understanding would look for and recognize the type of mathematical object involved in an algebra problem, while students with an operational understanding might have a tendency, when given an algebra problem, to rely on recalling the operations they had performed in ostensibly similar settings. Also, students with an operational understanding might recognize elements of a given type but not have an awareness of how those elements were related to each other or to other types of elements.

Not only have researchers identified different types of mathematical understanding, but they have offered ways to think about levels through which mathematical understanding develops. Dubinsky and colleagues (e.g., Arnon et al., 2014; Dubinsky, Dautermann, Leron, & Zazkis, 1994) described the evolution of understanding of a mathematical object through sequential levels: Action, Process, then Object.[1] APOS theory describes a progression of mathematical understanding, beginning with an *action* understanding in which students need to be cued to recall and carry out each step of a procedure, through a *process* understanding in which students can imagine and recall the sequence of steps of a mathematical procedure without actually carrying it out, and, finally, to an *object* understanding in which students can envision and operate on a mathematical procedure. The theory might be extended to imply that once the learner reaches an object-level understanding of a mathematical entity, in which he or she is able to operate on that entity, he or she can begin to recognize structures that involve that entity.

Whereas Sfard, Dubinsky, and colleagues dealt with levels of understanding of a single mathematical entity, Piaget and Garcia (1983/1989) described levels of understanding signified by the relationships among objects and coherence of schema. According to Piaget and Garcia, knowledge grows through a "transitional

[1]APOS theory also includes Schema as a collection of related Actions, Processes, and Objects and other sub-Schemas. To illustrate levels of understanding of a *single* mathematical entity, we adopt only Action, Process, and Object from APOS theory. To illustrate levels of understanding of more than one mathematical entity, we adopt Piaget and Garcia's triad.

mechanism... that leads from intra-object (object analysis) to inter-object (analyzing relations or transformations) to trans-object (building of structures) levels of analysis" (p. 28, parentheses in original). Specifically, the *intra-object* level is characterized by a learner's focus on the object in isolation from other objects and providing "only local and particular explanations" (p. 273). For example, a student may be able to describe and apply a single mathematical process for two different mathematical entities, but may view these processes as distinct and unrelated. As a schema develops, a new element is incorporated into the existing schema, followed by an assimilation or accommodation process. Piaget and Garcia's *inter-object* level is characterized by the learner coordinating an existing object with other similar, new ones and finding the relations among the objects. An organized grouping begins at this level through the coordination process but, at this level, the learner may not be able to explain yet how the composition of new and existing components can be seen as a whole. Finally, in the *trans-object* level, relations across the objects are established in the learner's mind and a collection of the object and others cohere and form a developed schema. According to Piaget and Garcia, this level can be characterized by synthesis across the objects that leads to "the building of structures" (p. 178). Theories advanced by the aforementioned researchers or research teams provide two distinct categories of the role of structure in mathematics learning: (1) descriptions of the level of understanding of a single mathematical entity (as in APOS theory or Sfard's operational and structural understanding) and (2) descriptions of different configurations of relationships among mathematical objects (as in Piaget and Garcia's levels).

Structural Perspective: An Overarching Construct in School Algebra and Abstract Algebra

The construct of structure is key to algebra. The ability to recognize elements of a given type and the ways in which like elements behave and relate to each other under certain operations is fundamental to both abstract algebra and the algebra that students encounter in the school mathematics curriculum. In the context of school mathematics, the types of algebraic elements that are operated on include expressions, equations, and inequalities, and the operations that are performed on them include multiplication, factoring, and applications of properties of equality (i.e., properties of the form $a = b \Rightarrow f(a) = f(b)$ for some function $f : A \rightarrow B$ and a, $b \in$ A). The study of abstract algebra can be characterized as the study of algebraic structures—systems of elements and their properties under certain operations. For example, a ring is a structure in which the elements in a set are related to each other under addition and multiplication,[2] according to a specific set of axioms. The

[2]Addition and multiplication can have different interpretations depending on how they are defined in the ring.

structure of a ring is embodied in specific instances in school mathematics, such as the set of integers and the set of polynomials and the typical operations defined on those elements.

Dreyfus (2014) defines structure as "elements and the relationships and connections between them" (p. 5). Building on Dreyfus's definition, we define a structural perspective as consisting of three components. The first component of structural perspective is *the ability to recognize and look for elements of a given type*. For example, at the elementary level, the first component would include looking for and using sets of ten when completing exercises involving the addition of whole numbers, and, at the secondary level, the first component would include recognizing that $64a^6 - 1$ is both a difference of squares and a difference of cubes.

The second component of structural perspective is *awareness of structure*—that is, *awareness of the defining features of a set of mathematical elements and the relationships and properties that relate the elements in the set to each other*. Secondary mathematics students would be showing an awareness of structure if they recognize that certain forms of algebraic expressions, such as differences of squares or cubes, facilitate the work of producing equivalent forms that reveal different information. An awareness of structure by students of school algebra includes knowing that properties of equality operate on algebraic equations and not on algebraic expressions. A teacher who has studied abstract algebra should understand that the properties of operations studied in school algebra constitute the properties of a field, suggesting an awareness of structure.

The third component of structural perspective is *the tendency, while doing mathematics, to attend to and look for structure*. Students studying calculus exhibit this component when they look for and recognize a Riemann sum structure in analyzing applications of calculus, such as determining volumes of solids of revolution, figuring out the average value of a function over an interval, calculating the present value of an income stream, or calculating consumer surplus.

Developing a structural perspective, which includes recognizing and having a tendency to look for mathematical objects, as well as knowing and using the relationships among mathematical objects, serves both students of school algebra and students of abstract algebra. A teacher who helps his or her students develop a structural perspective may enable them to consolidate their understanding of school algebra into a coherent whole.

The Role of a Structural Perspective in the Learning of School Mathematics

Examples that highlight the importance of a structural perspective in school algebra permeate the learning of school mathematics, from children's first formal experience with numbers to their introduction to calculus. As previously mentioned, in early school mathematics children are encouraged to recognize opportunities to rearrange the terms of a numerical expression in order to facilitate computation. In arithmetic exercises on which children are asked to work, the object under consideration is often a string of sums or products of numbers that need to be evaluated. Children learn to use a dual process of decomposing and reassembling to make calculations simpler. For example, decomposing 13 as the sum of 12 and 1 facilitates the following calculation: $59 + 13 = 59 + (1 + 12) = (59 + 1) + 12 = 60 + 12 = 72$. Later in their school mathematics experience, students learn to identify symbolic structures that are similarly amenable to decomposition, and which can make calculations easier (e.g., $99^2 + 99 = 99(99 + 1) = 99(100) = 9900$). Analogous processes performed in school algebra tasks require recognition of different types of mathematical objects. In school algebra, the objects under consideration include configurations of indicated sums or products of variable expressions. For example, students learn that $x^3 - 1$ can be decomposed into $(x - 1)(x^2 + x + 1)$ and that, conversely, the product $(x - 1)(x^2 + x + 1)$ can be simplified and represented as $x^3 - 1$. The perceptive algebra student recognizes the structure of a difference of cubes or squares as being special, and attends to such structures. The importance of looking for and capitalizing on structure continues beyond beginning algebra. As students advance in the study of school algebra, the phenomena they represent become more complicated and their work includes recognition of, and action on, increasingly complex algebraic and transcendental expressions (with various configurations of exponents, ratios, logarithms, and so on, in addition to sums and products). To help students develop a sufficiently robust structural perspective on algebra, teachers need to help their students develop a refined ability to recognize the types of mathematical objects with which they work, to see algebraic structure in the expressions with which they are dealing, and to have a tendency to look for that structure.

The tendency to look for algebraic structure does not always emerge naturally in college students of mathematics. This was evident in the first author's calculus class (a class of first-year students in a large U.S. university), when students were reviewing quiz problems on integration—a set of problems that required mixed approaches, including integration-by-parts, trigonometric substitution, and so on. To the instructor's surprise, when she asked students to identify the most difficult problem in the set of integral problems, they chose $\int \frac{e^{3x}-1}{e^x-1}dx$. This is despite the fact that some other problems required quite sophisticated techniques of integration. Their difficulty with the problem was even more surprising because they had recently studied a very similar problem, which asked them to calculate $\int \frac{x^3-1}{x-1}\,dx$. The problem they identified as most difficult could have been quite straightforward had students recognized the function being integrated as a difference of cubes divided by one of its factors.

The desired tendency to attend to and look for the relationship between elements (e.g., the relationship between $x^3 - 1$ and $x - 1$), and the flexibility to see an expression in either its original or an algebraically equivalent form, is one indicator of having a structural perspective. Teachers who approach algebra with a structural view are positioned to cultivate this perspective in their students.

Attention to structure is needed in the study of algebra at both micro- and macro-levels. On a micro-level, as shown in the previous examples, attention to structure may be regarding symbolic form. On a macro-level, attention to structure may involve more global features, such as the meaning of a familiar-looking operation on an unfamiliar set. Teachers with a sense of structure may recognize that although the notation for an arithmetic operation is familiar, the operation may behave differently on different sets (e.g., multiplication on the set of integers and on the set of polynomials). At the same time, these teachers recognize the ways in which the operation is the same from set to set. Teachers might draw on a sense of structure when they note that (1) students may not realize that operations are defined on sets, and that (2) the use of arithmetic operations when applied to new sets requires students to make a transition in the way they think about those operations. With each new structure, students need to be able to establish ways of operationalizing addition, subtraction, multiplication, and division on sets of new objects. In this significant transition, of applying seemingly familiar operations to unfamiliar objects (e.g., polynomials), teachers need to define what it means to add, subtract, multiply, or divide the unfamiliar objects, so that students recognize the difference in the old and new arithmetic operations, but also the "analogous" nature of the two systems (integers and polynomials, in this case).

The current chapter focuses on the structure of integers and polynomials[3] and, more specifically, on how integers and polynomials behave similarly under multiplication and division using the concept of factorization. Included in that discussion is how these structures can be understood coherently through the concept of integral domain. In what follows, we argue for the importance of a structural perspective and exemplify this view, specifically using the concept of factorization in the context of different (but closely related) structures, such as integers and polynomials, both of which are integral domains. The following section describes integral domain as a structure in which the construct of factorization plays a major role.

Integral Domain: A Structure in Which Factorization Matters

The connecting construct between abstract algebra and secondary mathematics that we articulate in this chapter is the concept of *factorization*. As a topic, factorization appears across elementary, secondary, and tertiary mathematics in different contexts. It is applied to the set of natural numbers at the elementary level, to the set of polynomials at the secondary level, and to integral domains at the tertiary level.

[3]Throughout this chapter, readers may assume polynomials are defined over $\mathbb{Q}$ (i.e., coefficients are rational numbers), unless otherwise stated.

Most of all, the mathematical importance of the construct of factorization cannot be overlooked, since factorization serves as a way of decomposing a mathematical object into a product of more fundamental, elementary objects within the structure. The decompositional nature explains why factorization provides a context for both the fundamental theorem of arithmetic and the fundamental theorem of algebra. In this chapter, we discuss why factorization matters in those particular sets (the set of natural numbers and the set of polynomials), but does not matter in others (such as in rational numbers), by linking factorization to the notion of divisibility. Then we define integral domain as an overarching structure that is exemplified by the set of integers and the set of polynomials. The aim of this analysis is to reveal how employing different contexts of factorization may be effective for developing a structural perspective.

The actions that students are taught to apply in the context of factorization vary across levels of mathematics. To be specific, at the level of elementary mathematics, students learn how to factor a given natural number, usually, a positive composite number, such as 24, by listing factor pairs or drawing factor trees. Students might be asked to write the factorization using appropriate exponents or to distinguish prime numbers from composite numbers. Factorization at this level is applied almost exclusively to positive integers. That is, students would rarely be asked to think about whether -2 can be considered a factor of 6, whether $\frac{1}{2}$ can be considered a factor of 6, or whether 6 can be considered a factor of $\frac{1}{2}$. Why do we not list $\frac{1}{2}$ as a factor of 6? Why do we never ask students to factor $\frac{1}{2}$? These are good mathematical questions for prospective teachers to consider because it leads to appreciating a feature of the set of integers that distinguishes it from other number systems, such as the set of rational numbers or real numbers. The answer involves an essential structural property of the integers: not every integer has a multiplicative inverse within the set of integers. In fact, two integers, 1 and -1, have multiplicative inverses within the set of integers, and no other element in the set of integers has its multiplicative inverse within the set. For example, there is no multiplicative inverse of 2 within the set of integers because $n \cdot 2$ does not result in a value of 1 for any integer n.

This means that within the system of integers, dividing by a number cannot, in general, be conceived of as equivalent to multiplying by its multiplicative inverse. Instead, division within the set of integers results in a quotient and a remainder; when the remainder is 0, we consider the divisor as a factor of the dividend. In other words, within the set of integers, not all divisions can be carried out with a remainder of 0. *Nonuniversal divisibility* is the phrase we use to refer to the property of a particular system, in which not all divisions within that system can be carried out with a remainder of 0.

In contrast, in the set of rational numbers, every nonzero element of the set has its multiplicative inverse within the set. Thus, divisibility is universal within the set of rational numbers since, for any nonzero rational number x, division by x is equivalent to multiplication by $\frac{1}{x}$. When divisibility is universal, consideration of factors is meaningless because, for a fixed rational number a, any nonzero rational number x would be a factor of a. Symbolically,

$$a = a \cdot 1 = a \cdot \frac{1}{x} \cdot x = \frac{a}{x} \cdot x$$

When any element of a set can be a factor, the notion of an element being a factor of another element loses its mathematical significance in that set. Because nonuniversal divisibility distinguishes some sets from others, it is useful to consider factorization within sets that have nonuniversal divisibility—in this case, within the set of integers.

In fact, integers and rational numbers are archetypal examples of two algebraic structures, *integral domain* and *field*, respectively. The key difference between the set of integers and the set of rational numbers—that is, nonuniversal versus universal divisibility—is what distinguishes the two abstract algebraic structures, *integral domain* and *field*. Integral domain is a structure in which factorization matters because of its nonuniversal divisibility.

An integral domain is, as the name suggests, a set of elements that behaves *similarly to the integers* under the operations governing the set. The set is closed, associative, and commutative under addition. It has an additive identity, and each element in the set has its additive inverse in the set. The set is also closed, associative, and commutative under multiplication, and it has a multiplicative identity. The distributive property under addition and multiplication also holds in the set, and it has no zero divisors (i.e., there are no a, b in the set for which $a \neq 0$, $b \neq 0$, and $a \cdot b = 0$).[4] In sum, an *integral domain* is defined as a commutative ring with a multiplicative identity in which the product of any two nonzero elements is nonzero.

When an integral domain satisfies an additional feature of containing a multiplicative inverse for each of the nonzero elements, we may call it a field. In other words, a *field* can be defined as an integral domain, in which every nonzero element has its multiplicative inverse. That is, a field is an integral domain with universal divisibility. Hence, the idea of factor or factorization loses its significance in a field, but stays meaningful in an integral domain that is not a field.

Integral domain is an important mathematical structure for secondary teachers because the structure of *integral domain*, coupled with the construct of *factorization* defined within an integral domain, may serve as a lens through which one can look at how the set of integers and the set of polynomials behave in consistent ways. Because the demands of tasks in which students are involved do not typically require recognition of the structural similarity between two sets, students are not often afforded the opportunity to attend to structural similarity. If prospective

[4]To say that a set has no zero divisor guarantees that, when two nonzero elements of the set are multiplied together, it will never result in zero. This may sound obvious, but the system of matrices provides examples of zero divisors. Two nonzero matrices can be multiplied together to give a zero matrix.

teachers are encouraged to consider the similarity in the structures and to observe parallel roles played by particular types of elements in the two structures (see Table 14.1), they might be able to restructure their prior understanding in ways that could allow them to take a unified approach in teaching factorization in school mathematics. They could use language that connects the mathematical ideas underlying factoring, rather than language that suggests factoring is a set of disconnected procedures; their students might focus on factoring as a decomposition of an algebraic expression rather than as a set of isolated procedures that apply to different symbolic expressions (with each procedure matched to a limited, specific case).

The juxtaposition of concepts in the set of integers and the corresponding concepts in the set of polynomials, as indicated in Table 14.1, highlights a correspondence between the two structures—one which supports a conceptualization of factorization that is independent of the object of factorization (an integer versus a polynomial). In both the integral domain of integers and the integral domain of polynomials, an element can be decomposed into fundamental building-blocks (called primes or irreducibles, respectively). When the decomposition is completed (called prime factorization or complete factorization, respectively), the form of factorization is unique up to reordering and multiplying by units.[5] In both structures, nonuniversal divisibility is a key (necessary, but not sufficient) property on which unique decomposition depends.

Understanding decomposition in the context of factorization requires recognition of the relationship among elements in a structure because it connects an element a in the set with other elements in the set, its prime factors, p_i's. Just as college students can easily recognize 12 as a multiple of 3 based on their arithmetic education, it should be natural for them to recognize $x^3 - 1$ as a multiple of $x - 1$, based on their education in school algebra. When the structural relationship of factorization of the difference of cubes can be recognized readily, students can also recognize $e^{3x} - 1$ as a multiple of $e^x - 1$ in the problems such as $\int \frac{e^{3x}-1}{e^x-1}dx$. However, students often consider the factorization of a polynomial solely as a tool for solving an equation, rather than as a way to relate different elements in the same structure through decomposition. School-algebra experience in factoring a polynomial that is restricted to equation-solving and finding the roots may be a barrier to developing a structural perspective. In what follows, we share episodes that help us to identify some of the difficulties that college students might encounter in their development of a structural perspective.

[5]This is not to say that any integral domain guarantees unique factorization for all elements. For example, $\mathbb{Z}\left[\sqrt{2}\right]$ is an integral domain in which elements do not necessarily have a unique factorization. For example, $7 = 1 \cdot 7 = \left(3 + \sqrt{2}\right)\left(3 - \sqrt{2}\right)$.

Table 14.1 The parallel structure of integers and polynomials

<table>
<tr><th>Within the set of integers . . .</th><th></th><th>Within the set of polynomials . . .</th></tr>
<tr><td>1 or −1
1 and −1 are factors of every element in the integers. In other words, for any integer n, $n = 1 \cdot n = (-1) \cdot (-n)$.
Hence, we consider this kind of factorization as trivial; 1 and −1 are trivial factors of any integer n.
NOTE: 1 is forced to be neither prime nor composite to conserve the simplicity of factorization or to avoid a trivial case for the prime factorization. If 1 were considered to be a nontrivial factor, there would be an infinite number of factorizations.</td><td>is parallel to</td><td>Units
A unit is an element that is a factor of every element in the set of polynomials. For example, 2 is a unit in the set of polynomials over $\mathbb{Q}$, because for any polynomial f, $f = 2 \cdot \left(\frac{f}{2}\right)$.
Hence, we consider this kind of factorization as trivial; units are trivial factors of any polynomial f.
NOTE: Unit is forced to be neither irreducible nor reducible to conserve the simplicity of factorization. If a unit were considered to be a nontrivial factor, there would be an infinite number of factorizations.</td></tr>
<tr><td colspan="3">Definition of unit that spans the integral domains of both integers and polynomials:
Unit has been defined as "[a]n element that is a factor of every element of a structure" (Sawyer, 1959, p. 87) or an element of a structure that "has a multiplicative inverse in [the structure]" (Fraleigh, 1994, p. 280).</td></tr>
<tr><td>Primes
A prime number is defined as a positive integer ($\neq 1$) that has no factors except trivial ones within the set of integers. Hence, the factorization $1 \cdot 5$ or $(-1) \cdot (-5)$ does not prevent 5 from being a prime number.
In other words, if n is a prime number, whenever n is written as the product of two integers p and q (i.e., $n = p \cdot q$), either p or q has to be 1 or −1.</td><td>is parallel to</td><td>Irreducibles
An irreducible polynomial is defined as a non-unit polynomial that has no factors except trivial ones within the set of polynomials. Hence, the factorization $2 \cdot \left(\frac{x-1}{2}\right)$ does not prevent $x - 1$ from being an irreducible polynomial.
In other words, if a polynomial f is irreducible, whenever f is written as the product of two polynomials p and q (i.e., $f = p \cdot q$), either p or q has to be a unit.</td></tr>
<tr><td colspan="3">Definition of irreducible that spans the integral domains of both integers and polynomials:
First, associates of an element refer to unit-multiples of the element. For example, $2(x - 2)$ is an associate of $x - 2$ in the set of polynomials over $\mathbb{Q}$. Irreducibles are the elements in the structure whose factors are only units and the associates of the element. So $x - 2$ as well as $2(x - 2)$ are irreducible elements in the set of polynomials over $\mathbb{Q}$, and 5 (whose only factors are 1, −1, 5, and −5) is an irreducible element in the set of integers. Note, in the set of polynomials over $\mathbb{Z}$, $x - 2$ is irreducible but $2(x - 2)$ is not irreducible because 2 is not a unit in the set of polynomials over $\mathbb{Z}$.</td></tr>
<tr><td>Composites
A composite number is defined as a positive integer ($\neq 1$) that has a nontrivial factorization.</td><td>is parallel to</td><td>Reducibles
A reducible polynomial is defined as a non-unit polynomial that has a nontrivial factorization.</td></tr>
<tr><td>Prime factorization
The act or result of decomposing a given integer into the product of primes.</td><td>is parallel to</td><td>Complete factorization
The act or result of decomposing a given polynomial into the product of irreducibles.</td></tr>
</table>

Difficulties Encountered by College Students in Their Development of a Structural Perspective

College students who are pursuing mathematics-intensive majors (including both secondary mathematics education majors and mathematics majors) do not necessarily develop a structural perspective, even in elementary contexts such as factorization. The following episodes illustrate ways in which seven students, who were pursuing mathematics-intensive majors at a large US university at the time of the study, dealt with structure in the context of factorization.

Participants and Background Information

These episodes were selected from a larger study, which focused on examining how students' mathematical knowledge construction at the collegiate level builds on their school mathematics understandings. The larger study, which is also the dissertation study of the first author, involved 40 interviews with college students and focused on their understanding of factorization. The three episodes shared in this chapter were selected based on the nature of student thinking related to factorization over different structures and, in particular, the comparison between the set of integers and the set of polynomials.

Participants were either secondary mathematics education majors (Jake, Lucy, and Sam) or mathematics majors (Calvin, Andy, Helen, and Jason). None of them were enrolled in or had yet completed a course in abstract algebra at the time of the interviews. Participation was invitation-based and voluntary. The final set of participants consisted of those volunteers who were available for interviews and who had demonstrated, in a prescreening survey, a solid background in school mathematics topics related to factorization (e.g., definition of polynomials, factoring a quadratic equation, various number systems).

In the following episodes, we present how the participants in this study (1) engaged with factorization in different structures, (2) dealt with a factorization situation when they viewed factorization as equivalent to finding roots, and (3) made sense of the structural similarities between the set of integers and the set of polynomials.

Episode 1: Jake's Operational Perspective on Factorization in Different Structures

In this episode, Jake treats factorization of polynomials and integers as the demand for isolated procedures that he is cued to carry out by the type of mathematical entity that is being factored. We describe how Jake thinks through questions related

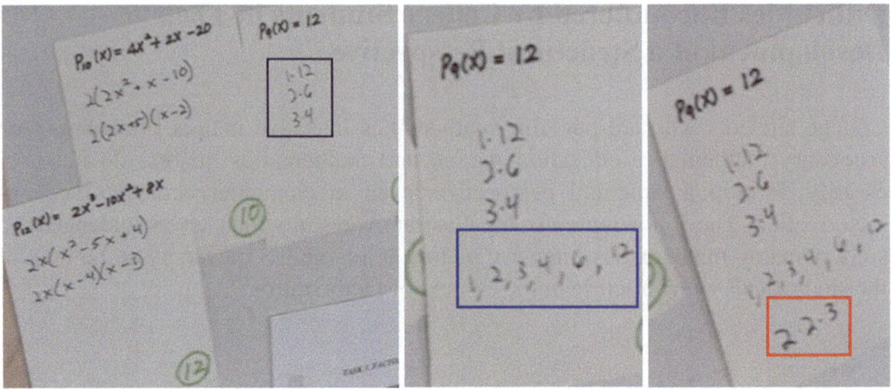

Fig. 14.1 Polynomial cards for p_9, p_{10}, and p_{12}, and Jake's answers on them

to factorization of three expressions, $p_9(x) = 12$, $p_{10}(x) = 4x^2 + 2x - 20$, and $p_{12}(x) = 2x^3 - 10x^2 + 8x$. When the interviewer asked him to factor these expressions, he answered as follows (see also Fig. 14.1a).

$$\begin{aligned} p_9(x) = 12 \rightarrow\ & 1 \cdot 12 \\ & 2 \cdot 6 \\ & 3 \cdot 4 \end{aligned}$$

$$\begin{aligned} p_{10}(x) = 4x^2 + 2x - 20 \rightarrow\ & 2\left(2x^2 + x - 10\right) \\ & 2\,(2x + 5)\,(x - 2) \end{aligned}$$

$$\begin{aligned} p_{12}(x) = 2x^3 - 10x^2 + 8x \rightarrow\ & 2x\left(x^2 - 5x + 4\right) \\ & 2x\,(x - 4)\,(x - 1) \end{aligned}$$

Jake's methods of factoring 12 and the other two nonconstant polynomials suggested that he was cued by the nature of the object being factored to treat factoring of polynomials as a different task from factoring integers. Jake seemed to think that the procedure for factoring a natural number required factor pairs, such as $1 \cdot 12$, $2 \cdot 6$, and $3 \cdot 4$, but he did not apply his finding-factor-pairs idea to the other two polynomials, $p_{10}(x)$ and $p_{12}(x)$. The difference in Jake's approaches to the factoring problems implies that the words, factor or factoring, were associated with different concepts and procedures for Jake, depending on the objects to be factored (i.e., integers versus nonconstant polynomials).

The interviewer probed Jake's reasoning by asking him to *list the factors* of the given polynomials, $p_9(x)$, $p_{10}(x)$, and $p_{12}(x)$. For the nonconstant polynomials, $p_{10}(x)$ and $p_{12}(x)$, Jake chose the components on the bottom line of his factorization to be listed as the factors of the original polynomial. For $p_9(x)$, he listed all the factors written in his factor pairs (also, see Fig. 14.1b).

$$p_9(x) = 12 \rightarrow 1, 2, 3, 4, 6, 12$$

$$p_{10}(x) = 4x^2 + 2x - 20 \rightarrow 2, (2x + 5), (x - 2)$$

$$p_{12}(x) = 2x^3 - 10x^2 + 8x \rightarrow 2x, (x - 4), (x - 1)$$

At that point, the interviewer brought up the difference in his way of listing factors. Considering that 6 was listed as a factor of 12, she asked if $x^2 - 5x + 4$ could also be listed as a factor of $p_{12}(x) = 2x^3 - 10x^2 + 8x$. Jake then introduced the notion of prime factorization to show 12 can be factored into $2 \cdot 2 \cdot 3$ (see Fig. 14.1c) and to explain why he did not have $x^2 - 5x + 4$ in his list of factors for $p_{12}(x)$. Jake explained $x^2 - 5x + 4$ is not a "prime factor" because it can be "broken down more into $2x + 5$ and $x - 2$." In this way, Jake gradually connected his responses on the nonconstant polynomials to his response in the case of the constant polynomial, and was able to address the inconsistency between his factoring of integers and of nonconstant polynomials.

However, it seemed clear in the beginning that what came to his mind first was not the same in the two cases—integers versus nonconstant polynomials—when he thought about factoring or listing factors. Making the connection between factorization of integers and that of polynomials did not seem to be a natural response for Jake. This episode suggests that factorization in a student's mind may be conceived of as procedures conducted differently depending on the context, not as a way of recognizing relationships between elements in a structure by decomposition. Although Jake could see their similarity when he was led to consider it, his compartmentalized operational perspective may have impeded him from spontaneously looking for structural similarity across the different systems.

Episode 2: Calvin's Procedural Equation-Solving Perspective on Factorization

In this episode, we describe how Calvin dealt with a discrepancy he noticed in his complete factorization[6] of a cubic polynomial, $p_3(x) = 4x + 16x^3$. His unsuccessful attempt to address the discrepancy suggests that his conception of factorization was

[6]In Episode 2, we consider polynomials over $\mathbb{C}$.

tightly connected with finding roots, in ways that perhaps impeded his attending to factorization as a decomposition. Calvin first started by factoring $p_3(x) = 4x + 16x^3$ into $4x(1 + 4x^2)$. Then he solved $1 + 4x^2 = 0$ for x by applying the square root function to both the right and left members of the equation. He concluded his factoring by writing $\left(x - \frac{i}{2}\right)\left(x + \frac{i}{2}\right)4x$, and noted the inconsistency between $4x(1 + 4x^2)$ and his final factored form of $\left(x - \frac{i}{2}\right)\left(x + \frac{i}{2}\right)4x$ (which would be equivalent to $4x\left(\frac{1}{4} + x^2\right)$). The following excerpts show how Calvin treated the discrepancy mainly by checking whether he found the correct zeros to the original function.

Calvin: I don't think this works.

Interviewer: Can you tell me why?

Calvin: If I… I need $4x^2$ somewhere, and I completely dropped that in my… So let me… I'll just leave that there. I'll use the quadratic equation I think where b is zero. [*Calvin performed an intermediate process using the quadratic formula which is not reported in this excerpt. Calvin applied the quadratic formula to* $1 + 4x^2 = 0$.] It will be negative 16 all over $2a$ which is 8. What? [pause] I get the same answer. 4 over $8i$, because… which is just the same thing. The roots are x is equal to plus or minus one half i. Maybe my transformation into the polynomial expression wasn't right. I've thoroughly confused myself.

When both the method of applying the quadratic formula to an equation related to the original expression and the method of applying the square root function to members of the equation gave him the same zeros, $\pm\frac{1}{2}i$, Calvin seemed to be quite confused. As a next step (Fig. 14.2), he evaluated the polynomial function at $x = \frac{1}{2}i$ to confirm that it was, in fact, a zero of the function.

Calvin: p_3 of one half i, I guess. That's equal to four times a half i… I'm just going to ignore this term just for the sake of brevity and room [*this term refers to* $4x$]. I'm just going to plug it in to one plus four, one half i, squared. Oh, what do you know? It will actually work I think. 'Cause you get one plus four times a quarter, then i squared is negative one…. That's one minus one equal zero, so it's a zero. Yeah. So I guess the way I would completely factor it is have it be $4x$… This would be the complete factorization, right here. [*Calvin points to the previous answer of* $\left(x - \frac{i}{2}\right)\left(x + \frac{i}{2}\right)4x$ *while saying "right here."*]

At this point, Calvin seemed to be convinced that the roots of $1 + 4x^2 = 0$ were indeed $\pm\frac{1}{2}i$ and expressed comfort with his original answer. Then, the conversation continued as follows.

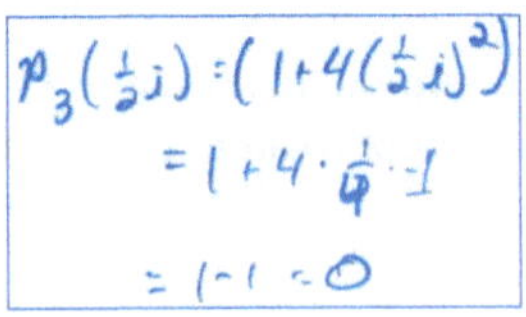

Fig. 14.2 Calvin evaluating the function value at $x = \frac{1}{2}i$

Interviewer: Can you tell me more about what troubled you?

Calvin: Oh hold on, no way. Okay. What troubled me was generally when you write the roots, if you were to FOIL it, you would get the same expression that you started with. I was FOIL-ing this expression. ["*This expression*" *refers to* $\left(x - \frac{i}{2}\right)\left(x + \frac{i}{2}\right)$.] I was not getting back to $1 + 4x^2$. For once, no, I definitely wasn't.

Interviewer: So what's your conclusion now then?

Calvin: My conclusion would be that... My tentative conclusion would be... I don't want to commit [to] that. It would be that this ["*this*" *refers to* $\left(x - \frac{i}{2}\right)\left(x + \frac{i}{2}\right)4x$] would be the complete factorization over complex [numbers]. And the way that I'm used to thinking about it with multiplying the roots together to get you back to your starting function—it's only consistent over $\mathbb{R}$. That would be my general conclusion.

Calvin's actions when he recognized an inconsistency between his goal (finding expressions whose product was $4x(1 + 4x^2)$) and the procedure he had learned for finding factors require some interpretation. When Calvin first discovered the mismatch between the expression to be factored, $4x(1 + 4x^2)$, and his own answer, $\left(x - \frac{i}{2}\right)\left(x + \frac{i}{2}\right)4x$, he responded by executing a different method of finding the roots, seemingly thinking that the discrepancy resided in his calculating initial roots incorrectly. The second method also gave him the same roots, $\pm\frac{1}{2}i$ and, when he evaluated the polynomial function at $x = \frac{1}{2}i$, he was able to confirm that $\frac{1}{2}i$ was, in fact, a zero of the function. In spite of spending a considerable amount of time trying to resolve the inconsistency, Calvin did not identify other sources that could have led to the mismatch between the two expressions. (The source of the mismatch was the leading coefficient's scalar role.) In the end, he made a "tentative conclusion" that things may behave differently over complex numbers than over real numbers, and this difference could be the reason why relationships that previously held did not work in the given situation.

One might wonder why it was so difficult for Calvin to address the discrepancy he noticed. Notably, the same result was generated by three other participants (Andy, Lucy, and Sam), none of whom seemed to notice the discrepancy. All four participants seem to be convinced that having the roots meant having the factorization. This phenomenon is too important to be "chalked up" to a lack of attention to precision. As one way to explain this phenomenon, it is reasonable to hypothesize that using the roots to factor a quadratic expression had always worked for the students and so their comfort with the routine of finding roots could have made them lose sight of factorization as a decomposition. It is also possible that they had not developed the flexibility to bring in procedures, as needed, and to return to thinking about the entities involved, thus allowing their procedural perspectives (i.e., performing a set of steps to get to the answer) to overpower a possible structural perspective (i.e., thinking about factorization as relating objects within the set).

Episode 3: Participants' Focus on Inscriptions Rather Than on Mathematical Objects

This episode describes how participants engaged with thinking about a CCSSM standard[7] on the similarities between systems of polynomials and integers. The following interview question was used to challenge them to draw on a structural perspective: "According to the Common Core State Standards, high school students need to *Understand that polynomials form a system analogous to the integers*. In what sense do you think they (i.e., polynomials and integers) are analogous?" In the attempts they made to relate polynomials and integers, most of the participants (Sam, Lucy, Calvin, Andy, and Helen) indicated that it made little sense to them that polynomials and integers "form analogous systems." (See Table 14.2 for their attempts to make sense of the statement "polynomials form a system analogous to the integers.") Instead of focusing on how elements in two systems related to each other in similar ways, they identified aspects of a polynomial (e.g., coefficient, exponent, degree, roots, and number of roots) in which integers might appear. Lucy, Helen, and Andy connected the degrees of the polynomials to integers, while Calvin connected the coefficients of the terms of polynomials to integers.

Only one participant (Jason) reasoned about how integers and polynomials are similar as structures, pointing out how factoring works in both structures and how they are both rings.[8] Jason first suggested, "[the Common Core] is probably talking about the factoring," potentially influenced by some of the preceding interview questions on factoring. He continued his reasoning as follows.

Jason: So, starting with integers, um. We have a nice theorem that says that we can factor an integer into powers of primes. That's ${p_1}^{n_1}{p_2}^{n_2}$ etc., to like ${p_k}^{n_k}$. And we can do that for any integer in a unique way. Um, similarly for polynomials, we can say that $a_nx^n + a_{n-1}x^{n-1}$ all the way down to $a_1x^1 + a_0$. And that equals x minus, uh, what do I think I want to use for root? Say r_0. Actually this is a degree and so it's going to have… $(x - r_1)$ … to $(x - r_n)$. So they're sort of similar.

In addition, he also noted how both the sets of integers and polynomials are "closed under various operations like, um, addition, subtraction, and multiplication." Then, he went on to say "you could consider this … I guess, really abstractly" and explained what he meant by that as follows.

[7]The standard in the Common Core State Standards for Mathematics [CCSSM] states, "Understand that polynomials *form a system analogous to the integers* [emphasis added], namely, they are closed under the operations of addition, subtraction, and multiplication; add, subtract, and multiply polynomials." (National Governors Association Center for Best Practices & Council of Chief State School Officers [NGA & CCSSO], 2010, p. 86).

[8]Although Jason had not taken an abstract algebra course by the time of the interview, he had been exposed to the notion of rings and groups in another class in which he was, at the time, enrolled. That class focused on complex numbers. The exposure occurred during the same semester as the interview.

Table 14.2 Participants' answers to the Common Core question

Participant	Answer
Sam	I'm not really sure what that means. I'm not sure how to interpret that standard.
Lucy	I mean . . . I guess with just like, x^2, you know you're going to have two roots. And like x^3, you know at max you're going to have three roots. So whatever the highest degree is *the highest amount of roots* you could possibly have. Maybe that's what they're trying to say. I don't know. I'm just trying to figure out what this means.
Calvin	In what sense are they analogous? Integers, just relating it to the power, *the exponent* has to be an integer in order to be a polynomial—Your *coefficients* can really be anything. Your *roots* can be anything depending on your coefficients, your powers.
Andy	I don't know how that would [be] analogous to integers in that sense because you would have a rational coefficient. You would have rational coefficients and you'd have rational roots with that, so I don't know how that relates to integers because it doesn't have much to do with integers in the piece. I don't know in what way they're trying to show that polynomials are analogous to integers . . . . I mean there's *the degrees* of polynomials are always integers, so maybe in that sense they're similar because every polynomial has form $a_0 + a_1x^1 + \ldots + + a_nx^n$. Every polynomial has that form by definition of a polynomial where the ns are all integers. You can't have an x raised to the negative—they're all strictly positive integers also.
Helen	In my opinion I don't think it's true though . . . . In this case I think they limit it to integer but I think polynomials they can be anything . . . . [High school students] should know that in the future when they learn it's going to be more complicated than integer such as like rational or irrational. It still can form polynomials. It can't be the *power* but *the number in front* can [be] something [that is] not [an] integer.

Jason: I'm trying to think about what I meant by that. Um. I mean. Let's see. Oh, they're both rings, right? Because . . . the set of integers, you can add them and with additions it's a group because you have your inverses and closed and all that. And then with multiplication, you can multiply them make some identity, but you can't do the inverse because it's 1 over 2. They're both integers, but that's not an integer [*that refers to 1 over 2*]. Similarly, with polynomials, I guess you can add them and subtract them. And you can multiply them. But you can't always divide them. Like $x - 1$ over $x - 2$, that's not the polynomial. That's just going to, a rational thing . . . . So I guess they are the similar structure in terms of the operations that you can do on them.

The way that Jason interpreted the statement "*polynomials form a system analogous to the integers*" was strikingly different from the way other participants interpreted the statement. The fundamental difference between Jason's approach and the approaches of other participants is the extent to which he paid attention to the relationships among the elements within each of the sets. Even though all participants were mathematics-intensive majors, who have, presumably, had ample experiences dealing with integers and polynomials, most of them seemed to lack a structural perspective in understanding similarities in the relationships among the elements in both the set of integers and the set of polynomials. In four cases, students identified the incorrect mathematical entity on which to operate (e.g., they looked for the factors of a term of a polynomial expression). That is, they looked for superficial similarities rather than structural ones.

Summary of Episodes 1–3

The episodes presented here illustrate three points related to a structural perspective. In the first episode, the student had a compartmentalized operational perspective. Jake did not have a tendency to look for structure across different systems. In the second episode, familiarity with procedure seemed to overpower a structural perspective. Calvin did not maintain an awareness of structure and what that allowed in this setting, moving rapidly instead to another structure. The focus of students in the third episode was on inscriptions, rather than on mathematical objects. Most students did not look for elements of a given type.

In all three episodes, a common phenomenon observed in the participants' mathematical responses was their difficulty in attending to structure. The tendency to attend to identifying the objects under consideration and to look for a structure that describes relationships among those objects was rarely witnessed. In most cases, an operational, rather than a structural, perspective seemed to dominate students' responses to the questions. The following section is an examination of this issue through the lens of knowledge construction, which suggests ways to engage prospective teachers in building toward a structural perspective in the context of factorization.

Discussion: A Categorical Framework for Reorganizing Existing Understandings of School Algebra (EDUS Framework)

Developing a structural perspective is not a simple task, partly due to the lack of opportunity to observe and to reconsider the boundaries one imposes on the objects and procedures with which one has been operating (e.g., Jake's boundary for applying factor-pairs was limited to integers and excluded polynomials) and to examine the relationships and properties that are used to connect objects within certain boundaries (e.g., Jason examined the relationships and properties that cut across integers and polynomials). In cases in which a prospective teacher's mathematical experiences are centered on operational understandings rather than structural understandings, the network of his/her mathematical knowledge, skills, and perspectives on certain topics (such as factorization) may be composed of isolated and disconnected components. A task of those who teach mathematics to prospective teachers is to help them restructure their networks of mathematics-related knowledge.

We suggest a restructuring that focuses on how different sets of mathematical elements form similar structures, which we believe could become powerful tools for prospective teachers' own understanding, and, by extension, for educating their future students.

School algebra students seem more likely to focus on symbolic routines than on the objects of algebra (Kieran, 1992). Consequently, they are more likely to develop an operational perspective on algebra than to develop a structural perspective. In order to make the transition from school algebra to the study of abstract algebra, students need to develop flexibility in their thinking about the objects of school algebra and the relationships among them. Students with an operational perspective on school algebra may find it difficult to recast their understanding to that end.

To understand how to better support the transition from understanding school algebra to understanding abstract algebra, we examine the transition through the lens of knowledge construction and introduce, in this chapter, a new, categorical framework for reorganization of existing mathematical understandings, which we call the EDUS framework.[9] The EDUS framework delineates ways in which university students' previous knowledge constructed in school algebra and their new knowledge acquired in abstract algebra can be developed into a coherent whole. In an analysis of literature on characteristics of mathematical knowledge, four components emerged that focused on qualities of mathematical knowledge (e.g., breadth, depth, coherence, and connectedness) and characteristics of progress in mathematical knowledge development (e.g., encapsulating process into object, developing schematic understandings). The characteristics that arose from the literature and seemed to capture the development of mathematical understanding are: Extending, Deepening, Unifying, and Strengthening. These characteristics formed the four categories of the EDUS framework.

To be specific, to smooth the transition from school mathematics to collegiate mathematics, university students might reorganize their existing mathematical understandings in the learning of collegiate mathematics by (1) *extending* the context in which a set of existing understandings are situated, (2) *deepening* the level of existing understandings of a certain, single mathematical object, (3) *unifying* existing understandings (that were previously unrelated by the student) under a specific overarching mathematical object, and (4) *strengthening* the links between existing understandings of more than one mathematical object.

Briefly, we give examples of the four categories. Students may *extend* the coefficient ring of polynomials, initially from $\mathbb{Z}$, to $\mathbb{Q}$, $\mathbb{R}$, and $\mathbb{C}$, and then to other fields, such as $\mathbb{Z}_7$. Students may *deepen* their understanding of factorization, initially by thinking of it as a step-by-step procedure, to thinking of it as a fundamental decomposition that is sometimes characterized by its uniqueness or completeness. Students may *unify* initially unrelated objects, such as the set of integers and the set

[9]EDUS framework was developed in the context of the first author's work on her dissertation, supervised by the second author, as a result of her analysis of literature on construction of mathematical knowledge.

of polynomials, by considering an umbrella notion of integral domain. Students may *strengthen* their understanding of connections between finding roots of a polynomial and factoring of the polynomial by considering the role of a unit in factorization.

To envision how one might reorganize his or her existing mathematical knowledge through the lens of the EDUS framework, we draw on levels of mathematical understandings from Piaget and Garcia's triad (Piaget & Garcia, 1983/1989) and from APOS theory (Arnon et al., 2014; Dubinsky et al., 1994). Three of these categories (extending, unifying, and strengthening) refer to the relationships between mathematical objects, and one (deepening) refers to the depth of knowledge of a single mathematical object. Recognizing this distinction, we employed a different, but related, theoretical basis for the deepening category than for the other three categories (extending, unifying, and strengthening).

Student's progress under *extending*, *unifying*, and *strengthening* categories can be understood from the perspective of Piaget and Garcia's (1983/1989) triad—that knowledge grows from intra-object (analysis of object) through inter-object (analysis of relations or transformations) to trans-object (construction of structures). Operating at the intra-object level regarding a mathematical entity is often a characteristic of students who have not yet developed a unified understanding of the entity, and operating at the inter-object level reveals the onset of a transition to a unified understanding. Jake's work (Episode 1), initially an example of an intra-object level of understanding factoring, illustrates the transition from a non-unified to a unified concept of factoring. Jake was able to describe and demonstrate factoring a whole number and factoring a polynomial separately, but his initial responses suggested that, to Jake, these two tasks required distinct and unrelated procedures that are performed in isolation from each other. With prompting from the interviewer, however, Jake appeared to come to one unified way of interpreting the task of factoring, which displayed an inter-object level understanding of factoring. This interview showed Jake engaging in the reorganizational act of unifying: he exhibited an overarching understanding of factoring that encompassed two previously unlinked processes.

In the trans-object level, relations across the objects are established in the learner's mind and a collection of the object and others cohere and form a developed schema. According to Piaget and Garcia, this level can be characterized by synthesis across the objects that leads to "the building of structures" (p. 178). Jason, operating at the trans-object level, demonstrated a unified understanding of factoring as he applied one overarching understanding to both the set of integers and the set of polynomials. This unified understanding was evident in Jason's interview (Episode 3) when he pointed out how factorization may lead to the decomposition of an element into a product of irreducible elements in both of the sets. He seemed to be operating at the trans-object level.

Student's progress on the *deepening* category can be understood from the perspective of APOS theory (Arnon et al., 2014). The first three constructs of APOS theory—action, process, and object—can be used to explain different levels at which a mathematical concept can be understood. We consider the progression from action level to process level to object level as *deepening* the level of existing understandings of a certain, single mathematical notion because it suggests increased quality and power of the learner's understanding of the same concept.

In the most basic level, the *action* level, a mathematical idea can be understood in terms of an action—"a repeatable physical or mental manipulation that transforms objects in some way" (Dubinsky et al., 1994, p. 270). The action level can be characterized by the learner's need for specific formulas, procedures, or external cues in order to interact with a mathematical entity. For example, a student with an action-level understanding of factorization might think factoring $x^3 - 1$ is about memorizing a specific formula (e.g., $a^3 - b^3 = (a - b)(a^2 + ab + b^2)$) and be unable to express it in factored form unless he or she was able to recall the formula.

Interiorizing the action level to the *process* level is thought to be accompanied by repeating and reflecting upon an action. At the *process* level, the action takes place in the mind of the learner and he or she does not need to perform the action in order to think about the idea. For example, a student with process-level understanding of factorization would be able to express the process of factoring in a generic way and explain what it means to factor without actually performing the action of factoring. Calvin, in Episode 2, seemed to demonstrate a process-level understanding by applying a generic approach (i.e., using the roots) to the factorization of $4x + 16x^3$. However, he failed to step back and see the big picture of factoring as decomposition, beyond applying the procedure of finding roots and rewriting the expression in terms of the roots.

At the *object* level, the student reflects on the process and thinks of it as a totality so that he or she can encapsulate the process of applying a transformation to an object, producing a resulting object. Then, the student can draw on the resulting object and apply it to take another action on the object. For example, a student with object-level understanding of factorization would be able to compare different factorizations (e.g., incomplete factorization versus complete factorization) and characterize a factorization as an entity (e.g., uniqueness, relativeness to the coefficient rings).

Building on the components of the EDUS framework—that is, *extending*, *deepening*, *unifying*, and *strengthening*—we suggest, in the next section, how mathematics instructors can engage prospective teachers in activities that promote development of these components. The development of these components is essential to the development of a structural perspective.

Developing a Structural Perspective in Abstract Algebra Through EDUS (Extending, Deepening, Unifying, and Strengthening)

In order for an abstract algebra course to be effective in secondary mathematics teacher preparation, it is essential that the prospective teachers' previous knowledge constructed in school algebra and new knowledge acquired in abstract algebra be developed into a coherent, connected entity at a higher level. Presenting formal definitions, propositions, and proofs in abstract algebra may simply result in reduced abstraction (Hazzan, 2001), in which university students only partially internalize the meaning of definitions, propositions, or proofs, and reduce the abstraction level of concepts by forcing them into a familiar context. Hazzan (2001) conjectured that this was probably so because, from the world of students' subjective knowledge, there was very little connection to the formally presented definitions, propositions, and proofs.

Alternatively, those who teach prospective teachers can help them reorganize their existing understandings of an algebraic entity through EDUS activities that, as described in the following sections, would contribute to the construction of a structural perspective. A singular goal for the study of abstract algebra is the development of an umbrella, structural understanding that would encompass students' understanding of both school and abstract algebra. Abstract algebra can provide a perfect opportunity for prospective teachers to view their existing concepts from a structural perspective, and their existing understandings may serve as building blocks for constructing this new knowledge. That new knowledge provides a new vantage point for students to consider the objects of algebra. As Freudenthal (1978/2004) noted, "Mathematics exercised on a lower level becomes mathematics observed on the higher level" (p. 61).

One way to engage prospective teachers with EDUS activities is to provide a problem context in which existing understandings of a subject, as they are, are not sufficient for the problem-solving or for justifying the subject's own claims. In such a context, existing understandings that were recalled and activated can be assembled and reflected on in a novel (to the subject) way (Dreyfus, Hershkowitz, & Schwarz, 2015). In what follows, we discuss the nature of activities that we believe can promote each of the four components in the EDUS framework that we suggest, in turn, can help prospective teachers develop a structural perspective in the context of factorization.

Developing a Structural Perspective Through *Extending*

Extending the contexts for existing concepts can help develop a structural perspective. Prospective teachers should be given chances to explore how the same notion (e.g., factorization) can be applied in a variety of contexts (e.g., polynomials over reals, polynomials over complex numbers) and examine how the relationship held within a structure may be extended to a broader structure (e.g., how the factorization of a polynomial over real numbers may be extended to the factorization of a polynomial over complex numbers). Prospective teachers in an abstract algebra course may study the fundamental theorem of algebra, for example, by extending the context of factorization. Prospective teachers may easily recognize, within the structure of polynomials over real numbers, that the decomposition of a polynomial into linear factors is not always possible without violating the condition that coefficients be within a particular set of numbers (e.g., factorization of $x^2 + 1$ over the real numbers). Essential to understanding the fundamental theorem of algebra is knowing what happens if one broadens the coefficient ring to be the complex numbers. Prospective teachers can be encouraged to reason about polynomials with complex coefficients (e.g., $x^2 + 1$, or $x^2 + (1 + i)x + i$) and factor them into the products of linear expressions. Within the broader structure of polynomials over the complex numbers, every polynomial can be decomposed into the product of linear expressions. Since the activity of extending encourages prospective teachers to look for and attend to relationships held between elements within a structure, it naturally contributes to a structural perspective in prospective teachers.

A prospective teacher who views an algebraic procedure (e.g., applying the quadratic formula) as applicable only to the set of polynomials over real numbers, for example, could be asked to apply the procedure to a larger set (e.g., polynomials over complex numbers) and to examine what parts of the procedure are essential to both the initial set and the targeted set. Making such comparisons by extending would help prospective teachers recognize how the elements in the two structures behave similarly or differently under a certain algebraic procedure.

Developing a Structural Perspective Through *Deepening*

Deepening the level of existing understandings from action level to process level to object level can help develop a structural perspective. Prospective teachers can be given chances to reflect on their actions and processes of applying a transformation (e.g., factorization), in order to examine how elements in a given set are related by the transformation. Such activities can promote an object-level understanding. Once the notion of factorization is understood at object-level, prospective teachers may use it to conceptualize a particular type of structure in which the complete factorization of any element is unique (i.e., unique factorization domain) in abstract algebra. Since the activity of deepening encourages prospective teachers to attend to elements related by a given transformation and examine the relationship between elements, it contributes to development of awareness of structure in prospective teachers.

Since many prospective teachers struggle to acquire object-level understanding, introducing any mathematical definition/theorem/proof that requires object-level understanding might need to be preceded by activities for deepening. A prospective teacher who views an algebraic object as a process could be asked to consider various applications of the process and reflect on the relationship between the activity and the result. After reflection on a number of such instances, the student could be asked to consider the entire process as an entity with which to reason, without thinking about a particular instance. For example, engaging with the activity of factoring the same set of polynomials over different domains and reflecting on the results may help them to deepen the process of factorization into an object (see Table 14.3).

Table 14.3 Activity for deepening

Please factor completely the same set of polynomials over each of the number systems, $\mathbb{Z}$, $\mathbb{Q}$, $\mathbb{R}$, and $\mathbb{C}$, and record your results in the following table.						
	$p_1(x) = x^4 + 12x^2 + 36$	$p_2(x) = x^5 - x^2$	$p_3(x) = 4x + 16x^3$	$p_4(x) = 12$	$p_5(x) = 4x^2 + 2x - 20$	$p_6(x) = 2x^3 - 10x^2 + 8x$
Over $\mathbb{Z}$						
Over $\mathbb{Q}$						
Over $\mathbb{R}$						
Over $\mathbb{C}$						
Based on your entries in this table, what can you say about factoring over different number systems? How can you make some general statement about your answers in this table?						

Developing a Structural Perspective Through *Unifying*

Unifying existing understandings (understandings that were previously unrelated by the student) under a broader, overarching idea can help develop a structural perspective. For example, prospective teachers' understandings of integers and polynomials may be unrelated initially, but a suitable context could allow them to compare and connect these two structures and gradually build an overarching construct, such as an integral domain, a unit, and the notion of irreducibles in abstract algebra. The activities of comparing and connecting different structures with which prospective teachers are familiar for the purpose of constructing a new structure in abstract algebra not only helps them to recognize the structures with which they are dealing but also encourages them to attend to defining features of the structures.

To relate different structures (e.g., integers and polynomials), one might use as a bridge a transformation (e.g., factorization) that applies to both structures. For example, prospective teachers in an abstract algebra class can be given chances to revisit how a mathematical concept (e.g., factorization) may be used in different situations, in order to construct a meaning of it that would generalize across all situations. A prospective teacher who views factorization as different in the context of integers and polynomials could be asked to construct a single definition of factorization that describes both types of factorization. Also, exploring unusual factorizations, such as $12 = 1 \cdot 2^2 \cdot 3$, $12 = (-1)^2 \cdot 2^2 \cdot 3$, or $x^2 - 1 = 2 \cdot \left(\frac{1}{2}\right) \cdot (x - 1) \cdot (x + 1)$, could lead prospective teachers to reflect on what would be considered redundant or inefficient in terms of factorization. The goal would be to recognize the particular role played by the units in the two structures. A similar exploration could be created for them to observe the parallel role played by primes and irreducibles, as discussed in Table 14.1. By identifying parallel roles in different settings, prospective teachers would be encouraged to recognize mathematical elements in different settings, and, in so doing, develop their structural perspective.

Developing a Structural Perspective Through *Strengthening*

Strengthening the links between existing understandings can help prospective teachers develop a structural perspective. To strengthen the links, prospective teachers can be asked to reexamine connections that they are frequently accustomed to making and taking for granted (e.g., ask them to consider the relationship between the roots of a polynomial and its factorization). They can be asked bidirectional tasks such as: determine the roots of a polynomial given its factorization, and determine the factorization of a polynomial given its roots. In so doing, it is necessary for prospective teachers to identify the entities they are connecting and to establish clear relationships among them. Doing so involves developing an awareness of a structure (i.e., set of mathematical objects and the relationships between them).

Also, by developing this bidirectional understanding of a mathematical relationship, prospective teachers can begin to transform their understanding of a mathematical entity from the operational to the structural.

For example, prospective teachers whose procedures for factoring override their conceptual understanding of the entities involved could be asked to explain the relationships between their procedures and the algebraic objects involved. In Calvin's episode, the case of factoring $p_3(x) = 4x + 16x^3$ provided a chance for him to revisit what he knew about factorization because his procedure for factoring (i.e., find the roots, r_i, of a polynomial equation, and rewrite that polynomial as the product of factors of the form, "$x - r_i$") did not result in what he expected. Although Calvin did not resolve the dilemma, groups of prospective teachers might engage in a meaningful discussion based on such an unexpected observation. Being able to recognize the role of unit (in this case, 4) in a factorization, and being aware that an equation has the same roots regardless of whether a unit is factored out, is evidence of a structural perspective that can be developed in the context of abstract algebra. The discussion might help prospective teachers recognize that the expression $4x^2 + 4x - 24$ is equivalent to the expression $4(x - 2)(x + 3)$, but not to $(x - 2)(x + 3)$, but that the polynomials $4(x - 2)(x + 3)$ and $(x - 2)(x + 3)$ are equivalent in the sense that they have the same roots. Understanding this difference would contribute to prospective teachers' awareness of structure.

Conclusion

In this chapter, we suggested a structural perspective as a central theme in both school algebra and abstract algebra, and we provided the EDUS framework to envision how one might engage prospective teachers in activities for developing a structural perspective in an abstract algebra course—a perspective that would encompass both abstract algebra and school algebra. We defined a structural perspective as having three interrelated components: (1) ability to recognize and look for elements of a given type, (2) awareness of structure, and (3) tendency to attend to and look for structure.

Typically, abstract algebra courses engage students in studying the elements of algebraic structures and the relationships among those elements—the first two components of our definition of structural perspective. The three-part definition of a structural perspective, however, may require the addition of a focus on the third component as an essential part of instructional goals in both school algebra and abstract algebra courses. Awareness of the elements and relationships within a structure may not necessarily require nor give rise to the tendency to attend to and look for structure. Prospective secondary teachers' development of this tendency is necessary in an abstract algebra course if we want them to model it in the context of teaching school algebra. Development of their students' awareness of structure, in addition, may contribute to their students' flexibility in applying operational and structural understandings, as appropriate. Connections between school algebra and abstract algebra may become more visible to prospective teachers who have developed a structural perspective, and abstract algebra ideas may be used by them to treat school algebra ideas in a connected and coherent way in their future teaching of school algebra.

We discussed our point of view using structures in which factorization matters (i.e., integral domains manifested in the set of integers and the set of polynomials) and employing empirical data gathered in the interviews with mathematics-intensive majors. In general, the context of factorization provides a pedagogically fertile setting because it allows prospective teachers to explore known (to them) structures, such as the set of integers, the set of rational numbers, or the set of polynomials from a new angle in abstract algebra—to think about integral domain, field, unit, irreducibles, etc.

The context of factorization also provided a fruitful setting for studying college students' understanding of their structural perspectives. The data in the study suggested that mathematics-intensive majors lack a structural perspective. During the subsequent interviews, however, some participants made meaningful progress in their ways of thinking about factorization of polynomials over a generic domain. We believe instruction in abstract algebra could take the form of having prospective teachers reflect on what they already know, and gradually transform their existing understandings in ways that extend, deepen, unify, and strengthen what already exists in their minds. By engaging in activities, such as extending, deepening, unifying, and strengthening, prospective teachers might be able to reorganize their understandings of familiar structures, such as the set of integers and the set of polynomials, and eventually arrive at the construction of abstract mathematical structures, while also developing the tendency to look for and attend to structures in the process of this construction.

References

Arnon, I., Cottrill, J., Dubinsky, E., Oktaç, A., Fuentes, S. R., Trigueros, M., et al. (2014). *APOS theory: A framework for research and curriculum development in mathematics education*. New York, NY: Springer.

Cuoco, A., Goldenberg, E. P., & Mark, J. (1996). Habits of mind: An organizing principle for mathematics curricula. *The Journal of Mathematical Behavior, 15*(4), 375–402.

Dreyfus, T. (2014). Abstraction in mathematics education. In S. Lerman (Ed.), *Encyclopedia of mathematics education* (pp. 5–9). New York, NY: Springer.

Dreyfus, T., Hershkowitz, R., & Schwarz, B. B. (2015). The nested epistemic actions model for abstraction in context—Theory as a methodological tool and methodological tool as theory. In A. Bikner-Ahsbahs, C. Knipping, & N. Presmeg (Eds.), *Approaches to qualitative research in mathematics education: Examples of methodology and methods, Advances in mathematics education series*. New York, NY: Springer.

Dubinsky, E., Dautermann, J., Leron, U., & Zazkis, R. (1994). On learning fundamental concepts of group theory. *Educational Studies in Mathematics, 27*(3), 267–305.

Fraleigh, J. B. (1994). *A first course in abstract algebra*. Reading, MA: Addison-Wesley.

Freudenthal, H. (1978/2004). *Weeding and sowing: Preface to a science of mathematical education*. Dordrecht, The Netherlands: D. Reidel.
Hazzan, O. (2001). Reducing abstraction: The case of constructing an operation table for a group. *Journal of Mathematical Behavior, 20*(2), 163–172.
Kieran, C. (1992). The learning and teaching of school algebra. In D. Grouws (Ed.), *Handbook of research on mathematics teaching and learning* (pp. 390–419). New York, NY: Macmillan Publishing Company.
Mark, J., Cuoco, A., Goldenberg, E. P., & Sword, S. (2010). Developing mathematical habits of mind. *Mathematics Teaching in the Middle School, 15*(9), 505–509.
National Governors Association Center for Best Practices & Council of Chief State School Officers. (2010). *Common core state standards for mathematics*. Washington, DC: Authors Retrieved from http://www.corestandards.org/Math/
Piaget, J., & Garcia, R. (1983/1989). *Psychogenesis and the history of science* (H. Feider, Trans.). New York, NY: Columbia University Press.
Sawyer, W. W. (1959). *A concrete approach to abstract algebra*. San Francisco, CA: W. H. Freeman and Company.
Sfard, A. (1991). On the dual nature of mathematical conceptions: Reflections on processes and objects as different sides of the same coin. *Educational Studies in Mathematics, 22*(1), 1–36.
Usiskin, Z. (1988). Conceptions of school algebra and uses of variables. In A. F. Coxford (Ed.), *The ideas of algebra, K–12. 1988 Yearbook of the National Council of Teachers of Mathematics* (pp. 8–19). Reston, VA: National Council of Teachers of Mathematics.

Chapter 15
Building a Coherent Research Program that Links Abstract Algebra to Secondary Mathematics Pedagogy via Disciplinary Practices

James Cummings, Elise Lockwood, and Keith Weber

Using Abstract Algebra to Teach Disciplinary Practices

In this commentary chapter, we discuss cross-cutting themes from the five chapters in this volume that explored approaches to connect engagement in abstract algebra to the development of mathematical disciplinary practices. Collectively, the authors of these chapters make a compelling case for rethinking the relationship between abstract algebra and the teaching of secondary mathematics. The majority of prospective secondary mathematics teachers will complete a course in abstract algebra. However, most of these teachers do not see how a course in abstract algebra (or advanced mathematics in general) can inform their teaching (e.g., Ticknor, 2012; Zazkis & Leikin, 2010). Thus, from the perspective of many teachers, completing an abstract algebra course is a requirement for teachers that they believe does not advance their professional goals. Given the strong links that we see between abstract algebra and secondary mathematics, this feels like a missed opportunity.

The chapters in this section each use a novel approach to link abstract algebra to teaching secondary mathematics, that we summarize as follows: In learning abstract algebra (perhaps in an ideal or an innovative environment), prospective teachers will be given the opportunity to engage in disciplinary practices, such as noticing,

J. Cummings (✉)
Department of Mathematical Sciences, Carnegie Mellon University, Pittsburgh, PA, USA
e-mail: jcumming@andrew.cmu.edu

E. Lockwood
Department of Mathematics, Oregon State University, Corvallis, OR, USA

K. Weber
Graduate School of Education, Rutgers University, New Brunswick, NJ, USA

N. H. Wasserman (ed.), *Connecting Abstract Algebra to Secondary Mathematics, for Secondary Mathematics Teachers*, Research in Mathematics Education,
https://doi.org/10.1007/978-3-319-99214-3_15

defining, or proving. The disciplinary practices that these prospective teachers engage in can also be used to complete some of the tasks of teaching secondary mathematics, such as designing lesson plans, responding to student questions, or simply making students aware of these disciplinary practices. Hence, by studying abstract algebra, prospective teachers can develop an increased awareness of or ability to engage in disciplinary practices that can productively inform their future teaching. As a further point, we presume that some of the disciplinary practices that teachers engage in are specific to abstract algebra. By this, we mean the disciplinary practices should be more prevalent in abstract algebra than in other advanced mathematical settings, or students would be engaging in these disciplinary practices in a different way in abstract algebra than they would in other settings. Hence, in learning abstract algebra, prospective teachers can engage in disciplinary practices that are useful for teaching that they were unlikely to encounter in other advanced mathematical settings. We elaborate on what we consider to be practices specific to abstract algebra later in this chapter.

The purpose of the chapters in this section is to explore how abstract algebra and secondary mathematics pedagogy can be connected via disciplinary practices. The authors of these chapters propose a number of interesting ideas that can motivate further research in this area. In Chap. 10, Zbiek and Heid describe their Situations project. A noteworthy contribution of their study is identifying which disciplinary practices are important by carefully investigating the actual practice of teaching. We think this is a valuable design heuristic. In Chap. 11, Baldinger examines how students' experience in their abstract algebra course influenced their problem solving behavior. The author has highlighted the importance of hypothesizing particular abilities in secondary mathematics that will improve as a result of taking an abstract algebra course, and she then empirically tested these hypotheses. In Chap. 12, Shamash, Barabash, and Even report on an innovative class structure in which a selective group of experienced Israeli mathematics teachers studied not only the core of modern abstract algebra, but also the historical arc from the subject's inception to cutting edge research in abstract algebra. A notable feature of their chapter is that their outcome measure was based on teachers' designing of lesson plans, which is clearly an important task of teaching. In Chap. 13, Lai and Donsig describe connections between linear algebra and geometric habits of mind, or productive ways of thinking that enable an individual to make sense of geometry. This chapter highlights an important aspect of abstract algebraic thinking—studying the structural properties that are preserved by mappings between algebraic structures—that is also useful in secondary mathematics. Lai and Donsig focus on the structure preserving properties of geometric transformations, but we believe this occurs in other areas of secondary mathematics as well. In Chap. 14, Lee and Heid illustrate how the search for structure in abstract algebra can potentially inform and unify the various types of factorization that are done in secondary mathematics.

The goals for the chapters in this section include exploring how links between abstract algebra and secondary mathematics can be generated, proposing interesting ideas for instruction and research, and stimulating further research in this area. We believe the authors of these chapters were successful in achieving these goals. Collectively, these authors established the potential of linking abstract algebra to secondary mathematics pedagogy via disciplinary practices, they proposed promising instructional methods that can inform both research and practice, and their work can inspire further research of this type. Even so, as we read the chapters, we found ourselves grappling with fundamental issues, including what each author meant by abstract algebra and pedagogy, as well as whether there was a shared understanding of these terms. In particular, the following questions emerged:

1. What constitutes a meaningful connection between abstract algebra and pedagogical practice?
2. What should the goals of teaching disciplinary practices from abstract algebra be? How can we determine if these goals are met?
3. What theoretical perspectives should we use to investigate these goals? What is the role of theory in conducting these investigations?
4. How do we handle contextual variation? How can researchers whose work is situated in different contexts learn from each other's work?

What we observed was that some chapters appeared to have inconsistent answers to these questions, while other chapters did not consider these questions at all. This is not a demerit of these chapters; it would certainly be unreasonable to expect the authors to reach a consensus on foundational issues in a field that is only just coming into existence. However, we do think these questions are necessary to consider in evaluating work of this type. We also believe that scholars need to reach some consensus on these foundational issues if they want to develop a coherent and consistent body of literature. In the next section, we describe why we believe these questions are important to address if researchers are to develop a research base where scholars can build upon each other's work. In the concluding section, we discuss how Sandoval's (2014) notion of "conjecture mapping" for conducting design research may potentially offer a language and a methodology for conducting research on the connections between abstract algebra and secondary mathematics teaching.

Questions Raised by the Chapters

Here we discuss four major questions that emerged for us from reading, comparing, and contrasting the chapters.

What Constitutes a Meaningful Connection Between Abstract Algebra and Pedagogical Practice?

From our perspective, a meaningful connection between abstract algebra and pedagogy should connect features that are *specific to abstract algebra* to a concrete improvement in mathematical instruction. We begin this discussion by highlighting what we think are fundamental content and ways of thinking that may be uniquely covered in an abstract algebra course.

In standard university curricula, the core content of an abstract algebra course is material about groups, rings and fields. Key topics include operations and their properties, identity and inverse elements, groups and subgroups, cosets, quotient groups, homomorphisms, rings and ideals, special classes of rings, prime factorization, rings of polynomials, and fields. In terms of ways of thinking, we view abstract algebra as the study of structures (i.e., a set of elements and operations on that set that satisfy certain properties) and structure-preserving mappings that can act on these structures. For example, in elementary group theory, we aim for students to recognize groups wherever they arise, to comprehend that two isomorphic groups are essentially the same, and to have a deep understanding of the importance of homomorphisms (for example in the context of normal subgroups, quotient groups and the First Isomorphism Theorem).

By contrast, there are other ways of thinking that are either universal in advanced mathematics (e.g., understanding through examples and counterexamples, certain ways of approaching problems) or are more salient in other parts of the advanced mathematics curriculum (e.g., combinatorial, geometric, probabilistic, or set-theoretic thinking). A meaningful connection between abstract algebra and pedagogy should ideally involve content and ways of thinking that are specific to abstract algebra.

If a student-teacher learned something about pedagogy in an abstract algebra setting, ideally, we would want the connection between abstract algebra and pedagogy to involve something *specific to abstract algebra* (and not a connection that a student could have made in any mathematics class). This applies not only to the content that is covered in an abstract algebra course, but also to the way that content is treated. Abstract algebra involves the study of prime numbers and polynomials, but so do some other advanced mathematics classes (e.g., number theory and real analysis, respectively). We would like any connection between abstract algebra and pedagogy regarding prime numbers and polynomials to involve concepts or ways of reasoning that are typical of abstract algebra, such as thinking of properties of prime numbers in Euclidean domains or splitting fields of polynomials.

Overall, we found that these chapters either did not specifically address the issue of why abstract algebra was particularly useful for the teachers, or we found many of the connections did not involve ways of thinking that were specific to abstract algebra. For instance, Baldinger's curriculum specifically concerned problem solving, but it was unclear how *specific aspects* of abstract algebra can improve teachers' problem-solving abilities. Plausibly, the participants could have gained as much from a course in real analysis or discrete mathematics. Lai and Donsig's chapter (Chap. 13) involves mathematical content (e.g., linear and affine transformations of the Euclidean plane) which is not traditionally part of abstract algebra, but instead tends to be covered in a beginning linear algebra course. We perceive that both Baldinger, and Lai and Donsig shifted the typical curriculum of abstract algebra to a focus on other topics, such as number theory (e.g., the Chinese remainder theorem) and linear algebra, respectively. These changes in curricula raise interesting questions: What content should be covered in an abstract algebra course for teachers? Do we need to substantially alter the content of an abstract algebra course if the course is to productively inform prospective teachers' future pedagogy? How much can a teacher or researcher alter the content of an abstract algebra course before the course is no longer focusing on abstract algebra?

Shamash et al.'s curriculum was innovative in that it clearly covered fundamental ideas of abstract algebra. However, the student-teachers' lesson plans that Shamash et al. used to illustrate the success of their curriculum generally did not involve these abstract algebra ideas. Rather, many of their lesson plans concerned the solving of cubic and quartic polynomials, which could be done without the study of the advanced abstract algebra or, indeed, even the core topics of an undergraduate abstract algebra course. Lee and Heid's article emphasized structure (in particular structural parallels between the ring of integers and the ring of polynomials). However, since the participants were interviewed before taking an abstract algebra course, it is hard to draw a direct connection between abstract algebra and the participants' performance in the tasks. Zbiek and Heid's article was innovative, and it was admirably specific in connecting specific topics in abstract algebra to specific tasks in pedagogy. Yet, even here, we wondered if the tasks actually were particular to abstract algebra, as some of the claimed connections seemed remote. For example, Example 1 (which involves viewing $f(x-k) = f(x+k)$ as associated pairs) seemed to be more about equivalence relations and equivalence classes than about cosets.

While overall the chapters suggested interesting ways to connect teachers' experiences in abstract algebra and their experiences teaching, the chapters left us still wanting clearer, more explicit connections between concepts and ways of thinking that are specific to abstract algebra with the pedagogical practices of preservice or inservice teachers. Providing these explicit connections could facilitate conversations between researchers who are interested in the connections between abstract algebra and pedagogy.

What Should the Goals of Teaching Disciplinary Practices from Abstract Algebra Be? How Can We Determine If These Goals Are Met?

At a high level, a goal of the research presented in this volume is that teachers' experience with abstract algebra will help them deliver better instruction and will improve students' mathematical learning. This is a lofty goal that is difficult to assess, as assessment would require gathering data in prospective teachers' future mathematics classrooms and quantifying the impact of the teacher's experience with abstract algebra on instruction and learning. If we view this overall goal as having two stages—first, examining the impact of experience with abstract algebra on the development of teachers' practices, and, second, examining the ways in which that learning affects the teachers' students—we can consider intermediate goals. In principle, these intermediate goals should be easier to assess. In particular, the chapters we summarize in this commentary focus on the first stage, without making claims about the impact on students.

We found that each of these chapters had quite different intermediate goals, as well as different strategies for assessing these goals. In particular, Shamash et al. sought to demonstrate the relevance of taking an abstract algebra course on experienced teachers' development. In this study, teachers designed lesson plans based on their experience in an abstract algebra course, and this lesson design formed one module in an enrichment program aimed at experienced teachers. These lessons were assessed as classwork for this program.

Baldinger aimed to show that an abstract algebra course should improve students' problem-solving abilities. She tested this hypothesis by assessing student teachers' problem-solving abilities with a short, written test before and after taking an abstract algebra course. Lee and Heid broadly sought to improve content knowledge for teaching among preservice teachers, and they focused on the role of structure and structural thinking. They conducted interviews with mathematics and mathematics education majors who had not taken an abstract algebra course, as they worked on tasks involving factorization of integers and polynomials.

Both Zbiek and Heid, and Lai and Donsig attempted to articulate connections between abstract algebra and secondary mathematics. Zbiek and Heid's goal was to describe a set of possible connections between abstract algebra and secondary mathematics that was based on the Situations project of Heid, Wilson, and Blume (2015). Lai and Donsig described a set of mathematical contexts in which geometry at the secondary level can be connected with algebra at the secondary and college level, particularly focusing on the relationship between abstract (linear) algebra and geometric transformations.

Although each of these respective intermediate goals may be appropriate for each book chapter, we feel that for future work in this area, it will be important for researchers to formulate appropriate learning goals and assessment strategies. This calls for a need for studies in this area to have learning goals and methods to assess them. It also calls for the field to think about what learning goals are appropriate within this context.

What Theoretical Perspectives Should We Use to Investigate These Goals? What Is the Role of Theory in Conducting These Investigations?

We observed that the chapters employ a very wide array of theoretical perspectives. Obviously, researchers have particular reasons for choosing their respective theoretical perspectives, and we have no doubt that each of these authors could clearly justify their choice of perspective. We are, nonetheless, struck by the sheer variation in these five chapters. Zbiek and Heid used MUST and Mathematical Activity perspectives (Heid et al., 2015), Baldinger drew on mathematical practices (CCSSI, 2010), Lai and Donsig framed their work in terms of (geometric) habits of mind (Cuoco, Goldenberg, & Mark, 1996), and Lee and Heid also used habits of mind, as well as operational versus structural understanding (Sfard, 1991), APOS (Dubinsky, Dautermann, Leron, & Zazkis, 1994), inter-object/intra-object understanding (Piaget & Garcia, 1983/1989), and they developed the Extending, Deepening, Unifying, and Strengthening (EDUS) framework. (A crucial difference to note is that the EDUS framework was an outcome, or contribution, of the chapter, not the analytical scheme used to interpret the data.)

The wide variety of theoretical perspectives is advantageous in that the authors are drawing on the collective work of many researchers. However, the use of such varied perspectives does have potential drawbacks, and we feel that the variation can hinder more coherent/cohesive progress toward the overarching goal of helping link abstract algebra to pedagogical practices. We discuss a couple of specific drawbacks to having such a wide array of theoretical perspectives. First, the variety of perspectives can inhibit progress by hindering productive communication between like-minded researchers with similar goals. We note that some of the perspectives seem analogous; for example, Mathematical Activity, mathematical practices, and habits of mind can perhaps be regarded as all trying to articulate the same phenomenon. Perhaps there could be more cohesion if more researchers chose (or agreed to choose) one perspective, rather than each employing a different viewpoint.

A second issue concerns which type of theoretical perspective researchers should be developing and applying. In particular, a key question is whether theoretical perspectives should be applied *top-down* (the researcher takes existing theories to frame her research questions, design her instruction, and analyze her data) or *bottom-up* (the theoretical frames emerge from the researchers' work). Ideally, there is a dialectical process in which the initial theory is modified and reframed in the light of the research results; hence, this way of categorizing theoretical perspectives should be viewed as a continuum rather than a dichotomy. From our reading of the chapters, it appeared to us as if the researchers have generally taken a top-down approach in choosing their theoretical perspective. As we noted, this is potentially advantageous in that the researchers are using the theoretical work of previous scholars to gain insight into the phenomena that they are studying. However, we suggest that an early commitment by each respective research team to a different theoretical perspective could potentially prevent researchers from noticing other phenomena of interest that are not privileged by the theoretical perspectives being applied. If we are not sure what a good contribution of an abstract algebra class can be for teachers, we may want to avoid focusing our attention too tightly too early on. A premature commitment to a theoretical perspective runs the risk of blinding us to important phenomena.

How Do We Handle Contextual Variation? How Can Researchers Whose Work Is Situated in Different Contexts Learn from Each Other's Work?

The chapters are concerned with various populations of teachers and students. Three of the articles involve data collected from work with groups of teachers, including a selective group of highly experienced teachers in Israel (Shamash et al. (Chap. 12)), preservice teachers with degrees in science and mathematics and varying levels of teaching experience in the USA (Baldinger (Chap. 11)), and mathematics and mathematics education majors in the USA (Lee and Heid (Chap. 14)). We acknowledge that differences between the populations being studied and the contexts in which they are studied are unavoidable. This is true for any body of mathematics education research, but especially studies in teacher education where curricula and teacher expectations vary greatly across institutions and across countries. Also, we note that the variation of contexts can potentially be viewed positively, because such variety has the possibility to demonstrate certain phenomena across a wide array of contexts, allowing researchers to distill what is important regardless of context. Further, if researchers in different contexts use the same instructional or methodological methods, but obtain different results, then this discrepancy can lead to the generation and evaluation of hypotheses for why these differences were observed. This, in turn, can lead to a greater understanding of what factors need to be in place for teacher education development to be effective.

However, we felt that it was unclear how we should generalize the results of these studies beyond the context in which each study took place. For instance, it is doubtful that the American mathematics education majors in Lee and Heid's study would have the same appreciation for cutting edge research in abstract algebra as the selective group of highly experienced Israeli teachers. Indeed, in the chapters in this section, we find that because the goals and theoretical perspectives are so varied (as we have indicated previously), it is difficult to distill common themes about the role of advanced abstract algebra on pedagogy in a meaningful way. We do not mean to be critical of the authors of the chapters, as their aim was to generate and explore promising ideas, rather than think about how their instruction and results would generalize across contexts. However, we do think that as research in this area moves forward, without a coherent perspective grounding their work, the contextual variation among the chapters could become a drawback rather than a strength. If researchers who work in this area could develop a shared understanding of pedagogical goals and a theory of how learning occurs, this research community could potentially actually leverage the contextual variation more effectively. What seems to be necessary is strong, unified theoretical frameworks that let us understand why effective instruction was actually effective, and shed light on what factors need to be in place for the instruction to be successful. This may include exploring questions like what background knowledge teachers need to have or what pedagogical settings or supports may be most beneficial.

Conjecture Mapping

The Need for a Theoretical Focus and Empirical Data on How Prospective Teachers Learn in Abstract Algebra

In the previous section, we discuss the importance of increased clarity and consensus on what the learning goals of abstract algebra should be for prospective teachers, how to determine if students' learning gains were due to abstract algebra, and what theoretical perspectives (and what types of theoretical perspectives) can be used to conduct research on how prospective teachers can benefit from abstract algebra. We also note the challenge that the research community faces in generalizing research findings beyond the contexts in which they occur. The chapters collectively provide an interesting theory on *what* prospective teachers can learn in an abstract algebra class and useful data illustrating *that* prospective teachers showed learning gains after completing an abstract algebra course. What we believe is also needed is both theory and data on *how* prospective teachers can and do learn from an abstract algebra course.

Regarding theory, we observe that although the authors use a wide variety of theoretical perspectives, these theoretical perspectives generally are not used to specify or analyze prospective teachers' learning processes in an abstract algebra class. For instance, MUST, mathematical activity, mathematical practices, and (geometric) habits of mind are useful in informing goals of instruction and identifying links between abstract algebra and secondary mathematics. Only the EDUS framework suggested specific *mechanisms* by which prospective teachers might better achieve the author's intended learning goals (in this case, perceiving mathematical objects and noticing and attending to mathematical structure).

Regarding the data that was presented, the two chapters with studies documenting student growth illustrate how prospective teachers improved their ability to solve problems (Baldinger (Chap. 11)) and that experienced teachers could design interesting lesson plans (Shamash et al. (Chap.12)). While both Baldinger and Shamash et al. give a detailed account of how the abstract algebra courses in their studies were taught, they do not present data on how students engaged in the course. For this reason, although we were persuaded *that* these abstract algebra courses were beneficial to the teachers, we do not have a sense of *how*, specifically, these abstract algebra courses helped (outside of the general course descriptions). As we previously noted, we were left with questions about how the abstract algebra content related to teachers' learning. Perhaps Baldinger's teachers' learning gains were not due to their abstract algebra class as all, but due to the other three mathematics courses that they were also taking at the time. Perhaps most of Shamash et al.'s experienced teachers' learning gains were entirely attributable to the historical discussion of the solutions to quadratic and cubic polynomials and the coverage of the advanced abstract algebra was irrelevant to the teachers' growth. Further, we also do not have a sense of how changing the curricula or the context of the abstract algebra courses would influence the learning gains that Baldinger and Shamash et al. reported. Knowing *how* teachers learn in their abstract algebra courses would go a long way in making the case that abstract algebra was pivotal in this learning and whether we should expect findings from one context to generalize to another context.

Design Research and Conjecture Mapping

We have argued that we need theory about and data pertaining to how prospective teachers learn in an abstract algebra classroom. A popular way that contemporary mathematics educators investigate how students learn is by design research (Cobb, Confrey, DiSessa, Lehrer, & Schauble, 2003). We think many aspects of design research are relevant to the research aims of the authors in this section. Design research is not characterized by adherence to a specific methodology, but rather by common epistemic commitments held by design researchers (Sandoval, 2014). We argue that some of these theoretical commitments can be useful for developing theory and practice about linking abstract algebra and secondary mathematics pedagogy. A first theoretical commitment is to simultaneously develop practical instruction that achieves the researcher's learning goals and a testable theory of how the instruction supported students' learning (Cobb et al., 2003; DiSessa & Cobb, 2004). A second theoretical commitment is to move beyond broad general learning theories to develop specific local instructional theories that explain how learning occurs in the specific situation being investigated (DiSessa & Cobb, 2004). A third theoretical commitment is to carefully analyze both the processes and the outcomes of learning to document how learning occurred or develop hypotheses about why learning did not occur (Cobb et al., 2003). (There are other tenets of design research, such as the need to engage in iterative design cycles that may also be useful for researchers to consider, but are less relevant to the aims of our commentary.)

Sandoval (2014) described a specific paradigm for conducting and reporting design research that he termed "conjecture mapping." Sandoval argued that this paradigm was flexible enough to account for the wide range of theoretical perspectives, learning goals, and instructional methods that different design researchers use while still providing a language and methodological structures to make comparisons across studies. We hypothesize that this paradigm might be particularly useful for researchers who wish to connect abstract algebra to secondary mathematics teaching, in that it focuses their attention to how teachers learn to improve their practice, how abstract algebra was critical in students' growth, and how theory can develop from data. We are not suggesting that researchers who are interested in the connections between abstract algebra and secondary mathematics need to adopt this paradigm; we are only suggesting this is one method that a researcher can use to simultaneously build upon the insights from the chapters in this volume, while addressing the questions we raised earlier in this chapter. In Sandoval's framework, when conducting research with conjecture mapping, the researcher engages in the following analyses:

- *High-level conjectures*: The researcher specifies a broad theory of how these learning outcomes can be supported by instruction.
- *Embodied instruction*: High-level conjectures on how to support learning are not sufficiently prescriptive to design specific instructional modules. The embodied instruction is a concrete instantiation of the high-level conjectures that explicates a wide array of features of instructional design. (This not only includes the content that is covered and the activities that the students will complete but also the resources that students will have available, how long students will be given to complete the task, whether the students will work collaboratively, what sort of scaffolding or support the instructor will provide, and so on).
- *Mediating processes*: The researcher predicts specific mediating processes that students will participate in as they engage in the embodied instruction, as well as a means of determining whether the mediating processes actually occurred. (The presence or absence of specific mediating processes can be measured by analyzing students' utterances or artifacts, such as their written solution to problems.)
- *Design conjectures*: Design conjectures are specific theories about which mediating processes specific instructional activities will elicit. Because the researcher specifies how to determine if the mediating processes occurred prior to instruction, these design conjectures are empirically testable.
- *Theoretical conjectures*: Theoretical conjectures are specific theories about how eliciting mediating processes will enable students to achieve the researchers' learning goals. As the research specifies how to determine whether mediating processes occurred and learning goals were achieved, the relationship between the presence of mediating processes and students' success in achieving their learning goals can be systematically analyzed.

Applying Conjecture Mapping to Teaching Productive Disciplinary Practices in Abstract Algebra

Sandoval (2014) presented conjecture mapping as a means for conducting design research in any context. Below we discuss what such research might look like when teaching disciplinary practices via abstract algebra to prospective secondary mathematics teachers and how conjecture mapping can at least partially address some of the concerns that we raised in the previous section.

- *Learning goals*: We believe the learning goals for teachers in an abstract algebra course should include the teachers being able to successfully engage in some activity directly related to teaching secondary mathematics. These activities can include lesson planning (as was done in the Shamash et al. chapter (Chap. 12)) or other activities, such as examining the correctness of students' work, providing explanations about or models for secondary mathematics concepts, or responding to hypothetical student questions (e.g., Wasserman, Fukawa-

Connelly, Villanueva, Mejia-Ramos, & Weber, 2017). This is why we found Zbiek and Heid's approach of identifying learning goals by inspecting the actual practice of teaching to be so promising. One aspect of learning goals that is challenging for researchers is to specify criteria for what constitutes successful teacher engagement in these pedagogical activities. For instance, in the Shamash et al. chapter (Chap. 12), we would be interested to know what criteria the authors used to determine if a lesson plan qualified as a "good" lesson plan. Such criteria can be difficult to generate, but are needed for researchers to evaluate the efficacy of a study or attempt to replicate one another's work.

Explicating learning goals addresses our second question about the researchers specifying what their learning goals are. Our suggestion for having the learning goals consist of teachers successfully engaging in some pedagogical activity provide clarity on what learning goals should be, from our perspective. Finally, we note that even if authors were in the preliminary stage of identifying potential connections between abstract algebra and secondary mathematics teaching (as Lai and Donsig (Chap. 13) and Lee and Heid (Chap. 14) were), the authors could still specify what specific activities teachers should be able to successfully engage in if they have become aware of these connections.

- *Mediating processes*: As the theme of this section involved connecting abstract algebra to secondary mathematics teaching via disciplinary practices, we assume that one of the mediating processes that the embodied instruction should elicit is teachers engaging in specific disciplinary practices. However, we expect that mere engagement will not be enough. We also expect that students will show an increased facility in, appreciation of, or propensity to engage in this disciplinary practice.
- *Design conjectures*: The design conjectures consist of mapping specific aspects of the instructional environment to the specific mediating processes that these elicit. As researchers in this volume are attempting to connect the specific topic of abstract algebra to secondary mathematics, it is incumbent upon the authors to specify how the specific abstract algebra content being addressed is important to elicit the desired mediating processes. This partially speaks to two of the concerns we raised earlier in the chapter. First, this highlights how the abstract algebra content was pivotal in helping students engage in specific disciplinary practices. Second, by identifying which aspects of instruction were necessary to elicit specific mediating processes, the author is implicitly providing hypotheses on what features of instruction would need to be in place for the instruction to work in other contexts.
- *Theoretical conjectures*: The theoretical conjectures consist of causal links between mediating processes and learning goals. Given the themes of this section, theoretical conjectures should include how teachers' increased facility in, appreciation of, or propensity to engage in this disciplinary practice allows them to succeed at specific pedagogical tasks. In tandem with the design conjectures, the theoretical conjectures can address a concern we raised earlier in the chapter—they allow the researcher to make a compelling case for how the

focus on abstract algebra elicited teacher engagement in disciplinary practices that consequently led to improved performance on pedagogical tasks.

- *High-level conjectures and local instructional theories*: Sandoval's conjecture mapping clarifies the role of top-down theories and bottom-up theories. Top-down theories inform high-level conjectures about how teacher learning can be supported in an abstract algebra environment and what appropriate learning goals of instruction might be. However, the ultimate outcome of the research consists of local instructional theories that are greatly informed by an analysis of students' work as they engage in the embodied instruction.

Of course, researchers need not use Sandoval's conjecture mapping to explore connections between abstract algebra and secondary mathematics teaching via disciplinary practices. Conjecture mapping is one way to encourage the researcher to engage in activities that we identified as worthwhile, including clarifying learning goals, establishing clear links between abstract algebra, disciplinary practices, and secondary mathematics teaching, building bottom-up theories that describe how teachers learn, and describing the aspects of instruction environments that need to be in place for instruction to be effective.

References

Cobb, P., Confrey, J., DiSessa, A., Lehrer, R., & Schauble, L. (2003). Design experiments in educational research. *Educational Researcher, 32*(1), 9–13.

Common Core State Standards Initiative. (2010). *Common core state standards for mathematics*. Washington, DC: National Governors Association Center for Best Practices and the Council of Chief State School Officers.

Cuoco, A., Goldenberg, E. P., & Mark, J. (1996). Habits of mind: An organizing principle for mathematics curricula. *Journal of Mathematical Behavior, 15*(4), 375–402.

DiSessa, A. A., & Cobb, P. (2004). Ontological innovation and the role of theory in design experiments. *Journal of the Learning Sciences, 13*(1), 77–103.

Dubinsky, E., Dautermann, J., Leron, U., & Zazkis, R. (1994). On learning fundamental concepts of group theory. *Educational Studies in Mathematics, 27*(3), 267–305.

Heid, M. K., Wilson, P. S., & Blume, G. W. (Eds.). (2015). *Mathematical understanding for secondary teaching: A framework and classroom-based situations*. Charlotte, NC: IAP.

Piaget, J., & Garcia, R. (1983/1989). Psychogenesis and the history of science (H. Feider, Trans.). New York, NY: Columbia University Press.

Sandoval, W. (2014). Conjecture mapping: An approach to systematic educational design research. *Journal of the Learning Sciences, 23*(1), 18–36.

Sfard, A. (1991). On the dual nature of mathematical conceptions: Reflections on processes and objects as different sides of the same coin. *Educational Studies in Mathematics, 22*(1), 1–36.

Ticknor, C. S. (2012). Situated learning in an abstract algebra classroom. *Educational Studies in Mathematics, 81*(3), 307–323.

Wasserman, N. H., Fukawa-Connelly, T., Villanueva, M., Mejia-Ramos, J. P., & Weber, K. (2017). Making real analysis relevant to secondary teachers: Building up from and stepping down to practice. *PRIMUS, 27*(6), 559–578.

Zazkis, R., & Leikin, R. (2010). Advanced mathematical knowledge in teaching practice: Perceptions of secondary mathematics teachers. *Mathematical Thinking and Learning, 12*(4), 263–281.

Part IV
Exploring Approaches to Secondary Teacher Education: Engaging in Abstract Algebra in Relation to Pedagogical Problems in Teaching Secondary Mathematics

Chapter 16
Exploring an Instructional Model for Designing Modules for Secondary Mathematics Teachers in an Abstract Algebra Course

Nicholas H. Wasserman and Patrick Galarza

Introduction

Many secondary mathematics teachers take advanced mathematics courses, frequently including a course in abstract algebra, as part of their preparation program. There are many reasons for doing so. One is that much of the content of secondary school mathematics is connected to and undergirded by structures studied in advanced mathematics (e.g., CBMS, 2012). In the case of abstract algebra, many school mathematics concepts themselves are instantiations of groups, rings, and fields. Yet secondary teachers regularly report that completing such courses provides little professional value and does not influence their subsequent instruction (e.g., Wasserman, Weber, Villanueva, & Mejia-Ramos, 2018; Zazkis & Leikin, 2010). This raises the challenging problem of how such advanced mathematics coursework might make more meaningful connections to secondary mathematics teachers' future professional work.

In this chapter, we explore the use of Wasserman, Fukawa-Connelly, Villanueva, Mejia-Ramos, and Weber's (2017) and Wasserman, Weber, and McGuffey's (2017) instructional model, previously used in the context of a real analysis course, to design modules for abstract algebra. First, we summarize some key aspects of the literature and recap the theoretical underpinnings of the instructional model. Second, we provide a detailed account of two modules designed for an abstract algebra course. The modules explore not just connections to secondary mathematics content, but also specific implications for teaching practice. Third, we elaborate on findings from a small-scale study that explored teachers' engagement with these two modules

N. H. Wasserman (✉) · P. Galarza
Department of Mathematics, Science and Technology, Teachers College, Columbia University, New York, NY, USA
e-mail: wasserman@tc.columbia.edu

N. H. Wasserman (ed.), *Connecting Abstract Algebra to Secondary Mathematics, for Secondary Mathematics Teachers*, Research in Mathematics Education,
https://doi.org/10.1007/978-3-319-99214-3_16

(and three other modules). Lastly, we reflect on some of the potential implications for the teaching of advanced content courses as they relate to secondary teacher education.

Literature and Theoretical Perspectives

In this section, we briefly recap some of the findings from the literature regarding two related questions: What are the potential benefits to secondary teachers in studying advanced mathematics? And, why are these benefits not necessarily being realized? Doing so provides theoretical motivation for the instructional model (Wasserman, Fukawa-Connelly, et al., 2017; Wasserman, Weber, et al., 2017) in this study. Notably, we provide only brief summaries here; for a more thorough discussion, see Wasserman, Weber, et al. (2017).

Essentially, there are two prominent arguments for secondary teachers studying advanced mathematics. The first is that the advanced content discussed often forms the foundation for mathematical ideas studied at the secondary level (e.g., CBMS, 2012). Felix Klein was one of the first to advocate for such an approach with teachers in his series *Elementary Mathematics from an Advanced Standpoint* (1932). These books elaborated on connections between the mathematics taught at universities, and the mathematics taught in secondary schools. More recent mathematics textbooks for secondary teachers explore similar connections (e.g., Bremigan, Bremigan, & Lorch, 2011; Sultan & Artzt, 2011; Usiskin, Peressini, Marchisotto, & Stanley, 2003). These examples suggest that advanced mathematics can serve to deepen and rigorously establish the mathematics that secondary teachers will teach. The second argument is that the study of advanced mathematics is also productive for inculcating teachers into what "doing mathematics" entails—that is, it is not just for acquiring some knowledge *of* mathematics, but also more broadly for acquiring knowledge *about* mathematics (e.g., Even, 2011). In elaborating on their provisional domain of "horizon content knowledge," Ball and Bass (2009) identified knowledge of key mathematical practices and of core mathematical values and sensibilities as important components of this domain. Both speak to the importance—for teachers—of understanding mathematics as a discipline. Rasmussen, Wawro, and Zandieh (2015) describe some of these practices, of defining, theoremizing, symbolizing, etc., as those that are core to the activity of professional mathematicians. The content of advanced mathematics courses often resembles such activities (by defining objects, making claims about those objects, providing proofs for those claims, etc.). The argument is that advanced coursework can be leveraged in such a way as to help teachers develop these disciplinary practices, so that they can engage their students in these practices as well.

Despite these arguably strong connections between advanced and secondary mathematics, there is little evidence that taking advanced mathematics courses influences secondary teachers' instruction (e.g., Goulding, Hatch, & Rodd, 2003; Monk, 1994; Wasserman et al., 2018; Zazkis & Leikin, 2010). There are multiple reasons why this may be the case. In accord with practice-based approaches to teacher knowledge (e.g., Ball, Thames, & Phelps, 2008; McCrory, Floden, Ferrini-Mundy, Reckase, & Senk, 2012; Rowland, Huckstep, & Thwaites, 2005), Wasserman (2016, 2018a) and Wasserman, Weber, et al. (2017) have argued that a tension exists with the connection not to the content of secondary mathematics, but to the teaching of secondary mathematics content. That is, just because one might improve his or her knowledge of secondary mathematics, this does not necessarily imply there will be a change in his or her teaching of secondary mathematics. Another reason is that although there are potential connections to secondary content and disciplinary practices, some scholars (e.g., Moreira & David, 2008; Deng, 2007) have argued that the same content and disciplinary practices appear quite differently in advanced mathematics and secondary mathematics contexts. Lastly, some have argued that secondary teachers' learning goals in an advanced mathematics course do not align with teacher educators' goals. Teachers may be trying merely to pass the course (e.g., Ticknor, 2012) or to look for ideas they can transport directly to their own teaching (e.g., Wasserman et al., 2018), or they may have a different view altogether about the nature of mathematical learning (e.g., Heid, Blume, Zbiek, & Edwards, 1998). These perspectives may fail to promote the kinds of desirable connections from advanced coursework.

A Novel Instructional Model

In this section, we look at instructional approaches in advanced mathematics courses. We contrast what appears to be a normative instructional model for teaching advanced mathematics courses with Wasserman, Fukawa-Connelly, et al.'s (2017) and Wasserman, Weber, et al.'s (2017) instructional model, which was designed specifically with secondary teachers in mind. This is not to say that the latter instructional model would not be productive for other students (e.g., mathematics majors), but rather, to clarify that the approach grew out of considering advanced courses in a secondary teacher education context.

Making connections to school mathematics content has been a primary model for teaching advanced mathematics courses. However, in this approach, there is an assumption that as a byproduct of learning advanced mathematics content, secondary teachers will better understand secondary mathematics content and will consequently respond differently to instructional situations in the future—a presumed "trickle down" effect (Fig. 16.1a). This sort of knowledge transfer (from the cognitive psychology literature, e.g., Perkins & Salomon, 1992) is notoriously difficult. Indeed, each of the three postulated explanations for why the potential benefits of studying advanced mathematics are not being realized makes the sort of transfer presumed in this model more difficult.

In Fig. 16.1b, we present Wasserman, Fukawa-Connelly, et al.'s (2017) and Wasserman, Weber, et al.'s (2017) alternative instructional model for teaching advanced mathematics. The model was first used in a real analysis course, but

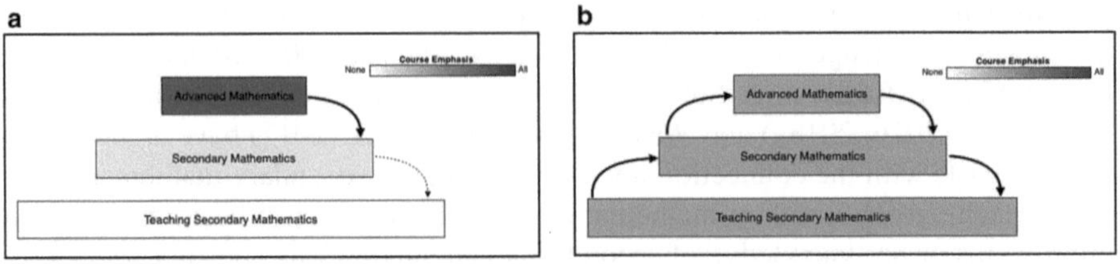

Fig. 16.1 (**a**) Implicit instructional model for advanced mathematics courses designed for teachers. (**b**) Wasserman, Fukawa-Connelly, et al.'s (2017) and Wasserman, Weber, et al. (2017) instructional model for advanced mathematics courses designed for teachers

was intended to be broad enough for thinking about instruction in other advanced mathematics courses. This model is composed of two parts: building up from and stepping down to practice.[1] To *build up from (teaching) practice,* the advanced mathematics content is preceded by a practical school-teaching situation. The building-up portion provides a context for the study of advanced mathematics in ways that are relevant to teachers' practices. The second part, *stepping down to (teaching) practice*, then uses the ideas acquired from advanced mathematics to reconsider the secondary mathematics and relevant pedagogical situations. Stepping down to practice explicitly clarifies the intended mathematical and pedagogical aims of the advanced mathematics content. In between building up from and stepping down to practice, the advanced mathematics topics are taught by the instructor in ways true to its advanced nature, with formal and rigorous treatment. Although we do not rehash here, Wasserman, Weber, et al. (2017) argued for why this instructional model might help reduce barriers to knowledge transfer in comparison with the first model (e.g., Barnett & Ceci, 2002) in light of the issues identified in the literature.

In what follows, we elaborate on two additional facets of this model.

Implications for Teaching

In addition to elaborating on the theoretical aspects of the instructional model, Wasserman, Weber, et al. (2017) reported on its use in a real analysis course. In a first iteration using these materials with 32 preservice and in service secondary teachers, the authors elaborated on the teachers' interactions with one particular module. This module attempted to make a specific disciplinary practice—described as "attending to the scope" of mathematical statements and justifications (e.g., clarifying that a statement or proof was only valid for a particular set of numbers)—more explicit in

[1]The phrase "stepping down to practice" does not intend to convey that teaching is somehow "lower" than advanced mathematics, but rather to suggest that the advanced mathematics has some utility in, or application to, the teaching situation.

terms of its connection to teaching. In classroom observations the following year, the authors found that the teachers, in their own secondary classrooms, engaged in actions that supported the module's pedagogical aims with respect to attention to scope. In later interviews, the teachers explicitly reported doing so because of their experiences in the real analysis course. That is, they reported their experiences in the real analysis course as being influential on their teaching of secondary mathematics. This provides some empirical support for the efficacy of the model in connecting learning in advanced mathematics to teaching in secondary mathematics.

Differentiating Types of Connections

As suggested in the literature, mathematical content and disciplinary practices are two primary ways the study of advanced mathematics might be connected to secondary mathematics. Essentially, both types of connections could be leveraged in the instructional model—the key difference is that the instructional model makes explicit not just the connections between the mathematics but also the implications for teaching mathematics. In addition to the distinction between mathematical content and disciplinary practices, we make one additional distinction between mathematical content connections. On the one hand, the connection could be that an idea in secondary mathematics serves as an example of an advanced mathematics concept; on the other hand, the connection could be that the advanced mathematics concept serves as an example of an idea in secondary mathematics. We refer to the first as a *generalization connection* and the latter as an *instantiation connection*, and elaborate further below. (See also Fig. 16.2.)

We call the connection between advanced and secondary mathematics a *generalization connection* when the advanced mathematics is a generalization of the secondary mathematics. That is, the secondary mathematics is an instance of the advanced mathematics—in some sense, a subset of the advanced mathematics. Consider, for example, the concept of an isomorphism as studied in an abstract algebra course. The logarithm, which is studied in secondary mathematics, is an example of an isomorphism—it is a function demonstrating the groups $(\mathbb{R}, +)$ and $(\mathbb{R}^+, \times)$ to be isomorphic: $\ln(ab) = \ln(a) + \ln(b)$. One might leverage this connection to the secondary teaching of logarithms since, for example, being an

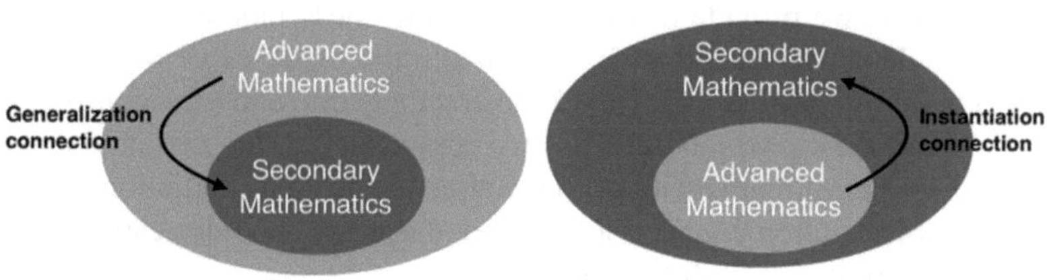

Fig. 16.2 Differentiating generalization and instantiation connections between advanced and secondary mathematics

isomorphism preserves the structure of the groups and can provide justification for some common properties of logarithms: $\ln(1) = 0$, $\ln\left(\frac{1}{a}\right) = -\ln(a)$, etc. In contrast, an *instantiation* connection is when the advanced mathematics is an example (or instance) of the secondary mathematics. That is, the advanced mathematics is a subset of the secondary mathematics. In this case, an isomorphism in an abstract algebra class could be framed as one example of a function (a broader secondary mathematics topic), and this connection could be leveraged to think about the teaching of functions in secondary mathematics.

We make one additional comment about the distinction between instantiation and generalization connections. Instantiation connections likely belong in particular advanced mathematics courses: studying isomorphisms (in abstract algebra) is important to be able to recognize the logarithm as one such example. Generalization connections, however, may be useful in several advanced courses. Although an isomorphism is an example of a function, there are also other advanced mathematics courses in which functions are studied, and these contexts could be used similarly to connect to the teaching of function in secondary mathematics.

Two Modules for an Abstract Algebra Course

As part of a larger project, a research team engaged in designing five modules, in accord with Wasserman, Fukawa-Connelly, et al.'s (2017) and Wasserman, Weber, et al.'s (2017) instructional model, that could be used in an abstract algebra course. The design process incorporated the following tasks: looking through abstract algebra textbooks to identify potential aspects of abstract algebra content that might be connected to secondary teaching, elaborating the kinds of connections being made to secondary mathematics and to secondary teaching, and articulating a rationale for how knowledge of this abstract algebra content could subsequently influence instruction in secondary school mathematics. This third aspect involved generating hypotheses for how secondary teachers might initially respond to particular teaching situations, as well as how the study of abstract algebra content might influence and shape their pedagogical responses. This constituted the first step in developing hypothetical trajectories for how teachers might come to acquire an understanding of these connections via the specific questions and tasks in the module.

In this section, we provide details about two of these modules. We discuss the tasks and problems associated with each of the three phases of the instructional model: (1) building up from practice; (2) abstract algebra; and (3) stepping down to practice. We use these two modules for two reasons. First, the two modules exemplify different types of mathematical connections—one an instantiation connection, the other a generalization connection. Second, these two modules capture the most salient findings from across all five modules.

Module 3: Functions

Module 3 was designed around the content of group isomorphisms in abstract algebra. The module aimed to use group isomorphisms as an example of a function—an instantiation connection—in order to discuss functions between more abstract sets of objects. That is, the connection between the advanced mathematics and secondary mathematics was not about isomorphisms, per se, but rather it used the context of group isomorphisms (a more abstract functional relationship) as an instantiation of a function. The overarching premises of the module were that many secondary teachers only consider functions on $\mathbb{R} \times \mathbb{R}$, and that a broader understanding of functions could shape the types of examples teachers use in their secondary teaching. More diverse examples of functions could help exemplify nuances within and boundaries around the idea of functions as more than merely symbolic rules.

Building Up from Practice

To begin the module, prospective and practicing teachers (PPTs) are given a classroom situation, depicted in Fig. 16.2. In addition to playing out the scenario—i.e., providing a hypothetical response by Mrs. Y—PPTs are asked, "If you were introducing a unit on functions, what definition and examples would you use? What ideas would you emphasize? Explain your reasoning." The expectation was that PPTs would provide fairly normative definitions of a function (i.e., not the definition in the classroom situation), and might emphasize things like the vertical line test, a black-box analogy, or modifiers like "each" or "unique." In addition, we posited that most of their examples would be of functions on $\mathbb{R} \times \mathbb{R}$, and, moreover, ones that had identifiable mathematical formulas.

After their discussion of the teaching situation, PPTs were asked to consider the various ways in which one can choose to express a mathematical function—with equations often being a very concise way of representing the relationship pairs (but not the only way). PPTs considered this for the function mapping in the teaching situation (Fig. 16.3). We expected PPTs to gravitate toward trying to identify an equation—which is challenging in this case—and ultimately struggle to find a concise representation. The primary point of the exercise is that, although there is a formula that would produce this finite set of ordered pairs ($f(x) = \frac{11}{15}x^2 - \frac{392}{15}x + \frac{1194}{5}$), there are a variety of ways that we might choose to express the mapping that a mathematical function represents.

Abstract Algebra

The abstract algebra portion began with a brief two-part definition of a function: the first part is the definition of a relation (as a subset of $A \times B$), with classification

Classroom Scenario: A high school teacher, Mrs. Y, is continuing her unit on functions. She is on day two – which has the goal of analyzing various functions using different representations.

Mrs. Y: Does the following relationship between A and B represent a function?

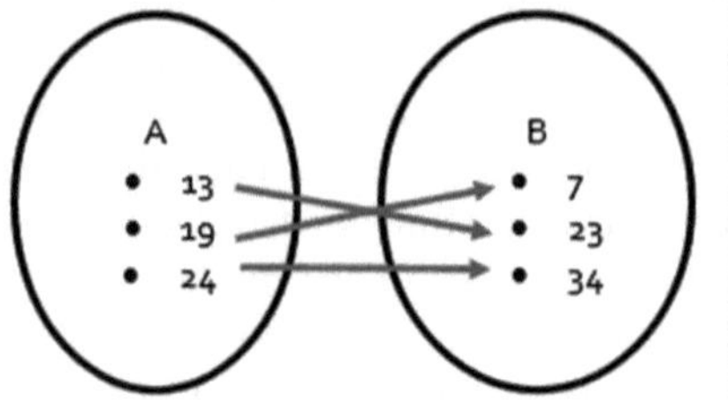

Evan: But your lesson from yesterday said $y = x$ is a function; and I don't see any equation and I don't see any y and x with those numbers. I am confused.

Evan: Also, you asked us to write down the definition from the textbook, which stated, "An expression where y is connected to the variable, x, is called a *function*. For every function, one can draw a diagram..." I don't see an expression. Are the ovals and arrows supposed to be the diagram?

Mrs. Y: Yesterday, we didn't speak about all the possible ways of representing functions—

Evan: Oh, Mrs. Y... I think I got it...I would say $y = x + 10$, if you think of A as x and B as y.

Mrs. Y: ...

Fig. 16.3 Teaching situation for the Function Module

as one-to-many, one-to-one, or many-to-one; the second part is the definition of a function as a relation with each element in A being mapped to exactly one element in B (i.e., *not* a one-to-many relation). The next activity in the abstract algebra portion built on a concept PPTs were expected to be familiar with from a previous module: "Consider the binary operation table below [addition modulo 12] (in Fig. 16.4a), and describe the function (i.e., mapping) that this binary operation table represents. What are some elements in the domain of this function? What are some elements in the range of this function? Express the function in the most concise way you can." Given that we had defined a binary operation as a function from $A \times A \rightarrow A$ in a previous module, this activity asked PPTs to recognize and express this function mapping: for $a, b \in A = \{0, 1, 2, \ldots 11\}$, $(a, b) \rightarrow (a + b)$[mod 12]. Our sense was that PPTs might be familiar with how to use an operation table to determine the outcome of an operation, but less familiar conceptualizing this operation table as a function, in part because it has an ordered pair (not a singular value) as its input.

After PPTs worked on the binary operation task, the instructor provided an elaboration on a group table based on the Western 12-tone music system (Fig. 16.4b). The example is a "Distance from *C*" operation, which combines the distance of two notes (+ or − from middle *C*), and which yields the note corresponding to that combined distance from middle *C*. In addition to conceptualizing this group table, the instructor also pointed out that some notes (such as *F* and *G*), combined with themselves many times, can create all 12 tones—effectively acting as generators of the group. The purpose in introducing these two specific examples (addition modulo 12 and distance from *C*) is the opportunity to discuss the structural similarities between them—they are isomorphic groups. The realization of this structural similarity leads to providing a definition of two groups being isomorphic. The definition of group isomorphism was scaffolded, beginning with a less formal

a

+	0	1	2	3	4	5	6	7	8	9	10	11
0	0	1	2	3	4	5	6	7	8	9	10	11
1	1	2	3	4	5	6	7	8	9	10	11	0
2	2	3	4	5	6	7	8	9	10	11	0	1
3	3	4	5	6	7	8	9	10	11	0	1	2
4	4	5	6	7	8	9	10	11	0	1	2	3
5	5	6	7	8	9	10	11	0	1	2	3	4
6	6	7	8	9	10	11	0	1	2	3	4	5
7	7	8	9	10	11	0	1	2	3	4	5	6
8	8	9	10	11	0	1	2	3	4	5	6	7
9	9	10	11	0	1	2	3	4	5	6	7	8
10	10	11	0	1	2	3	4	5	6	7	8	9
11	11	0	1	2	3	4	5	6	7	8	9	10

Addition modulo 12 operation table

b

Example. An Example from Music—Distance from C

Western music is based on a 12-tone system, namely the set of notes, $M = \{A, A\#, B, C, C\#, D, D\#, E, F, F\#, G, G\#\}$. Each note can be thought of in relation to middle C. We will define the operation $\oplus_C$ ("C-sum") to be combining the distances of two notes (+ or – in relation to middle C), which yields the corresponding note (that distance in relation to middle C).

$\oplus_C$	A	A#	B	C	C#	D	D#	E	F	F#	G	G#
A	F#	G	G#	A	A#	B	C	C#	D	D#	E	F
A#	G	G#	A	A#	B	C	C#	D	D#	E	F	F#
B	G#	A	A#	B	C	C#	D	D#	E	F	F#	G
C	A	A#	B	C	C#	D	D#	E	F	F#	G	G#
C#	A#	B	C	C#	D	D#	E	F	F#	G	G#	A
D	B	C	C#	D	D#	E	F	F#	G	G#	A	A#
D#	C	C#	D	D#	E	F	F#	G	G#	A	A#	B
E	C#	D	D#	E	F	F#	G	G#	A	A#	B	C
F	D	D#	E	F	F#	G	G#	A	A#	B	C	C#
F#	D#	E	F	F#	G	G#	A	A#	B	C	C#	D
G	E	F	F#	G	G#	A	A#	B	C	C#	D	D#
G#	F	F#	G	G#	A	A#	B	C	C#	D	D#	E

"Distance from *C*" operation table

Fig. 16.4 (**a**) Addition modulo 12 operation table. (**b**) "Distance from *C*" operation table

characterization ("...if they are the same groups but with different names of elements"), progressing to a middle-ground definition ("...if there exists a one-to-one function from G to H such that if $a \rightarrow x$ and $b \rightarrow y$, then $a * b \rightarrow x \cdot y$"), and ending with the more formal and standard definition ("...if there exists a one-to-one function ϕ from G to H such that $\phi(a * b) = \phi(a) \cdot \phi(b)$ for all $a, b \in G$"). PPTs were then tasked with identifying the isomorphic mapping ϕ between the two groups. Notably, this mapping is not a straightforward mapping between the two tables in Fig. 16.4 (0 maps to C, not A). Creating the mapping was intended to serve two purposes: (1) to reinforce the notion of a mapping between two sets that are more abstract; and (2) to discuss the preservation of the structure of the set and the operation across this mapping—i.e., the identity element maps to the identity element, generators map to generators, etc.

Stepping Down to Practice

After engaging with the abstract algebra, PPTs are asked to consider some mappings from secondary mathematics. That is, prior to stepping down to the teaching situation, PPTs are asked first to consider some ideas from secondary mathematics. The first problem was to identify a mapping depicting the "derivative" relationship (i.e., some function $D : f \rightarrow f'$). At this stage, we expected PPTs to be more familiar with functional mappings between abstract sets of objects (e.g., differentiable functions), expressing such mappings in a variety of ways, and describing their properties (e.g., many-to-one). The second was an open-ended question for PPTs to come up with some other interesting examples of functions (mathematical and non-mathematical), which included a prompt: "You might consider, for example, the distance formula, $d = \sqrt{(x_2 - x_1)^2 + (y_2 - y_1)^2}$, as a possible function, or other topics studied in secondary mathematics." Again, the purpose here was for PPTs to recognize various examples of functions, some of which might be related to familiar secondary content, but which, traditionally, may not be thought of as a function. Drawing from the earlier prompt, some PPTs might view the distance formula as a function: $d : (x_1, y_1, x_2, y_2) \to \sqrt{(x_2 - x_1)^2 + (y_2 - y_1)^2}$. The third problem asked PPTs to consider $(\mathbb{R}, +)$ and $(\mathbb{R}^+, \times)$, and the function, $\phi : a \to e^a$, as one function that demonstrates these two groups as being isomorphic. PPTs were tasked with investigating an inverse function (for ϕ), $\varphi : a \to \ln(a)$, that would be a one-to-one mapping in the reverse direction. The exercise is structured so that PPTs recall the general isomorphism property ($\ln(ab) = \ln(a) + \ln(b)$), identify the identity elements of each set (which necessitates the mapping, $\ln(1) = 0$), and prove that $\ln\left(\frac{1}{a}\right) = -\ln(a)$. PPTs were to recognize that because logarithms represent exponents, multiplying in the first set is equivalent to adding in the second. The questions allowed PPTs to revisit mathematics that they are already familiar with and reflect on relationships they may not have had the machinery to describe prior.

Finally, returning to the initial secondary teaching situation, PPTs were asked to discuss, again, the definitions, examples, and key ideas that they (now) reflect on when considering how they might prepare to teach an introductory unit on functions. We anticipated that new function examples would be more abstract (i.e., include non-numerical, multivariable, and/or non-mathematical examples). We further anticipated that PPTs would have a stronger emphasis on and recognition of a function's domain, a function's outputs, and a function's utility and potential application for thinking about structural preservation.

Module 5: A k-Product Property

Module 5 was designed around some of the properties of rings and fields in abstract algebra. In particular, the module aimed to consider the role of the additive identity, 0, in rings and fields and the implications this element has for solving equations. This includes multiplication by 0, the zero-product property, and that these properties are unique to 0 (i.e., there cannot be a k-product property for $k \neq 0$). The solving of equations in arbitrary rings and fields is a generalization of the solving of equations in secondary mathematics, which often presumes the field of real numbers—an example of a generalization connection. The overarching premises of the module were that many secondary teachers would not recognize the inherent structure in nonstandard student approaches and, when confronted with nonstandard approaches, would not clarify the implicit assumptions and limitations in students' statements or arguments, instead proposing approaches that do not build on student reasoning.

Building Up from Practice

To begin the module, PPTs are given a classroom situation (depicted in Fig. 16.5) in which a student, Evan, has incorrectly solved a quadratic equation.[2] They are prompted to "Discuss Evan's solution approach. What would you do to help Evan avoid this approach in the future? Explain your reasoning." The anticipated response was that PPTs would recognize that Evan cannot solve the equation this way (with 4 instead of 0) and get the correct answer. We also anticipated that PPTs would not justify why that was the case, nor recognize Evan's approach as an (incorrect) alteration of the zero product property. We anticipated that their way to help Evan would be to tell him to multiply the binomial and set the equation equal to zero before factoring—which constitutes a more standard approach. However, such an approach does not build on Evan's reasoning, nor does it discuss why one of Evan's solutions ended up being correct.

After PPTs considered the teaching situation, they were asked to find solutions to various modifications of a quadratic equation—but on $\mathbb{R}$, $\mathbb{Z}$, and $\mathbb{N}$. The purpose in doing so was to emphasize the importance of the domain on whether quadratic equations had 0, 1, or 2 solutions.

Abstract Algebra

The abstract algebra portion began by defining a ring and defining a field (each with a few relatable examples and non-examples). One of the examples was to determine whether addition modulo 9 and multiplication modulo 9 is a ring or a field (or both). A definition for a ring was provided in the module, and a discussion of the many intermediate stages between a ring and a field (ring with unity, commutative ring, etc.) was omitted to avoid distracting from the module's primary goal. The bulk of the abstract algebra content consisted of a sequence of four proofs about the

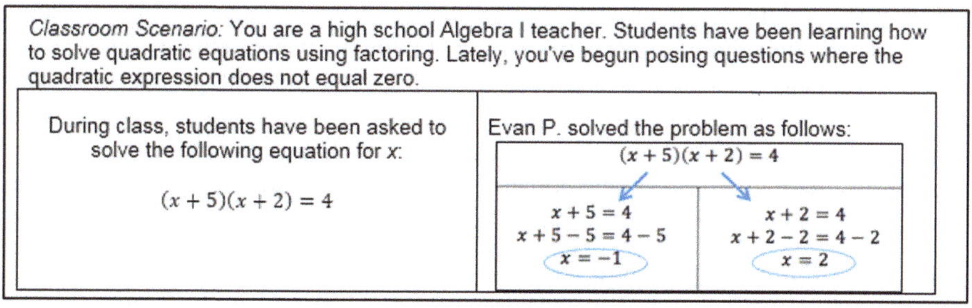

Fig. 16.5 Teaching situation for the *k*-Product property Module

[2]Heid, Wilson, and Blume (2015) discuss a similar teaching situation prompt.

properties of rings and fields—ones that particularly consider the role of the additive identity, 0. The first two are reasonably familiar and are as follows:

Theorem 1 (Multiplication by Zero) In a ring $R(+, \times)$, $0 \times b = 0$ (and $b \times 0 = 0$) for all $b \in R$.

Proof $0 \times b = (0 + 0) \times b = 0 \times b + 0 \times b$ (additive identity and distributive properties). Since $R(+)$ is a group, the cancellation laws hold for addition, which yields: $0 = 0 \times b$. The second proof, that $b \times 0 = 0$, is entirely similar to the first. QED.

Theorem 2 (Zero Product Property) In a field $F(+, \times)$, $\forall a, b \in F$, if $a \times b = 0$, then $a = 0$ or $b = 0$.

Proof Let $a, b \in F$, and let $a \times b = 0$. If $a = 0$, then we are done, so suppose $a \neq 0$. Then $a' \in F$ (with a' being the multiplicative inverse of a). Since $a \times b = 0$, then $a' \times (a \times b) = a' \times (0)$, which means $(a' \times a) \times b = a' \times 0$ (associative property) and $b = a' \times 0$. And by Theorem 1 (which holds in fields as well as rings) we get $b = 0$ (which means $a = 0$ or $b = 0$). QED.

The rationale for these two theorems was to claim that, in a ring (and a field), if 0 is one of the factors, then the product will also be 0, and in a field (but perhaps not a ring), whenever the product is 0, it must be the case that either a is 0 or b is 0—i.e., there are no other possible ways for a product to be 0. Both address the potential uses of 0 in solving equations. In the next task, PPTs were given the addition and multiplication modulo 9 tables (the ring, $\mathbb{Z}_9(+_9, \times_9)$) and were asked to solve the following two equations: (1) $3x = 0$; and (2) $(x + 5)(x + 2) = 0$. The purpose in doing so was to differentiate rings from fields, especially in relation to zero-divisors. That is, if $x = 0$ in equation (1), it is still the case that we know the product will be 0 (Theorem 1 holds), but it is no longer the case that we can be certain there are no other solutions (Theorem 2 does not hold). We anticipated that PPTs might initially miss some of the solutions, so we made sure to prompt them to "use the table" to help identify all solutions. The realization from solving equation (1) that there are nonzero solutions makes the solution method for equation (2) more difficult. This part requires finding the solution $x = 1$, which makes neither factor 0 but, instead, happens to make the two factors 6 and 3, for which the product is 0. That is, unlike the solution approach the PPTs were likely to suggest in the pedagogical situation to Evan, in this example, we cannot apply the zero-product property to reduce the quadratic to two linear equations. The third and fourth theorems, perhaps less familiar, solidify the uniqueness of the additive identity in these two properties.

Theorem 3 (No multiplication by k property for $k \neq 0$) In a ring, $R(+, \times)$, we know that $0 \times b = 0$ for all $b \in R$ (Theorem 1). In a sufficiently large ring (at least two elements), $R(+, \times)$, that property is *unique* to the additive identity 0—that is, there can be no nonzero $k \in R$ such that $\forall x \in R$ we have that $k \times x = k$.

Proof Suppose there were such a k, with $k \in R$ and $k \neq 0$, where $k \times x = k$ for all $x \in R$. Since $0 \in R$, then, by assumption, $k \times 0 = k$. But by Theorem 1, $k \times 0 = 0$. This is a contradiction. QED.

Theorem 4 (No k-product property for $k \neq 0$) In a field, $F(+, \times)$, we know that if $a \times b = 0$, then $a = 0$ or $b = 0$ (Theorem 2). In a sufficiently large field (at least four elements), $F(+, \times)$, that property is *unique* to the additive identity 0—that is, there can be no nonzero $k \in F$ such that $\forall a, b \in F$, if $a \times b = k$, then $a = k$ or $b = k$.

Proof Suppose there were such a k, with $k \in F$ and $k \neq 0$, where $\forall a, b \in F$ such that $a \times b = k$, then $a = k$ or $b = k$. Since F has at least four elements, $\exists a \in F$ such that $a \neq k$, $a \neq 1$, $a \neq 0$. So a' exists (since $a \neq 0$) and $a' \neq 1$ (since $a \neq 1$). Let $b = a' \times k$ (which means $b \in F$). Since $a' \neq 1$, we know $b \neq k$. However, $a \times b = a \times (a' \times k) = k$, but neither $a = k$ nor $b = k$. This is a contradiction. QED.

Stepping Down to Practice

The purpose in proving Theorems 3 and 4 becomes clearer as the PPTs return to considering equations on real numbers. PPTs were asked to give a statement of the zero-product property on $\mathbb{R}$, and then to discuss: "Could it be the case to have a "k-Product property" (for $k \neq 0$ and $k \in \mathbb{R}$) such that $\forall a, b \in \mathbb{R}$, if $a \cdot b = k$, then $a = k$ or $b = k$? Explain your reasoning and how it relates to the theorems proved." Essentially, we anticipated PPTs would recognize that since $\mathbb{R}$ is a sufficiently large field (with at least four elements), all of the preceding theorems are applicable. The aim of the exercise is to highlight the uniqueness of zero in a k-product property on a field, which means that the algebraic methods often used in high school cannot be used exactly as they are with any nonzero value, as Evan did.

As a final exercise, PPTs reflected on their original response to Evan and whether their response to Evan would be different in any way. The idea was that PPTs might recognize that Evan appeared to use a k-product property by setting $(x + 5) = 4$ and $(x + 2) = 4$. Now, if it were true that a product of 4 could only be achieved by having one of the other factors be 4, then this would be okay; however, this approach cannot work because the real numbers form a (sufficiently large) field. However, that is not to say that Evan's approach cannot be used—in fact, his approach identified one of the solutions. We anticipated, at this stage, that PPTs might point out why one of Evan's solutions was correct—which has to do with the fact that when $(x + 5) = 4$, it is also true that $(x + 2) = 1$, and thus $(x + 5)(x + 2) = 4 \cdot 1 = 4$. Indeed, PPTs themselves had no other choice but to engage in this sort of reasoning when they were solving $(x + 5)(x + 2) = 0$ with modulo 9 arithmetic. In fact, in Evan's situation, the equation could be solved by setting the two factors of the expression equal to two factors of 4 (whose product is 4). Unlike the zero-product property solution, however, this must be understood as the conjunction of two equations, and there is no guarantee that such a solution to this system of equations exists (as when the two linear factors are both set to equal 2). In a more general case, the solution may not even have integer factors, which means that such a method is not necessarily exhaustive. In sum, we anticipated that PPTs would emphasize the unique role that 0 plays in solving equations, and that the study of 0 in relation to properties of rings and fields might be influential for deepening their own understanding about various properties and approaches for solving equations.

Observing Changes in Secondary Teaching Responses

The design of five abstract algebra modules according to the instructional model was part of a larger research study. The larger study used a design research perspective (e.g., Cobb, Confrey, diSessa, Lehrer, & Schauble, 2003), treating participants' engagement with the materials as a focal point. In this chapter, we report on two students (PPTs) from a program in secondary mathematics teacher education and their participation in a total of five sessions, one for each module. One was a preservice teacher (Pam), the other an in-service teacher (Irene) with 5 years of experience teaching secondary mathematics (but not currently teaching). We collected and analyzed two sources of data: (1) a (transcribed) video-recording of PPTs' collaborative engagement with the materials, which also included all of their written work; and (2) an (transcribed) audio-recording of a post-teaching-experiment semi-structured interview.

In this chapter, we will focus our analysis on these two PPTs' responses to the teaching situation at the beginning of the module in comparison to their responses at the end of the module (not in comparison to our researcher-hypotheses about their developmental trajectory within the module, which was part of the larger study). In this regard, the instructional model, which began and ended with discussing a teaching situation, also served the research design. The analysis we focus on in this chapter developed in two stages. First, we characterized PPTs' responses to the secondary mathematics teaching situation at the beginning and end of each module (referred to as *characterizations of initial/end teaching response*, respectively). Each characterization of their teaching responses—either the initial or the end—was supported by several instances in the transcript in which the PPTs attended to this aspect in their responses. These characterizations allowed us to identify observed differences (or similarities) in PPTs' responses pre- and post-module to the secondary teaching situation. We refer to these differences as *Δ-teaching characterizations*. Second, for each *Δ-teaching characterization,* we then analyzed PPTs' engagement with all phases of the module and their post-interviews, in order to identify instances in which their thinking appeared to shift in relation to the *Δ-teaching characterization*. That is, the second phase of analysis intended to reveal why such instructional changes may have transpired in relation to the module—although we were not always able to locate such shifts from their work during the module. We aimed to triangulate as much evidence as possible to support our answers, which included the following sources: (1) PPTs' self-reports during the post-interview; and (2) PPTs' dialogue and written work on different tasks from the module. For each instructional change (i.e., *Δ-teaching characterization*), we also considered what phases of the module—building up, abstract algebra, stepping down—appeared to be most associated with the change. Third, we then looked at all the changes identified across all five modules and used thematic analysis (e.g., Creswell, 2012) to characterize the different types of instructional changes we observed.

Findings from the Functions Module

In the Functions module, PPTs responded to a dialogue between a student and a teacher about the meaning of function, and in the process, were also asked to identify the definition(s), examples, and key ideas they would emphasize in introducing a secondary unit on functions. From the first stage of the analysis, we identified five Δ-teaching characterizations based on their initial and end responses to the teaching situation. (Table 16.1 provides a summary.) Consider, for example, the second Δ-teaching characterization, "Emphasize more abstract examples of function, not just numerical examples." In their initial response, the PPTs' examples were pictorial mappings, which demonstrated the idea of uniqueness with an example from $\mathbb{R} \to \mathbb{R}$, $\{(1, 1), (2, 2), (3, 3)\}$, and a non-example from $\mathbb{R} \to \mathbb{R}$, $\{(1, 1), (2, 2), (3, 3), (3, 2)\}$. Further examples included a step function (from $\mathbb{R} \to \mathbb{R}$), which was a function, but "if you had two closed circles (at the same x-value), then it would not be a function." Their initial teaching response was characterized by their use of examples of functions from $\mathbb{R} \to \mathbb{R}$. Their end response was different:

Researcher:	Uh, talk about, maybe some of the things you might do, uh, definitions and examples you might use or see with students like Evan.
Pam (talking to Irene):	So, the birthday?
Irene:	Yeah, and I liked, I liked the piano, or anything that's, you know, not so mathy [sic], I guess.

The PPTs went on to discuss other examples of functions they would use, all of which were examples that used more abstract sets of objects (not just real numbers) in the mapping. Each *Δ-teaching characterization* (in Table 16.1) was supported by several such instances in the PPTs' dialogue and written work.

In the second part of the analysis, we triangulated from all available sources to attempt to understand more about why such instructional changes may have transpired. Consider again the second Δ-teaching characterization. While discussing examples at the end of the module (e.g., "birthday" and "piano" mappings), PPTs said the following:

Researcher: The birthday, piano, real world, so why that?
Pam: I just think they help them connect, like, what the idea of a function is.
Irene: Yeah, and sometimes I feel that in math, you have to do...

Table 16.1 Characterizations of PPTs' responses to the teaching situation for the Function Module

Characterizations of initial teaching response	Characterizations of end teaching response	Δ-Teaching characterizations
Give "every input has unique output" definition of function	Give two-part definition sequence: (1) define relations; and (2) "every input has unique output" definition of function	Δ1. Use relation definition to introduce function definition
Examples of functions were numerical (from $\mathbb{R} \to \mathbb{R}$)	Examples of functions were abstract (from more abstract sets, $A \to B$)	Δ2. Emphasize more abstract examples of function, not just numerical examples
Emphasized multiple representations of functions (e.g., equations, tables, graphs)	N/A	Δ3. Did not emphasize multiple representations
Emphasis on uniqueness aspect of function	Emphasis on uniqueness and mapping aspect of function	Δ4. Additional emphasis on mapping aspect of function
N/A	Have students look at functions via isomorphisms	Δ5. Discuss isomorphisms and their properties

Pam: Only numbers.
Irene: Yeah, like add, subtract, multiply, and divide, yeah, numbers, and . . . there's this idea a function is outside of just add, subtract, multiply, and divide . . . it helps identify the idea that the function is just some mapping we describe by however we want.

During the interview after the module, we probed further into some of their thinking. Here, they mentioned part of the rationale for doing so was "just to give [students] other examples of things that are functions, besides what we traditionally talk about in an algebra classroom." The primary activity in the module the PPTs identified during the interview as productive was the binary operation table task during the abstract algebra portion: "That's where I feel we, at least for me, I turned the corner about thinking about a function outside of just some linear situation . . . The fact that your domain can actually be an ordered pair . . . " Their work on this task indicated four important conceptual shifts—equation-view, mapping-view, multivariable-view, and dependent-view—as the PPTs came to view the binary operation table as a function (Wasserman, 2018b). In the interview, Irene indicated the following:

Irene: I have other examples of things that are functions now that I didn't have before . . . maybe there's another mathematical thing that I could show them outside of the traditional $y = x + 3$ . . . that is a function that's not normally something we would talk about as a function.

One such example the PPTs discussed was the quadratic formula, i.e., the function, $(a, b, c) \to \left(\frac{-b+\sqrt{b^2-4ac}}{2a}, \frac{-b-\sqrt{b^2-4ac}}{2a}\right)$, which they proposed previously

Table 16.2 Aspects of the Function Module influencing each Δ-teaching characterization

Δ-Teaching characterizations	Building up from practice tasks	Abstract algebra tasks	Stepping down to practice tasks
$\Delta 1$. Use relation definition to introduce function definition		– Sequence used for defining function	
$\Delta 2$. Emphasize more abstract examples of function, not just numerical examples		– Binary operation table task – "Piano" (i.e., "Distance from *C*") example	– Identify other interesting examples of functions in secondary mathematics
$\Delta 3$. Did not emphasize multiple representations	*Not in relation to module; identified as a consequence of* $\Delta 2$		
$\Delta 4$. Additional emphasis on mapping aspect of function		– Binary operation table task	– Identify other interesting examples of functions in secondary mathematics
$\Delta 5$. Discuss isomorphisms and their properties		– Isomorphism mapping between "Piano" example and addition modulo 12	

when asked to identify interesting examples of functions in secondary mathematics. Thus, their work on the abstract algebra and stepping-down-to-practice portions of the module broadened their example-space of functions in ways that appeared to influence the examples they would use with students. Table 16.2 provides a summary of why each Δ-teaching characterization appeared to have transpired in relation to the different aspects of the module.

Findings from the k-Product Property Module

In this module, PPTs responded to a particular student's solution approach for solving a quadratic equation and indicated what they might do to help that student adopt mathematically appropriate approaches. From the first stage of analysis, we identified three Δ-teaching characterizations (see Table 16.3). Consider, for example, the first Δ-teaching characterization, "Emphasize justification of ZPP approach." In their initial response, PPTs were uncompromising; they felt Evan's method—which did not utilize the ZPP—could not possibly be a reasonable means of approaching the question. However, PPTs had no argument justifying their choice to use the ZPP besides operating out of habit:

Table 16.3 Characterizations of PPTs' responses to the teaching situation for the *k*-Product Property Module

Characterizations of initial teaching response	Characterizations of end teaching response	Δ-Teaching characterizations
Emphasize identifying approaches as correct (ZPP) or incorrect (student)	Emphasize justifying ZPP approach as more efficient	$\Delta 1$. Emphasize justification of ZPP approach
Emphasize the student's misunderstanding of a procedure (ZPP)	Emphasize the student's procedure being a generalization of the ZPP to nonzero numbers	$\Delta 2$. Emphasize uniqueness of zero in ZPP approach
Change Evan's response to the more normative ZPP approach	Use Evan's approach to compare and contrast his solution method with more normative ZPP approach	$\Delta 3$. Leverage student's approach to compare and contrast potential solution methods

Pam (talking to Irene): So…for example…[Evan] didn't factor anything. It's already there for him, but…the whole idea that when you're solving, you use the zero-product property."

Later, their discussion continued:

Irene (talking to Researcher): [Pam] said that we could point out the [ZPP] and why it works…if one [of the factors] is zero, it doesn't matter what the other one is."

In sum, their initial teaching response did not explore the student's chosen method or provide justification for why their method was better. However, their end response revealed a newfound awareness:

Pam (talking to Irene):	But I think where you're even saying that you had the semi-ah-ha moment [was when] you had a concrete example [solving equations in ring, $\mathbb{Z}_9(+_9, \times_9)$] in front of you, you recognized that [the factors] could be equal to so many different things…and so, when we looked at this example again, we were like 'oh, we could set these factors equal to not only 4 and 1, but 1 and 4, negative 1 and negative 4, and so on.'
Irene:	…it just helped me think about it more carefully, maybe, than we thought about it the first time… like, wait, we could have all these other cases as well, and oh my goodness, this is very complicated.
Pam:	Right.
Irene:	And maybe there's an easier way, and, oh, why is 0 the easier way to do the problem?

The PPTs went on to detail how they would introduce and invalidate the notion of a "*k*-Product property" (for nonzero $k \in \mathbb{R}$), and then juxtapose the complexity of checking different numbers' factor-pairs with the simplicity of the zero-product approach.

In the second part of the analysis, we triangulated from three sources to attempt to understand more about why such instructional changes may have transpired.

Consider again the first Δ-teaching characterization. PPTs recognized their own development in this aspect:

Researcher:	Reflecting on the overall module, do you feel that... you learned something from this module?
Pam:	Yeah... normally you just teach ZPP and normally you don't explain why we use the ZPP, and now, I know that I should kind of show, like, a counter example. Like, you could use the *k*-Product property—well, not really, but like...
Irene:	A version of it....
Pam:	Right, but there are just so many... different factors that can work out, and it's hard to go through all of them, and that's why we use the ZPP.

The primary activities in the module that the PPTs identified as contributing to their understanding were found in the abstract algebra section, while exploring the differences between rings and fields:

Irene (talking to Pam):	Yeah, Evan. Cause remember you said he just memorized the rules, so now he thinks that if something times something is 4, then one has to be 4, the other has to be 4...
Pam:	That's only if you're in a ring. That could work if you're in a ring, right?
Irene:	It doesn't work on the reals. We know that.
Researcher:	Why not?
Pam:	Because it's a field.

It was from the exercises in this section that PPTs began realizing the importance of the uniqueness of the additive identity within a field, and the impact that it has when considering a product of real numbers. One example PPTs decided would be ideal for invalidating the (nonzero) *k*-Product property—and highlighting the weaknesses of *not* using the ZPP—was $(x + 5)(x + 2) = 48$; PPTs argued that students would be overwhelmed by the perceived tedium of checking for factor combinations that could possibly solve the equation. Likely, students also would only consider integer combinations (which are insufficient to solve the example). Although it was a variant of the one given in the classroom scenario, this example was designed and selected based on PPTs' experiences in the abstract algebra and stepping-down-to-practice sections of the module. Thus, we see the activities in the module as contributing to a deepening of PPTs' understanding about how quadratic equations are solved and, in particular, why they are solved frequently by way of the ZPP. Table 16.4 provides a summary of why each Δ-teaching characterization appeared to have transpired in relation to the module.

Synthesis Across Modules

Lastly, we share some of the findings across all five modules. In this analysis, we considered both *what* the characterization of the instructional change was, as well as *why* that instructional change appeared to have occurred. Of the 17 Δ-teaching characterizations we identified across all five modules, 15 appeared to have clear and specific reasons for the changes. (The remaining two were "coincidental"—that is, the difference identified appeared to be due to circumstantial aspects that were independent of the module and independent of other teaching changes.) For these 15, we characterized five different types of instructional changes the PPTs made in relation to the teaching situations. We elaborate briefly on each type of change.

Directly Incorporating from the Module

Some instructional changes ($n = 3$) were the PPTs *directly incorporating ideas from the module* into their teaching response. That is, after engaging with materials from the module, the PPTs demonstrated wanting to utilize these activities with their students. Consider, for instance, the "discuss isomorphisms and their properties" change from the Functions Module. After having created an isomorphic mapping between two groups during the abstract algebra section, the PPTs wanted to incorporate something like this into their own teaching of functions to secondary students. Two of these three changes were taking ideas from the abstract algebra content; the other was taking an idea that was present in activities from each of the three phases (building up from practice, abstract algebra, and stepping down to practice).

Table 16.4 Aspects of the k-Product property Module influencing each Δ-teaching characterization

Δ-Teaching characterizations	Building up from practice tasks	Abstract algebra tasks	Stepping down to practice tasks
$\Delta 1$. Emphasize justification of ZPP approach		– Solve linear and quadratic equations on the ring, $\mathbb{Z}_9(+_9, \times_9)$.	
$\Delta 2$. Emphasize uniqueness of zero in ZPP approach		– Use Theorems 3 and 4	– Discuss whether a k-Product property could exist on $\mathbb{R}$.
$\Delta 3$. Leverage student's approach to compare and contrast potential solution methods	*Not in relation to module; identified as a consequence of $\Delta 1$ and $\Delta 2$*		

Further Highlighting Content

Other instructional changes ($n = 3$) were the PPTs *further highlighting and emphasizing specific mathematical ideas* in their teaching response. That is, their initial responses indicated that they would include some specific mathematical idea in their teaching, but at the end, they placed significantly greater emphasis on this mathematical idea in their response. The "Additional emphasis on mapping aspect of function" from the Function Module was one such change. Here, their initial teaching responses included examples that incorporated the mapping aspect of function (e.g., $\{(1, 1), (2, 2), (3, 3)\}$); however, their examples were not intended to emphasize mapping, per se, but rather used as a means to emphasize the uniqueness aspect of a function. The idea of mapping was present in their initial response, but it was not emphasized. Their responses at the end, however, placed much more emphasis on the mapping aspect of function, at least in part due to the realization that the elements being mapped, themselves, can play a role in whether or not a mapping fulfills the uniqueness criteria of a function.

Further Highlighting Justification

Another instructional change ($n = 2$) was *further highlighting and emphasizing mathematical justification* in their teaching response. In these instances, although their initial responses made some attempt at justification, at the end, justification became a primary emphasis of the PPTs' response to the teaching situation. In the k-Product property Module, "Emphasize justification of ZPP approach" was one of the two examples. Here, while their initial responses made attempts to justify why the more normative ZPP approach was correct and why the student's approach was incorrect, their responses at the end placed much more emphasis on justification. In particular, their response at the end was not simply justifying why one was correct, but rather, justifying why one approach was more efficient in terms of determining solutions. That is, the justification changed somewhat, and, as a result, the notion of justification was much more prominent in their teaching response at the end.

Leveraging a Previously Unidentified Connection

In contrast to "further emphasis" on something (as in the previous two changes), other instructional changes ($n = 3$) *leveraged a previously unidentified connection* in their teaching response. That is, the PPTs' initial response did not incorporate some particular idea, but, at the end of the module, the PPTs sought to use that idea in their explanations. For instance, "Emphasize uniqueness of ZPP approach" was one such change. In their initial response, the PPTs did not explicitly recognize the student's solution approach as a variant of the zero-product property—they simply indicated the student's approach was incorrect and the ZPP was correct. At the end, the PPTs viewed the student's approach as trying to apply a k-Product property for a nonzero value of k. The module's emphasis on the unique role of zero in the field of real numbers contributed to this realization and its inclusion in their teaching response.

Consequences of Other Instructional Changes

Lastly, some ($n = 4$) teaching changes were not identified as being related to PPTs' work during the module (i.e., we did not identify shifts in their thinking about these Δ-teaching characterizations from their work on the module). Rather, these changes appeared to be *consequences of other instructional changes.* Consider, for example, "Did not emphasize multiple representations" from the Functions Module. This change was not identified in relation to particular aspects of the module. However, the examples the PPTs used for functions at the end of the module were characterized as being more abstract, not just functions from $\mathbb{R} \rightarrow \mathbb{R}$. Notably, more abstract examples of functions, such as the "piano" mapping, are not regularly associated with "multiple representations" in the way that we frequently associate, for example, graphs, tables, and equations in functions from $\mathbb{R} \rightarrow \mathbb{R}$. We do not typically see "equations" that capture these functional relationships or use "graphs" to depict them—although we could. Thus, since one of the instructional changes was to use more abstract examples, the non-emphasis on multiple representations was viewed as a potential consequence of this other change.

Discussion

In this chapter, we have looked at one approach to designing modules for an abstract algebra course with secondary mathematics teachers—building up from and stepping down to teaching practice. We shared in detail the primary components of each phase of the instructional model for two modules. The purpose was to provide specific examples of modules for use in an abstract algebra course, which were designed in accord with this instructional model. In addition, we have shared some of the findings from having two PPTs engage with and reflect on these materials. In this section, we elaborate on a few points of discussion.

The Instructional Model

The primary aim of the instructional model was to help make the study of topics in abstract algebra more relevant to the work of teaching secondary mathematics. Modules were designed accordingly. Evident from this study with two PPTs, this model has promise in terms of effecting change on secondary teachers' instructional responses. In four of the five modules, the PPTs indicated at least one way in which their response to the teaching situation would be different after having engaged with the content in the module.[3] This suggests that the model has promise in terms of making the study of advanced mathematics more pertinent to secondary teachers for their future work in teaching. And although there are certainly differences between responses to teaching situations and what PPTs actually might do as the teacher in a classroom, we regard having identified potential instructional changes at this stage as promising.

We elaborate on one facet of the model. As mentioned, the phases of the instructional model (i.e., "building up from a teaching situation" and then "stepping back down") were useful for research purposes. That is, the instructional model provided us with pre- and post-module data about PPTs' responses to a teaching situation for the purposes of comparison. However, although this was certainly the case, the actual purpose for this in the instructional model was about making connections to teaching, not research. By building up from practice, the instructional model aimed to set the stage for why the study of abstract algebra might be pertinent. The teaching situation at the beginning of the module was intended to be motivational; it also had the advantage of contextualizing the study of abstract algebra in relation to ideas about secondary mathematics. As Pam reflected after the Function Module: "[The teaching situation] put my brain in function mindset... Like, I just was thinking about functions, so everything that I know about functions was in my brain." Responding to the teaching situation again at the end of the module was intended to help the teachers think about how the abstract algebra content might be related to situations in teaching. That is, it aimed to help the PPTs consolidate their sense for how the abstract algebra content might affect their instructional practices. These aspects of the instructional model were intended to alleviate the three concerns identified in the literature: to make more explicit the links to secondary teaching, to acknowledge and address potential differences in abstract algebra and secondary mathematics, and to orient the goals of teachers by making the abstract algebra aims related to teaching situations.

The Abstract Algebra Modules

Designing five modules provided us with an opportunity to look more broadly at the instructional model; yet it also allowed us to gain specific insights about each module individually. Based on the two PPTs' interactions with and reflections on the content of the modules reported in this chapter, we have a better sense as to which aspects appeared to be particularly influential.

[3]In the one module where there were no Δ-teaching characterizations, the initial response from the PPTs incorporated the aims of the module, meaning that there was very little room for change in teaching; additionally, the alignment between the abstract algebra and secondary content was not as strong. Because of this, none of the five identified types of change were all that feasible.

Part of the larger research project aimed at studying and modifying the modules based on comparing PPTs' responses to our researcher-hypotheses. However, the analysis described in this chapter about the comparison of PPTs' initial/end teaching responses was also instructive. In the Functions Module, for example, although we were focused mostly on the abstract algebra content of isomorphisms, it was the binary operation table task that was the primary catalyst for PPTs' shifts in thinking. What we felt might be a relatively easy task turned out to be challenging—a place where the PPTs struggled, but in productive ways, and ultimately ended up deepening their sense of function. This analysis served a dual-purpose with respect to reflecting on, and revising, the modules. On the one hand, it provided insight into *how* particular instructional changes might be accomplished. For example, the emphasis on justifying the ZPP approach for solving quadratics not just as a "correct" approach but as an "efficient" one appeared to be related to solving equations on the ring, $\mathbb{Z}_9(+_9, \times_9)$. This abstract algebra activity necessitated a "look-at-each-factor-individually" approach to determine whether the resultant product matched. On the other hand, the analysis also identified *what kinds of instructional changes* might be reasonably related to particular abstract algebra activities. Solving equations on a ring with zero-divisors, such as $\mathbb{Z}_9(+_9, \times_9)$, is a widely used task in abstract algebra classes. Linking it to a discussion about the ZPP in terms of its efficiency in solving secondary quadratic equations would be one way to connect this activity to secondary teaching.

Types of Instructional Change

Of the five identified "types" of instructional changes, three of them seem to align with what many might regard as desirable goals for teaching in studying advanced mathematics. Having secondary teachers further highlight mathematical content and mathematical practices, such as justification, are clearly desirable results. That is, one purpose for studying advanced mathematics would be to help teachers identify and emphasize particularly meaningful mathematical ideas in their instruction. In addition, leveraging unidentified connections would similarly be desirable. We want teachers to study these advanced topics, which are new, and then have these new ideas shape what they attend to in their own teaching. As a part of studying these modules, we saw such changes in PPTs' instructional responses, which is encouraging. There were two other types of changes that were less expected. We elaborate on their significance for thinking about mathematics teacher education.

In their end-instructional responses, we saw the two PPTs from this study incorporate exact ideas from the modules. This is a phenomenon that we have observed before (Wasserman et al., 2018), wherein the utility of advanced mathematics is related to whether or not a teacher could "transport," as is, the advanced mathematics materials to their own teaching. Yet it still catches us off guard. Part of the problem is that when teachers take, in their exact form, materials that were intended to be part of abstract algebra instruction or secondary mathematics teacher education, then they are using materials with their students that were neither designed nor intended for secondary students. For example, the PPTs' impulse to incorporate isomorphic mappings as an example of functions seems to push the boundary for their students beyond secondary content. Such a result, while not inherently bad, may not be desirable. For us, this brings up two consequences. First, this tendency echoes a need to be more explicit with PPTs about the kinds of instructional changes that might be the aim from studying advanced mathematics. We, as a field and as teacher educators, need to articulate and exemplify the kinds of implications on teaching that would be especially beneficial. Second, this also indicates the need to acknowledge and address PPTs' beliefs and expectations about the relationship between advanced mathematics studies and secondary teaching, and, in particular, to help them transcend their transport model tendencies.

Finally, we saw some changes that were consequences of other instructional changes. In an analogy to chaos theory, this finding reminds us that in complex systems (such as secondary teaching), changes in one thing are often accompanied by unexpected changes somewhere else. For example, in PPTs' use of more abstract examples to exemplify the concept of function, we had not foreseen the potential consequence of a de-emphasis on utilizing multiple representations of functions. Of course, this is sensible in some regards, as abstract examples may be harder to capture in the same variety of representational forms. But it points out some unintended consequences that merit further exploration. For example, it made us consider whether we should explicitly incorporate multiple representations of abstract functions into the module—i.e., graphing them on a Cartesian plot of $A \times B$. Additionally, it made us appreciate further some of the novel mathematical examples (such as the quadratic equation), which frequently had multiple numerical inputs, rather than other abstract sets, because of the more immediate ability to write equations and draw a graphical mapping—of, say, points in 2-space to points in 1-space or 3-space, etc. However, while there were situations in which these unanticipated consequences may have been less than ideal, there were others in which they were positive. In various modules, PPTs had shown the proclivity to leverage students' approaches in their instructional responses. But in the k-Product Property Module, their initial response avoided this. It was only after they recognized the student's ideas in terms of efficiency that they were really able to then use and build off the student's approach in their instructional response. In this way, the module appeared to help the PPTs' understand and legitimize the hypothetical student's reasoning in this particular teaching situation. We regard such a change positively. In sum, changing instruction may have unforeseen consequences that are both positive and negative. This reminds us to think cautiously about the scope and design of teacher education, as it is teachers who are ultimately responsible for navigating the complex environments we call classrooms.

References

Ball, D. L., & Bass, H. (2009). *With an eye on the mathematical horizon: Knowing mathematics for teaching to learners' mathematical futures*. Paper presented at the 43rd Jahrestagung der Gesellschaft fur Didaktik der Mathematik, Oldenburg, Germany.

Ball, D. L., Thames, M. H., & Phelps, G. (2008). Content knowledge for teaching: What makes it special? *Journal of Teacher Education, 59*(5), 389–407.

Barnett, S. M., & Ceci, S. J. (2002). When and where do we apply what we learn?: A taxonomy for far transfer. Psychological Bulletin, 128(4), 612.

Bremigan, E. G., Bremigan, R. J., & Lorch, J. D. (2011). *Mathematics for secondary school teachers*. Washington, DC: Mathematical Association of America.

Cobb, P., Confrey, J., diSessa, A., Lehrer, R., & Schauble, L. (2003). Design experiments in educational research. *Educational Researcher, 32*(1), 9–13.

Conference Board of Mathematical Sciences (CBMS). (2012). *The mathematical education of teachers II*. Providence, RI and Washington, DC: American Mathematical Society and Mathematical Association of America.

Creswell, J. W. (2012). *Educational research: Planning, conducting, and evaluating quantitative and qualitative research* (4th ed.). Upper Saddle River, NJ: Pearson.

Deng, Z. (2007). Knowing the subject matter of a secondary-school science subject. *Journal of Curriculum Studies, 39*, 503–535.

Even, R. (2011). The relevance of advanced mathematics studies to expertise in secondary school mathematics teaching: Practitioner's views. *ZDM, 43*(6–7), 941–950.

Goulding, M., Hatch, G., & Rodd, M. (2003). Undergraduate mathematics experience: Its significance in secondary mathematics teacher preparation. *Journal of Mathematics Teacher Education, 6*(4), 361–393.

Heid, M. K., Blume, G. W., Zbiek, R. M., & Edwards, B. S. (1998). Factors that influence teachers learning to do interviews to understand students' mathematical understandings. *Educational Studies in Mathematics, 37*(3), 223–249.

Heid, M. K., Wilson, P. S., & Blume, G. W. (Eds.). (2015). *Mathematical understanding for secondary teaching: A framework and classroom-based situations*. Charlotte, NC: Information Age Publishing.

Klein, F. (1932). *Elementary mathematics from an advanced standpoint: Arithmetic, Algebra, Analysis* (E. R. Hedrick & C. A. Noble, Trans.). Mineola, NY: Macmillan.

McCrory, R., Floden, R., Ferrini-Mundy, J., Reckase, M. D., & Senk, S. L. (2012). Knowledge of algebra for teaching: A framework of knowledge and practices. *Journal for Research in Mathematics Education, 43*(5), 584–615.

Monk, D. H. (1994). Subject area preparation of secondary mathematics and science teachers and student achievement. *Economics of Education Review, 13*(2), 125–145.

Moreira, P. C., & David, M. M. (2008). Academic mathematics and mathematical knowledge needed in school teaching practice: Some conflicting elements. *Journal of Mathematics Teacher Education, 11*(1), 23–40.

Perkins, D., & Salomon, G. (1992). *Transfer of learning, International encyclopedia of education* (2nd ed.). Oxford, UK: Pergamon Press.

Rasmussen, C., Wawro, M., & Zandieh, M. (2015). Examining individual and collective level mathematical progress. *Educational Studies in Mathematics, 88*(2), 259–281.

Rowland, T., Huckstep, P., & Thwaites, A. (2005). Elementary teachers' mathematics subject knowledge: The knowledge quartet and the case of Naomi. *Journal of Mathematics Teacher Education, 8*(3), 255–281.

Sultan, A., & Artzt, A. F. (2011). *The mathematics that every secondary school math teacher needs to know*. New York, NY: Routledge.

Ticknor, C. S. (2012). Situated learning in an abstract algebra classroom. *Educational Studies in Mathematics, 81*(3), 307–323.

Usiskin, Z., Peressini, A., Marchisotto, E. A., & Stanley, D. (2003). *Mathematics for high school teachers: An advanced perspective*. Upper Saddle River, NJ: Pearson.

Wasserman, N. (2016). Abstract algebra for algebra teaching: Influencing school mathematics instruction. *Canadian Journal of Science Mathematics and Technology Education, 16*(1), 28–47.

Wasserman, N. (2018a). Nonlocal mathematical knowledge for teaching. *Journal of Mathematical Behavior, 49*(1), 116–128.

Wasserman, N. (2018b). Exploring the secondary teaching of functions in relation to the learning of abstract algebra. In A. Weinberg, C. Rasmussen, J. Rabin, M. Wawro, & S. Brown (Eds.), *Proceedings of the 21st Annual Conference on Research in Undergraduate Mathematics Education (RUME)* (pp. 687–694). San Diego, CA: RUME.

Wasserman, N., Fukawa-Connelly, T., Villanueva, M., Mejia-Ramos, J. P., & Weber, K. (2017). Making real analysis relevant to secondary teachers: Building up from and stepping down to practice. *PRIMUS, 27*(6), 559–578.

Wasserman, N., Weber, K., & McGuffey, W. (2017). Leveraging real analysis to foster pedagogical practices. In A. Weinberg, C. Rasmussen, J. Rabin, M. Wawro, & S. Brown (Eds.), *Proceedings of the 20th Annual Conference on Research in Undergraduate Mathematics Education (RUME)* (pp. 1–15). San Diego, CA: RUME.

Wasserman, N., Weber, K., Villanueva, M., & Mejia-Ramos, J. P. (2018). Mathematics teachers' views about the limited utility of real analysis: A transport model hypothesis. *Journal of Mathematical Behavior, 50*(1), 74–89.

Zazkis, R., & Leikin, R. (2010). Advanced mathematical knowledge in teaching practice: Perceptions of secondary mathematics teachers. *Mathematical Thinking and Learning, 12*(4), 263–281.

Chapter 17
Groups to the Rescue: Responding to Situations of Contingency

Rina Zazkis and Ofer Marmur

Does Advanced Knowledge Matter?

"Is there a need for advanced mathematics studies in the professional education and development of secondary school mathematics teachers?" (Even, 2011, p. 942). While we, as many of our colleagues in mathematics education, are convinced that a teacher's knowledge of mathematics beyond what is taught is essential for effective teaching, this idea is not widely shared by teachers. For example, in a study by Zazkis and Leikin (2010), secondary school teachers were asked to determine the extent to which they use their "advanced mathematical knowledge" (AMK) in teaching, where "advanced mathematical knowledge" was defined as the knowledge acquired in their tertiary education. Many teachers reported *rarely* using AMK in their teaching, or even regarded it as *non-essential.*

Wasserman (2016) described how knowledge of "advanced mathematics may positively impact instruction" (p. 29). "Advanced mathematics" here refers to topics beyond school curriculum, noting significant similarities in curricula across the world. Wasserman focused on topics and ideas of abstract algebra, and demonstrated not only how these are connected to school mathematics, but also how they can shape or alter the teaching of school mathematics. Arithmetic operations and their properties were among the identified topics for which abstract algebra can impact teaching. Explicit examples of planned teaching-activities were included, in which experience with abstract algebra transformed teachers' perceptions and, ultimately, their teaching.

More recently, Wasserman (2018) introduced the topological metaphor of *mathematical landscape*. He considered the *local mathematical neighborhood* to be

R. Zazkis (✉) · O. Marmur
Faculty of Education, Simon Fraser University, Burnaby, BC, Canada
e-mail: zazkis@sfu.ca

N. H. Wasserman (ed.), *Connecting Abstract Algebra to Secondary Mathematics, for Secondary Mathematics Teachers*, Research in Mathematics Education,
https://doi.org/10.1007/978-3-319-99214-3_17

the mathematics being taught, and the *nonlocal mathematical neighborhood* as consisting of "ideas that are farther away" (p. 118). He suggested that this division "intends to tackle the notion of mathematical knowledge *outside the scope* of what one teaches more directly" (p. 118). These metaphors are linked to the notion of *knowledge at the mathematical horizon* (Ball, Thames, & Phelps, 2008; Hill, Ball, & Schilling, 2008), focusing on teachers' horizons (rather than students' or curricular horizons), which Zazkis and Mamolo (2011) reconceptualized as advanced mathematical knowledge used in teaching.

Wasserman (2018) distinguished between three perspectives on the significance of exposure to advanced nonlocal mathematics: advanced mathematics for its own sake, advanced mathematics as connected to school mathematics, and advanced mathematics as connected to the teaching of school mathematics. He asserted that teachers' development of and understandings about nonlocal mathematics must not only relate to the *content of school mathematics* but to the *teaching of school mathematics content*. The premise is that exposure to advanced mathematics can help teachers acquire *key developmental understandings* (Simon, 2006), which affect content perceptions and mathematical connections, and, consequently, impact the teaching itself.

However, as noted, many teachers perceive their knowledge of advanced mathematics as irrelevant to their teaching practice. We believe that one of the reasons for this perception is lack of explicit connections between the ideas of advanced mathematics and—what Wasserman (2018) referred to as—the local mathematical neighborhood. By this we mean that such connections between advanced mathematical knowledge and school curriculum are rarely emphasized in university courses and are not self-evident to teachers. Zazkis and Leikin (2010) noted that even teachers who believe that their advanced mathematical knowledge is "extremely important" in teaching and is used "all the time," had difficulty in exemplifying where and how such knowledge is used. Rather, these teachers mostly provided meta-mathematical examples that emphasize the importance of aesthetic considerations in mathematical activity, the use of rigorous mathematical language, or student exposure to the history of mathematics. The advantage of AMK was seen as contributing to personal confidence in problem-solving, rather than connected to particular curricular content. While such an advantage is indisputably important, it describes only part of the potential benefits for teaching that stem from exposure to advanced mathematical knowledge.

AMK in Support of Situations of Contingency

In this chapter, we demonstrate how an understanding of advanced mathematical topics can be useful in instructional situations that do not involve advanced mathematics. However, while Wasserman (2016) concentrated on the use of AMK in *planning* for instruction, we supply an additional perspective by focusing on the benefits of AMK in *situations of contingency*. We borrow the notion of contingency from the Knowledge Quartet framework (Rowland, Huckstep, & Thwaites, 2005), in referring to classroom events that are unforeseen and therefore cannot be planned for. In the sections reported below, situations of contingency refer to real classroom events in which unexpected student questions, comments, or approaches arose. As suggested by Rowland et al. (2005), such events require teachers to be able to "think on their feet," respond to student inquiries, and be willing to deviate from the original planned agenda. We suggest that AMK can serve as a valuable resource in such unexpected situations, whether utilized consciously or subconsciously as a guide. While Rowland et al. described situations of contingency in school classrooms, we acknowledge that such situations also present themselves in the context of teacher education courses, as well as in undergraduate mathematics. Accordingly, this chapter focuses on situations exemplifying how introductory concepts of *group theory* can "come to the rescue" in various contexts of teaching.

Group theory possesses implications in a wide variety of domains, including geometry, physics, chemistry, and modern cryptography. It is closely related to the notion of symmetry, which, for example, is demonstrated by the use of group theory to describe and solve the well-known Rubik's Cube. Group theory has even served as the mathematical basis for the plot of one of the episodes of the TV show "Futurama." In this episode, a machine was built to switch between two people's minds, yet could only be operated once for the same two bodies, leaving them unable to switch back without the help of a theorem from group theory. However, even though groups relate to a wide range of interesting topics, these are not typically part of the school curriculum. Moreover, learning group theory is considered an extremely difficult topic for many undergraduate students, which in turn negatively impacts their attitudes towards abstract algebra (Dubinsky, Dautermann, Leron, & Zazkis, 1994; Weber & Larsen, 2008). Consequently, it is reasonable to assume that future teachers studying at university are not likely to focus on group theory as an important topic in support of their future career. That is, unless their attention is drawn to the contrary.

We focus on group theory as supporting instructional situations of contingency based on the *abstract* nature of groups. Dubinsky et al. (1994) claimed that "an individual's knowledge of the concept of group should include an understanding of various mathematical properties and constructions independent of particular examples" (p. 268). While abstraction is regarded as a source of difficulty, it nonetheless possesses the potential to make the algebraic construct relevant and applicable for a large variety of concrete examples and mathematical topics. As such, it is the abstractness of a group that may make it useful for teachers in a wide range of unexpected mathematical situations.

Responding to Situations of Contingency

In what follows, we exemplify several instructional situations in which teachers' knowledge of group theory can shape their responses to unexpected student questions or approaches. Furthermore, we demonstrate how task design can be subsequently informed or altered in order to enrich learners' understanding of the underlying mathematical concepts or ideas.

The first two sections ("Order of Operations" and "Superscript (-1)") are situated in mathematics education courses for prospective teachers. The subsequent three sections ("Seemingly Invertible Functions," "On Glide Reflection," and "Is 14/2 a Prime Number?") are situated in various undergraduate mathematics courses. Note that the meaning of "learners" varies according to the context, and refers to either school students, undergraduate students, or prospective teachers.

Order of Operations

It may appear surprising that a conversation on the order of operations in arithmetic can be related to abstract algebra. After all, the topic of order of operations belongs to the elementary school curriculum, where initially students learn that the operations of multiplication and division have priority over addition and subtraction, unless brackets suggest otherwise. The idea is later extended in middle and secondary school to involve exponents. The current discussion, however, focuses on the order of and relationship between division and multiplication.

Consider the following task given to a group of prospective teachers:

> Robin was given a worksheet with ten calculation problems, which involved multiplication and division only. The problems were in the spirit of the following examples, all alternating multiplication and division, where a, b, c, etc., were replaced by nonzero numbers:
>
> $$a \times b \div c$$
>
> $$a \div b \times c \div d$$
>
> $$a \div b \times c$$
>
> $$a \times b \div c \times d \div e \times f$$
>
> In each case, Robin performed division before multiplication, regardless of the order in which the operations appeared. This means, for example, that she treated the problem $a \times b \div c$ as if it were $a \times (b \div c)$, even though there were no brackets.
>
> What do you think of her solutions? How do you imagine you would have responded to such a scenario in class if you encountered this mathematical behavior from one of your students?

This task, presenting an imaginary situation of contingency, was designed based on prior explorations related to the order of operations with secondary school teachers (Zazkis, 2017) and with prospective elementary school teachers (Zazkis

& Rouleau, 2018). Following the reported disagreement among participants on the order in which operations should be performed, we formulated the task as a response to an approach carried out by a student.

The task was presented to several groups of prospective teachers (PTs), and the responses varied significantly. Some claimed that Robin's answers were all wrong. Others claimed that Robin's answers were all correct, assuming there were no computational errors. However, the majority of respondents suggested that the correctness of Robin's answers depends on the specific numbers in the computational exercises.

After providing initial responses individually, the PTs engaged in a conversation on the correctness of Robin's approach. The first type of claim some of the PTs conveyed was that: "You should do the operations in order of appearance if they are on the same level." The word "level" was explained in terms of hierarchy, where multiplication and division are on a higher level than addition and subtraction (see also Ameis, 2011). The counterclaim was: "If Robin is doing everything right and makes no mistakes, then her answers are correct." Upon elaboration, "doing everything right" referred to Robin's method of performing division before multiplication regardless of the order of appearance (without any calculation errors).

At this point, we would like to take a detour from the presentation of the classroom scenario in order to situate Robin's method in the discussion on order of operations. The method of giving priority to division over multiplication was unexpected for us and served as a situation of contingency (Rowland et al., 2005). This led to a reflection on the conventional order of operations, and, in fact, inspired the design of this task in the context of a teacher education course. However, the method may not appear strange to teachers in Canada, where the task was carried out. The reason for this is the acronym BEDMAS—interpreted as Brackets, Exponents, Division, Multiplication, Addition, Subtraction—that accompanies the teaching of the order of operations in many Canadian schools, and is used extensively by teachers. Though this acronym is not found in any of the recent textbooks we examined, the mnemonic is passed down from teacher to student as a memory-assisting device. The counterpart in the USA is PEMDAS—Parentheses, Exponents, Multiplication, Division, Addition, Subtraction. The phrase "Please Excuse My Dear Aunt Sally" is frequently invoked to assist students in memorizing the acronym, which is in turn used to assist in memorizing the order of operations. No such commonly used phrase exists for BEDMAS.

Typically, while there is a presumed "order of operations" for addition and multiplication (i.e., multiplication first, addition second), the same is not true for multiplication and division (i.e., we simply go left to right). Note that while Parentheses (P in PEMDAS) and Brackets (B for BEDMAS) are synonyms, the order of Division (D) and Multiplication (M) is switched in the two acronyms. This obviously does not imply there are different ways of performing sequences of arithmetic operations. However, both PEMDAS and BEDMAS *seem* to indicate an order—and, in particular, a *different* order for multiplication and division. In fact, it is not uncommon for learners to interpret the acronym as a reference to a "rigid" order, which leads to mistakes when multiplication is performed out of order before division (e.g., $10 \div 2 \times 5 \neq 10 \div (2 \times 5)$), as implied by PEMDAS (Glidden, 2008). However, what about division before multiplication as implied by BEDMAS? Does it also lead to mistakes when compared with the normative approach of going from left to right?

We return now to the classroom scenario and the significant disagreement amongst the PTs about Robin's approach. The disagreement on what is the correct or proper order of operations—where some PTs referred to the division-first order in BEDMAS and others preferred left-to-right order for operations on the same "level"—quickly turned to a search for examples and counterexamples. Examples of alternating multiplication and division expressions were sought and considered by those who supported Robin's approach, as well as those who rejected it. Each group attempted to find a counterexample, that is, a particular exercise in which one of the methods would lead to a mistake. PTs who claimed that the correctness of the approach depended on specific numbers were also engaged in considering multiple examples in an attempt to find a particular combination of numbers in which the two methods, left-to-right in order and division-first out-of-order, led to different results. The initial examples, such as $5 \times 10 \div 2$—in which both methods lead to the same result, i.e., $5 \times 10 \div 2 = 5 \times (10 \div 2)$ —were considered unsatisfying, and the search continued with examples involving longer sequences of computations, fractions, and negative numbers. This tendency was observed and described by Zazkis and Chernoff (2008), who noted that different counterexamples have a different convincing power in disputing mathematical claims.

After a long search, in which the occasional "Eureka!" assertions were dismissed as computational mistakes, no desired example was found. This led to a conjecture, which was initially extremely dissatisfying, that both approaches are correct.

With some prompting by the instructor, the conversation turned toward an attempt to understand and explain why both approaches lead to the same numerical result and to prove that this is indeed the case for all choices of numbers. Some participants in the discussion suggested that "it works" because "division is the inverse of multiplication." "It works" here referred to the fact that division-first out-of-order leads to the same result as by following the written order. However, based on a similar line of thought, this suggestion was followed by a "devil's advocate's" counterclaim that "multiplication is the inverse of division," and "it doesn't work." This is since multiplication-first out-of-order leads to an error. For example: $10 \div 2 \times 5 \neq 10 \div (2 \times 5)$.

Significant prompting was necessary to direct the PTs' attention to how multiplication and division differ from one another, and more specifically to the fact that multiplication, unlike division, is associative. In order to address the question at hand, it was also essential to reformulate the claim "division is the inverse of multiplication" into "division by K can be replaced by multiplication by $\frac{1}{K}$."

That is, the proof that both approaches for division are correct relies on several observations related to properties of real numbers. To elaborate,

$$(a \times b) \div c =$$

$$\left[\text{interpret division as multiplication by inverse (reciprocal)}\right]$$

$$= (a \times b) \times \frac{1}{c} =$$

$$\left[\text{invoke associativity}\right]$$

$$= a \times \left(b \times \frac{1}{c}\right) =$$

$$\left[\text{replace multiplication with division}\right]$$

$$= a \times (b \div c)\,.$$

The above proof does not make any explicit references to group theory, and could be achieved without the use of advanced mathematical knowledge. However, the above short and simple proof encompasses three properties that relate to the definition of a mathematical group: (1) a *single* operation (that is, thinking in terms of multiplication only, without division); (2) the existence of an *inverse* element; and (3) *associativity* of the operation. Notably, the division-first out-of-order approach draws on the associativity of multiplication; the multiplication-first out-of-order approach invokes the *non*-associativity of division, which explains why doing so can lead to an error. Accordingly, we argue that having such immediate access to knowledge of the algebraic structure of real numbers may aid a teacher in evaluating Robin's approach more quickly and easily. We emphasize that unifying multiplication and division into a single operation of multiplication highlights the abstract nature of the group structure. With this in mind, there is no ambiguity regarding the precedence of order with respect to division and multiplication.

While it is understandable that the mathematics taught at school does not allude to group theory when discussing order of operations, mathematics textbooks for teachers also typically address associativity without any explicit connection to groups. For example, in considering properties of operations with natural numbers, Musser, Burger, and Peterson (2010) mention that addition and multiplication are associative and commutative. These properties are further extended to rational and real numbers. However, the issue with associativity is that it is often mentioned in tandem with commutativity, since the basic four arithmetic operations possess either both or neither (Hadar & Hadass, 1981; Larsen, 2010; Zaslavsky & Peled, 1996). Exposure to group theory makes the associative property meaningful and disassociated from the commutative property.

The lengthy discussion that was needed to resolve the "division first" property suggests that attention to associativity was not in the teachers' active repertoire of knowledge. Zazkis and Rouleau (2018) and Zazkis (2017) have offered several theoretical explanations for this phenomenon. In particular, they described the prevalence of "met-before" ideas (Tall, 2013) and the resulting cognitive conflict when these ideas are challenged. Moreover, using Wasserman's (2018) terminology, it was noted that the notion of associativity, as an abstract idea, had not shaped teachers' understanding of associativity in the local mathematical neighborhood. In the context of the current chapter, the discussion of the "division first" property highlights how proficiency with properties of groups can indeed be helpful in unexpected instructional situations and in the evaluation of unusual student solutions.

Superscript (–1)

"$5^{-1} = \frac{1}{5}$, but f^{-1} refers to the inverse of a function f, and usually $f^{-1} \neq \frac{1}{f}$. So, have they run out of symbols, or what?"

This student question inspired an investigation into prospective secondary teachers' (PSTs) understanding of the appearance of (-1) as a superscript after a number or a function symbol. The detailed results of this investigation are reported in Zazkis and Kontorovich (2016) and Kontorovich and Zazkis (2017). Here, we provide a brief summary of the previously reported results, emphasizing the connection to group theory.

Mathematically, the symbol (-1) in a superscript points to an inverse. That is, a character followed by a superscripted (-1), such as in 5^{-1} and f^{-1}, represents an inverse element in a group structure. To recall, a group element (a^{-1}) is considered to be an inverse of another group element (a), if a binary operation ($\cdot$) involving the two elements results in an identity element (e). Symbolically, this relationship is given as $a \cdot a^{-1} = a^{-1} \cdot a = e$. However, the two cases, 5^{-1} and f^{-1}, differ in terms of the set and the binary operation. For the set of rational numbers without zero, the implied binary operation is multiplication; whereas for the set of bijective functions, the binary operation is the composition of functions.

Although we have used the term "inverse" so far (with respect to different operations), a common terminology in English is to refer to 5^{-1} as the "reciprocal" of 5. Zazkis and Kontorovich (2016) suggested that this difference in terminology is unhelpful for teachers in realizing the common features of the two cases. In fact, when engaged in a task around the student question regarding the same symbol use for the two cases, the majority of the PSTs highlighted the *differences* between the two cases. Their central claim was that it is the *context* that determines whether the exponent of (-1) should be interpreted as inverse (in the case of functions) or reciprocal (in the case of numbers).

On the other hand, there were several PSTs who highlighted the notion of "opposite," "undoing," or "returning to the starting point," when explaining the common features of 5^{-1} and f^{-1}. Though not explicitly stated by the PSTs, this notion is related to the property of an inverse element of an element in a group, where the operation between the two results in the identity element.

In line with our previous arguments, we suggest that knowledge of group theory may not be essential for recognizing the similarity in the two appearances of the exponent (-1). However, understanding the theoretic group notion of "inverse element with respect to an operation" brings the discussion to a higher level of abstraction, where different ideas exemplify the same structure. This in turn helps in reconciling the "curious appearance" of the symbol (as referred to by Zazkis & Kontorovich, 2016). Moreover, it helps in addressing a situation of contingency raised by a student question, or even deviating from the agenda by initiating a conversation with the learners on the chosen symbol in the context of the inverse property.

Seemingly Invertible Functions

Could there be functions that appear to be the inverse of each other, yet are not? The following case is presented in two parts. The first part describes an original formulation of a task in an undergraduate Real Analysis course, which evoked a situation of contingency caused by an unexpected student question. The second part describes the adaptation made to the lesson design for new groups of undergraduate students, which in turn, triggered new situations of contingency as a result of new student questions. In both the original and adapted version of the task, basic properties of group theory served as a guide for the instructor's responses and as a basis for further mathematical inquiry. While Wasserman (2017) has explored teachers' understanding of inverse functions, as related to the group structure, we here demonstrate how group structure guides an instructor of undergraduate mathematics in a lesson focused on properties of functions.

While different mathematical sources suggest various definitions for an *invertible function* (with respect to function composition), the following definition was regularly used in the aforementioned course:

Let $f : D \to E$. We say f is invertible
if there exists a function $f^{-1} : E \to D$
such that $f^{-1}(f(x)) = x$ for every $x \in D$
and $f(f^{-1}(y)) = y$ for every $y \in E$.

One can find equivalent—though differently formulated—definitions for an invertible function, such as the following definition given by Devlin (1981, p. 52):

Let $f : A \to B$. We say f is invertible
if there exists a function $g : B \to A$
such that for all $a \in A$, $b \in B$,
$f(a) = b \iff g(b) = a$.

We emphasize the fact that the domain and codomain are an integral part of these equivalent definitions, meaning that the invertibility property of a function relies on the chosen sets for the domain and codomain. We further note that according to the above definition(s), a function $f : A \to B$ is invertible if and only if it is both injective and surjective (see Devlin, 1981).

Part 1: The Original Task

Consider the following task:

Let $f, g : \mathbb{R} \to \mathbb{R}$ be functions such that $f(g(x)) = x$ for every $x \in \mathbb{R}$.

1. Prove that $g(x)$ is injective (one-to-one)
2. Prove that $f(x)$ is surjective (onto)

The above task was given as part of an undergraduate Real Analysis lesson on the topic of functions. It was presented after the students had already encountered explicit examples of injective and surjective functions, as well as non-examples of functions that did not fulfill the conditions of injectivity or surjectivity. Additionally, the students had already worked with invertible functions at that stage of the lesson (i.e., both injective and surjective). The teaching goal of the presented task was to have the students prove injectivity and surjectivity when the functions are given in an abstract manner and not by an explicit formula.

When faced with the first part of the task, one of the students wondered: "What is there to prove? Doesn't the condition $f(g(x)) = x$ for every $x \in \mathbb{R}$ already imply $g(x)$ is *invertible* and hence *one-to-one*?" This feeling that there is no "real work" required in the problem was echoed by other students, and similar arguments were also heard regarding $f(x)$ in the second part of the task. Essentially, the students considered the functions $f(x)$ and $g(x)$ to be each other's inverses.

From a purely mathematical perspective, the question at hand is whether it is sufficient to verify only one-sided composition in order to determine that a function is invertible. In other words, do we really need to verify the opposite-order composition $g(f(x)) = x$ for every $x \in \mathbb{R}$ *in addition* to the given condition that $f(g(x)) = x$ for every $x \in \mathbb{R}$? The answer to this question is "yes, we do," as explicitly exemplified in Part 2 below. Nonetheless, referring to the group definition may already serve as a guide for the answer. In a group, for every element $a \in G$ there exists an element $b \in G$ (denoted by a^{-1}) such that $a \cdot b = b \cdot a = e$. Meaning that, by *definition*, it is required for an inverse element in a group to satisfy *both* orders of operation $a \cdot b$ and $b \cdot a$, and hence it is required to verify them both. Note that verifying only one-sided condition $a \cdot b = e$ is sufficient in commutative groups; however, function composition is not commutative.

In fact, upon close examination, the set $A = \{f : \mathbb{R} \to \mathbb{R} \mid f \text{ is invertible}\}$ with the operation $\circ$ (composition) is indeed a group. We acknowledge that when faced with a situation of contingency like the one described above, an instructor does not necessarily need to carry in mind the whole perception of the set of invertible functions as a group. We accept that such overarching connections take time to be built. Nonetheless, an internalized recognition of the existence of an algebraic structure of mathematical objects (including objects outside the algebraic domain) may aid teachers in their thought process. In this case, considering an inverse function as an instance of an inverse element in an abstract group structure, can serve as a guideline for clarification and subsequent task design, as detailed below.

Part 2: Changes in the Task Design

The classroom situation described above led to instructional adjustments in the task design. In the following semester, the task was modified to include an additional third part:

3. Prove or Disprove: $g(x)$ is Invertible

The rationale underlying this change was to give educational space for the question that previously seemed present in the students' minds. Furthermore, the educational-mathematical goal was to enable student exploration of whether one-sided composition is sufficient for determining that a function is invertible.

It should be noted that the common example of $g(x) = \sqrt{x}$ and $f(x) = x^2$ does not meet the requirements of the task. While it is true that $f(g(x)) = \left(\sqrt{x}\right)^2 = x$, this only holds for all $x \geq 0$, and not for all $x \in \mathbb{R}$ as stipulated in the task. Furthermore, the function $g(x)$ itself is defined only for $x \geq 0$, while the given condition $g : \mathbb{R} \to \mathbb{R}$ implies the domain of $g(x)$ to be all real numbers.

The counter-example ultimately arrived to in class was:

$$g(x) = 2^x, \quad f(x) = \begin{cases} \log_2 x, & x > 0 \\ 1, & x \leq 0 \end{cases}.$$

Notice that $g(x)$ is not invertible as a function from $\mathbb{R}$ to $\mathbb{R}$ since it is not surjective (the codomain $\mathbb{R}$ does not equal the image of the function, which is $(0, \infty)$). Also $f(x)$ is clearly not invertible because it is not injective. Nevertheless, $f(g(x)) = f(2^x) = \log_2 2^x = x$ for all $x \in \mathbb{R}$.

However, the new task design invited new student questions that were not raised in the former lesson. After having finished the solution of all three parts of the task, one of the students asked: "If we were given that $g(x)$ was *invertible*, in addition to the given condition $f(g(x)) = x$ for every $x \in \mathbb{R}$, would *that* now imply that $g(f(x)) = x$ for every $x \in \mathbb{R}$? (See also Koichu, 2008, for a discussion on how situations of "if not, what yes?" promote meaningful learning opportunities for students.)

Also in this new situation of contingency, an understanding of group properties can "come to the rescue" in order to determine that the student's suggestion is correct. A theorem in group theory (that typically appears in textbooks immediately after the definition of a group) asserts that an inverse element is *unique*. Meaning that for a given element g, if an element f satisfies $fg = gf = e$, then necessarily

$f = g^{-1}$. In relation to the student question, the uniqueness property of the inverse element guided the instructor in responding to the unexpected student question. The explanation of uniqueness from group theory can easily be communicated as applicable to the situation explored in class: $f(x) = f(g(g^{-1}(x))) = g^{-1}(x)$ for all $x \in \mathbb{R}$ (which shows that f is the inverse of g).

This, however, was not the only new question raised as a result of the changes in the task design. An additional student question was: "But why isn't $g(x)$ invertible? I remember learning that exponential functions such as $g(x) = 2^x$ are invertible, and that the inverse is $\log_2 x$." In relation to this question, another student inquired: "Why is $g(x) = 2^x$ a function from $\mathbb{R}$ to $\mathbb{R}$, and not from $\mathbb{R}$ to all *positive numbers*?"

Indeed, if we consider $g(x) = 2^x$ as a function $g : \mathbb{R} \to (0, \infty)$ then $g(x)$ is not only injective but also surjective, and hence invertible (the inverse being $\log_2 x$). Based on our personal experience as teachers, we find that students typically find the distinction between $2^x : \mathbb{R} \to \mathbb{R}$ and $2^x : \mathbb{R} \to (0, \infty)$ to be puzzling at the very least, if not even an incomprehensible and artificial distinction. We suspect the underlying reason to be insufficient instructional attention given to the role of the *codomain* of a function (i.e., the set of *possible* values the function may obtain), and how this differs from the *image of the function* (i.e., the set of *actual* values obtained by the function). We recall that any one-to-one function can be made invertible when limiting the codomain to be the image of the function. Therefore, lack of explicit attention into how the codomain is defined in a specific problem might imply that students would inaccurately regard any injective function as bijective, i.e., invertible. Furthermore, an additional contributing factor to possible student confusion may be the inconsistent terminology used in relation to functions. We found that the term *range* is sometimes used to refer to the codomain of a function (e.g., Childs, 2009, p. 140), whereas at other times it is identified with the image of the function (e.g., Dummit & Foote, 2004, p. 2; Stewart, 2008, p. 11). This ambiguous terminology further blurs the much-needed distinction between the different terms.

The formulation of the task, in which a generic function is defined to have the same set as a domain and codomain, in this case $\mathbb{R} \to \mathbb{R}$ (as is also the case in many textbooks), can also be explained when considering group properties. That is, such a definition ensures *closure* under an operation (composition), as well as attributes to *inverse functions* the same meaning as *inverse elements* (in the group $\{f : \mathbb{R} \to \mathbb{R} \mid f \text{ is invertible}\}$ with functional composition). Concretely speaking, in the above counterexample of $g(x) = 2^x$, it is perhaps not surprising that students restrict the codomain to $(0, \infty)$, since g belongs to a well-known family of functions—exponential functions—with well-known inverses of logarithmic functions. However, it is possible to construct an example in which a similar restriction does not apply. For example, consider the additional counterexample in Fig. 17.1 for $f(g(x)) = x$ not implying invertibility of g.

In the example in Fig. 17.1, of a less familiar function $g(x)$,

$$g(x) = \begin{cases} x & -1 \le x \le 1 \\ x+1 & x > 1 \\ x-1 & x < -1 \end{cases},$$

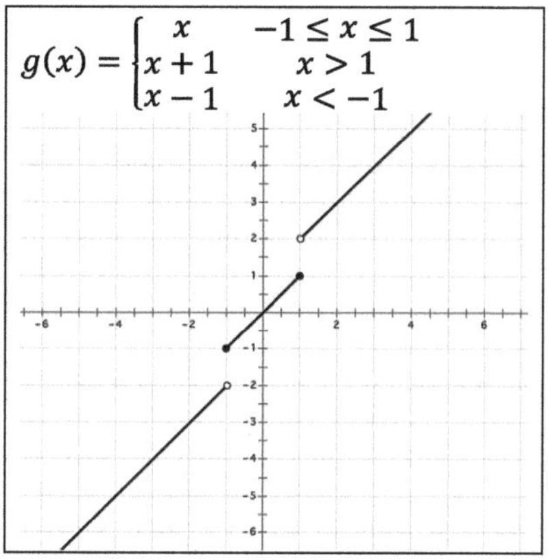

Fig. 17.1 Another counterexample for ***f*(*g*(*x*)) = *x*** not implying invertibility of ***g***

the image of the function is not a single interval (such as in the previous example), but rather the union of several disjoint intervals. Accordingly, we believe it is more likely that students would consider $g(x)$ to be a function from $\mathbb{R}$ to $\mathbb{R}$, rather than from $\mathbb{R}$ to $(-\infty, -2) \cup [-1, 1] \cup (2, \infty)$. Especially given that the former option seems to be the simpler choice, not requiring any computations. As such, $g(x)$ is injective and not surjective, and therefore not invertible.

In order to construct $f : \mathbb{R} \to \mathbb{R}$ that satisfies the condition $f(g(x)) = x$ for every $x \in \mathbb{R}$, we need to "undo" what the function g has done for all elements in the image of g. Additionally, for every $x \in \mathbb{R}$ that is not in the image of g, we may simply assign any random value (for example: 17). We thus obtain the following function:

$$f(x) = \begin{cases} x & -1 \le x \le 1 \\ x - 1 & x > 2 \\ x + 1 & x < -2 \\ 17 & 1 < x \le 2 \text{ or } -2 \le x < -1 \end{cases}.$$

Notice that, indeed, $f(g(x)) = x$ for all $x \in \mathbb{R}$:

$$f(g(x)) = \begin{cases} f(x) = x & -1 \le x \le 1 \\ f(x+1) = f(t > 2) = t - 1 = (x+1) - 1 = x & x > 1 \\ f(x-1) = f(t < -2) = t + 1 = (x-1) + 1 = x & x < -1 \end{cases}.$$

However, in the opposite compositional order we get, for example, $g(f(2)) = g(17) = 18 \neq 2$.

To conclude the current section, we call attention to the above constructed example as supplying us with an *example space* (Watson & Mason, 2005) for functions that "appear" to be invertible (i.e., they satisfy $f(g(x)) = x$ for all $x \in \mathbb{R}$), yet they are not. One can take any injective function $g : \mathbb{R} \to \mathbb{R}$ that is not surjective, and subsequently construct $f : \mathbb{R} \to \mathbb{R}$ to "undo" g for all $x \in g(\mathbb{R})$, while assigning any random value for all $x \notin g(\mathbb{R})$. This example space also relates to the discussed property of an inverse element in a group as operating on *both* sides to obtain the identity element.

Furthermore, the above discussion can serve as a basis for enhancing mathematical understanding by making connections to other domains in mathematics. For example, suspected inverses appear with matrices in the context of a linear algebra course, where $AB = I$ does not necessarily imply $BA = I$. This is exemplified in the choice of $A = \begin{pmatrix} 0 & -1 & 3 \\ 0 & 1 & -2 \end{pmatrix}$, $B = \begin{pmatrix} 12 & 17 \\ 2 & 3 \\ 1 & 1 \end{pmatrix}$, which yields $AB = \begin{pmatrix} 1 & 0 \\ 0 & 1 \end{pmatrix} = I$, although $BA = \begin{pmatrix} 0 & 5 & 2 \\ 0 & 1 & 0 \\ 0 & 0 & 1 \end{pmatrix} \neq I$.

This example of making connections between different mathematical domains also relates to our earlier claim that the abstract nature of the group concept makes it applicable and useful for a wide variety of mathematical situations. Furthermore, the understanding acquired in the context of a specific problem could, in turn, serve as a basis for understanding a new problem, even when situated in a different domain.

On Glide Reflection

"Why is there a *glide reflection*?"

This was a question from a student in an undergraduate Geometry course for PSTs, who was surprised that the textbook listed four isometries or rigid motions: translation, rotation, reflection, and glide reflection. The student wondered: "We do not have a special name for the composition of a translation and a rotation, so why do we need a special name for the composition of a translation and a reflection?" The perceived discord is further strengthened when considering that translation, rotation, and reflection are accompanied with the more familiar terms of slide, turn, and flip, respectively. But, there is no analogous non-mathematical term for glide reflection.

In fact, this question also bothered the authors when each was first introduced to transformations on a plane. The mystery is unveiled when one considers a group structure: for isometries to form a group it is necessary to add glide reflection to the list. With the addition of glide reflection, the set of isometries is *closed* under composition of transformations. Alternatively put, every rigid motion on a plane can be described by a *single* isometry, one of the four. Furthermore, we argue that thinking of isometries in the plane in terms of groups is more likely to evoke exploratory variations and adaptations to the teaching of the topic, which, in turn, invites deeper student learning of the structure of the set of different isometries.

For example, the following question may be posed for a skeptic learner who is not convinced of the need for a glide reflection since this is a composition of two other isometries: "Are all other isometries in fact also 'needed'?" Meaning, is it possible to obtain all isometries in the plane with repeated compositions of only *one* of them? Such a notion of using a single element to generate all others is in fact rooted in group theory. It is the fundamental idea behind cyclic groups, which, for example, is the case for "1" in the additive group of integers. In relation to the current discussed topic, we note that all rigid motions may indeed be described as compositions of *reflections* only.

The above realization ultimately leads to a different question than the one that originally initiated the discussion: Considering that rotation, translation, and glide reflection can all be represented as compositions of reflections only, could there be an *additional* single isometry (meaning not any of the above) that is a composition of reflections? Differently put, is there a rigid motion on a plane that can be described by a *single* isometry, which is not any of the four?

In such sequencing of questions, the instructor's use of advanced mathematical knowledge of group structure, even though not necessary, is not only helpful, but also exposes learners to real exploratory working methods of mathematicians towards mathematical discoveries (see also Hadamard, 1945; Polya, 1981).

Is $\frac{14}{2}$ *a Prime Number?*

For those of us who took an introductory programming course at university, some might recall sleepless nights trying to figure out why the program was not "behaving properly," even though there did not appear to be any mistakes in the code. Finding programming bugs was, at times, a "mission impossible."

Several years ago, an undergraduate student was writing a MAPLE program that was supposed to output the list of prime factors of its input. The student approached the instructor for help, as the program she had written was producing unusual results. The approach taken by the student, though not efficient, was to determine factors of the number and then check if a factor was prime. However, despite what looked like a correct sequence of commands, the program did not function as expected.

In MAPLE, the command *isprime* tests for whether the input is prime. That is, it returns *true* if the input is prime and *false* if the input is composite. The following results (from one of the previous versions of MAPLE) exemplify what was problematic with the program:

$$isprime(7)$$
$$true$$
$$isprime(3+4)$$
$$true$$
$$isprime\left(\frac{14}{2}\right)$$
$$false$$

14 divided by 2 is definitely 7, and 7 is definitely a prime number. How can this strange behavior of MAPLE be explained?

In this case, the perceived discrepancy was due to a choice made by the software programmers: the function *isprime* required the input (the argument) to be an integer. Even though 14 divided by 2 is 7, in MAPLE, the result of division was considered by the program to be a rational number, and rational-number 7 was not equated with integer-number 7. To elaborate, the result of $\frac{14}{2}$ was stored as 7.0, and *isprime*(7.0) resulted in *false* since the input was not an integer.

Fortunately, this choice was reconsidered by the programmers in subsequent versions of MAPLE (versions greater than or equal to 12), where the response of the program clarifies the situation:

$$isprime\left(\frac{14}{2}\right)$$

true

$$isprime(7.0)$$

Error, (in *isprime*) argument must be an integer

$$isprime\ (3.5 \cdot 2)$$

Error, (in *isprime*) argument must be an integer

What does all this have to do with group theory? We believe that the notion of *closure* is central in explaining the strange behavior of the program. Integers are closed under the operation of addition. Integers are not closed under the operation of division, but rational numbers (without zero) are. As such, any result of division between two integers was considered to be a rational number. Furthermore, the program cannot operate on numbers of a different "kind." As such, in $3.5 \cdot 2$, for example, both numbers are considered rational, as well as their product (7.0).

In fact, and on a more general note, the idea of closure under an operation is the primary force behind extending different number sets. That is, while the set of natural numbers is closed under addition and multiplication, seeking closure under subtraction and division results in introducing integers (including negatives) and rational numbers, respectively.

As for the student's MAPLE program, the instructor need not be familiar with group theory in order to be able to help the student isolate the problem. Knowledge of how the *isprime* command functions would be sufficient in this case. Nevertheless, awareness of the *closure* property in relation to different number sets and operations can serve as a guide that is also applicable in other programming cases. It is often required that operations be defined based on the types of objects for which they are valid, and that operations result in an object of the same type—since this simplifies some of the complexities of programming. That is, closure is not only a mathematical idea of import, but also an idea that has implications in applied mathematics settings like programming.

Discussion

"You don't have to be tall to play basketball, but it sure can help."

The approach we have taken throughout this chapter might be described as pragmatic. We have admitted that in all presented examples, basic knowledge of group theory is in fact neither necessary nor obligatory for addressing the (more elementary) mathematics. Nevertheless, and as suggested by the basketball analogy, it can be helpful. As has been demonstrated through the different examples, alluding to the (nonlocal) abstract structure of (local) mathematical objects, and more specifically to the defining properties of a group (i.e., closure, associativity, identity element, and inverse element), can serve as a mathematical guide for teachers in situations of contingency.

However, as opposed to a fixed height of an adult who wishes to play basketball, mathematics teachers *can* "grow taller" throughout their life span as teachers. We suggest not only that the classroom situations presented in this chapter could benefit from the teacher's use of advanced mathematical knowledge, but also that this, in turn, may support the teacher's professional development. This is applicable to prospective and practicing school teachers, teacher educators, and university instructors alike. In fact, as we planned the structure of this chapter, we noted that all the stories we wished to share referred to something that John Mason calls an "experience of disturbance."

> What triggers a new phase of personal development? Most frequently there is some form of disturbance which starts things off. It may be a surprise remark in a lesson, a particularly poor showing on a test, something said by a colleague, something asserted in a journal or book, or a moment of insight. (Mason, 2002, p. 10)

Disturbance, according to Mason (2002), may serve as a catalyzer for learning and for professional development. The term "is intended to signal simply that some action is initiated" (p. 139). The presented situations of contingency, which were a result of learners' questions, comments, or approaches, served as experiences of disturbance that initiated deviations from the agenda in the immediate timeframe, and resulted in new task designs in the long term. We believe these foster the mathematical development of teachers, and, in turn, their readiness for situations of contingency in the future.

However, and as argued by Mason, experiences of disturbance may be perceived as positive or negative—as opportunity or pressure. We suggest that the knowledge resources teachers have may play a role in responding to situations of contingency. More specifically, and as made evident throughout the presented examples, possessing advanced mathematical knowledge can serve as a "crutch to lean on" when dealing with such situations, which subsequently supports the likelihood of the disturbance becoming an opportunity for mathematical growth.

References

Ameis, J. A. (2011). The truth about PEDMAS. *Mathematics Teaching in the Middle School, 16*(7), 414–420.

Ball, D. L., Thames, M. H., & Phelps, G. (2008). Content knowledge for teaching: What makes it special? *Journal of Teacher Education, 59*(5), 389–407.

Childs, L. N. (2009). *A concrete introduction to higher algebra* (3rd ed.). New York, NY: Springer.

Devlin, K. J. (1981). *Sets, functions and logic: Basic concepts of university mathematics*. New York, NY: Springer.

Dubinsky, E., Dautermann, J., Leron, U., & Zazkis, R. (1994). On learning fundamental concepts of group theory. *Educational Studies in Mathematics, 27*(3), 267–305.

Dummit, D. S., & Foote, R. M. (2004). *Abstract algebra* (3rd ed.). Hoboken, NJ: John Wiley and Sons.

Even, R. (2011). The relevance of advanced mathematics studies to expertise in secondary school mathematics teaching: Practitioners' views. *ZDM Mathematics Education, 43*(6–7), 941–950.

Glidden, P. L. (2008). Prospective elementary teachers' understanding of order of operations. *School Science and Mathematics, 108*(4), 130–136.

Hadamard, J. (1945). *The psychology of invention in the mathematical field*. Princeton, NJ: Princeton University Press.

Hadar, N., & Hadass, R. (1981). Between associativity and commutativity. *International Journal of Mathematical Education in Science and Technology, 12*(5), 535–539.

Hill, H. C., Ball, D. L., & Schilling, S. G. (2008). Unpacking pedagogical content knowledge: Conceptualizing and measuring teachers' topic-specific knowledge. *Journal for Research in Mathematics Education, 39*(4), 372–400.

Koichu, B. (2008). If not, what yes? *International Journal of Mathematical Education in Science and Technology, 39*(4), 443–454.

Kontorovich, I., & Zazkis, R. (2017). Mathematical conventions: Revisiting arbitrary and necessary. *For the Learning of Mathematics, 37*(1), 29–34.

Larsen, S. (2010). Struggling to disentangle the associative and commutative properties. *For the Learning of Mathematics, 30*(1), 37–42.

Mason, J. (2002). *Researching your own practice: The discipline of noticing*. London, UK: Routledge Falmer.

Musser, G. L., Burger, W. F., & Peterson, B. E. (2010). *Mathematics for elementary teachers: A contemporary approach* (9th ed.). Hoboken, NJ: Wiley.

Polya, G. (1981). *Mathematical discovery: On understanding, learning, and teaching problem solving (Combined)*. New York, NY: Wiley.

Rowland, T., Huckstep, P., & Thwaites, A. (2005). Elementary teachers' mathematics subject knowledge: The knowledge quartet and the case of Naomi. *Journal of Mathematics Teacher Education, 8*(3), 255–281.

Simon, M. A. (2006). Key developmental understandings in mathematics: A direction for investigating and establishing learning goals. *Mathematical Thinking and Learning, 8*(4), 359–371.

Stewart, J. (2008). *Calculus: Early transcendentals* (6th ed.). Belmont, CA: Thomson Learning.

Tall, D. (2013). *How humans learn to think mathematically: Exploring the three worlds of Mathematics*. New York, NY: Cambridge University Press.

Wasserman, N. H. (2016). Abstract algebra for algebra teaching: Influencing school mathematics instruction. *Canadian Journal of Science, Mathematics and Technology Education, 16*(1), 28–47.

Wasserman, N. H. (2017). Making sense of Abstract Algebra: Exploring secondary teachers' understandings of inverse functions in relation to its group structure. *Mathematical Thinking and Learning, 19*(3), 181–201.

Wasserman, N. H. (2018). Knowledge of nonlocal mathematics for teaching. *Journal of Mathematical Behavior, 49*, 116–128.

Watson, A., & Mason, J. (2005). *Mathematics as a constructive activity: Learners generating examples*. Mahwah, NJ: Lawrence Erlbaum Associates.
Weber, K., & Larsen, S. (2008). Teaching and learning group theory. In M. P. Carlson & C. Rasmussen (Eds.), *Making the connection: Research and teaching in undergraduate mathematics education* (pp. 139–151). Washington, DC: Mathematical Association of America.
Zaslavsky, O., & Peled, I. (1996). Inhibiting factors in generating examples by mathematics teachers and student teachers: The case of binary operation. *Journal for Research in Mathematics Education, 27*(1), 67–78.
Zazkis, R. (2017). Order of operations: On conventions, mnemonics and knowledge-in-use. *For the Learning of Mathematics, 37*(3), 18–20.
Zazkis, R., & Chernoff, E. J. (2008). What makes a counterexample exemplary? *Educational Studies in Mathematics, 68*(3), 195–208.
Zazkis, R., & Kontorovich, I. (2016). A curious case of superscript (-1): Prospective secondary mathematics teachers explain. *Journal of Mathematical Behavior, 43*, 98–110.
Zazkis, R., & Leikin, R. (2010). Advanced mathematical knowledge in teaching practice: Perceptions of secondary mathematics teachers. *Mathematical Thinking and Learning, 12*(4), 263–281.
Zazkis, R., & Mamolo, A. (2011). Reconceptualising knowledge at the mathematical horizon. *For the Learning of Mathematics, 31*(2), 8–13.
Zazkis, R., & Rouleau, A. (2018). Order of operations: On convention and met-before acronyms. *Educational Studies in Mathematics, 97*(2), 143–162.

Watson, A., & Mason, J. (2005). *Mathematics as a constructive activity: Learners generating examples*. Mahwah, NJ: Lawrence Erlbaum Associates.
Weber, K., & Larsen, S. (2008). Teaching and learning group theory. [illegible]
[illegible]
[illegible]
[illegible]
[illegible]
[illegible]
[illegible]

Chapter 18
Some Applications of Abstract Algebra to High School Teaching: Task Design and Polynomial Interpolation

Al Cuoco

Introduction

As an undergraduate, I had no intention of becoming a teacher. I took a straight mathematics major that was preparation for graduate school—analysis, algebra, topology, and probability. I loved it. But I saw little connection to my high school studies. Then, in my senior year, I took a graduate course that used Lang's *Algebra* (Lang 1967), and, while no one said anything about precollege mathematics, I began to see some connections, connections that would resurface later when I found myself teaching high school.

This was in the 1970s. For personal and political reasons, I took a high school teaching job and planned to hide out for a few years (this would be impossible now, teaching with no experience in mathematics education, but there was a teacher shortage in those years). A few years turned into two decades, interrupted by a leave to return to graduate school, where I worked in number theory. This serendipitous confluence of experiences, working in high school mathematics and in mathematics research, has led to a career-long personal agenda to find ways to close the (huge) gap between school mathematics and mathematics as it is practiced by mathematics professionals.

Author Note: Many of the ideas in this chapter were hammered out in collaboration with Bill McCallum, Paul Monsky, and Joe Rotman. These folks have been inspirations.

A. Cuoco (✉)
Education Development Center, Waltham, MA, USA
e-mail: acuoco@edc.org

N. H. Wasserman (ed.), *Connecting Abstract Algebra to Secondary Mathematics, for Secondary Mathematics Teachers*, Research in Mathematics Education,
https://doi.org/10.1007/978-3-319-99214-3_18

What I've learned from these experiences has implications for the mathematical preparation of teachers. This paper concentrates on some of these implications that live at the intersection of abstract algebra and number theory. More precisely, I'll discuss:

(1) the use of algebraic methods to design tasks for students,
(2) applications of classical methods in algebra to the problem of fitting a polynomial function to a set of data.

While these examples are aimed at content courses for prospective mathematics teachers, the second one, with suitable modifications, can (and has) been made tractable for high school students (EDC 2014). These ideas are developed more extensively in a recent text that I co-authored with the late Joseph Rotman (Cuoco & Rotman 2013).

The focus of this chapter is on the profession-specific mathematics that many high school teachers (including me) have found useful in their careers. I hope it will be of interest to those who teach content courses for practicing and prospective teachers, to mathematicians who teach mathematics courses for undergraduates, and to teachers who look for mathematics that they can use in their work. I deal only incidentally with the crucially important issues of pedagogy and teaching practice, leaving those discussions for others. Instead, I present some examples of what I consider to be *applied mathematics*—mathematics applied to the profession of mathematics teaching.

The Mathematics of Task Design

In Cuoco (2000), I conjectured that a great deal of modern algebra was developed to solve problems that arise in the design of mathematical problems for students that "come out nice." For example, when launching the Law of Cosines in a second year high school algebra class, finding the size of $\angle Q$ in ΔUQS where $UQ = 5$, $SQ = 8$, and $SU = 7$ is a better choice than choosing a triangle whose side-lengths are 5, 9, and 7, because the measure of $\angle Q$ in the (5, 8, 7) triangle is 60°, whereas, in the $(5, 9, 7)$ case, $\angle Q$ has measure $\cos^{-1} \frac{19}{30}$.

We certainly don't want students to expect that every problem they encounter has simple-looking solutions; messy numbers are part of life. But most of the teachers I know prefer to *launch* an idea with a numerical example where the solution is integral (or at least rational), saving the more general examples for when the basic ideas are nailed down. One reason for this is that computational overhead can cloud the underlying method. Another is that such problems provide students with feedback that they are on the right track. These reasons are closely related to Wasserman's notion of concealing complexity (or micro-level trimming), as described in Wasserman (2015).

I have another reason, connected to the design of courses for prospective and practicing teachers: The problem of finding a scalene triangle with integer side-lengths and a 60° angle is an example of the kinds of "meta problems" that come up in teachers' rooms all over the country, and the mathematical tools that can be used to design such problems sit squarely inside the abstract algebra and number theory that undergraduates take. Yes, examples of problems that have integer solutions can be found on the web, but the benefits a teacher gets from developing general-purpose methods to design some particular class of tasks transcends the immediate goal of finding such problems. The methods provide

- some general purpose tools that can be applied to a whole variety of situations related to the mathematics used in teaching,
- some real, profession-specific applications of abstract algebra to the profession of teaching mathematics,
- examples of a certain class of habits of mind that leads one to look into modern mathematics as a source for connections useful in lesson design and planning

Some meta-problems that have come up over the years with teachers I know include

- How can you generate cubic polynomial functions with integer coefficients whose zeros, critical points, and inflection points all have integer coordinates? Example: $x^3 + 9x^2 - 120x - 128$.
- How can you design *Heron triangles*—triangles with integer side-lengths and integer area? Example: Side-lengths 13, 14, and 15.
- How can you find rectangles with integer side-lengths so that when you cut congruent squares out of each corner and fold up the sides to make a box, the size of the cutout that maximizes the volume is rational? Example: Length 7, width 15.
- How can you find triangles in the coordinate plane that have lattice point vertices and integer side-lengths? Example: $(3, 2)$, $(-189, 258)$, and $(-57, 34)$

The list goes on, and, in fact, once teachers get in the game, new examples start flowing. Speaking of flowing, here's a typical-looking algebra word problem that was devised by teachers at the Park City Mathematics Institute some years ago:

> A boat is making a round trip, 135 miles in each direction. Without a current, the boat's speed would be 32 miles per hour. However, there is a constant current that increases the boat's speed in one direction and decreases it in the other. If the round trip takes exactly 9 hours, what is the speed of the current?

Solve it, and you'll see the expert design; the application of algebra to its design is described in Kerins (2003).

Pythagorean Triples

Surely, one of the oldest meta-problems involves the search for Pythagorean triples—triples of positive integers (a, b, c) that are sidelengths of a right triangle, so that

$$a^2 + b^2 = c^2.$$

Diophantus of Alexandria developed, around 250 CE, a geometric method for generating such triples. Stated in modern language, he realized that a rational point on the unit circle (the graph of $x^2 + y^2 = 1$), when written in the form $\left(\frac{a}{c}, \frac{b}{c}\right)$, produces a Pythagorean triple.

$$\left(\tfrac{a}{c}\right)^2 + \left(\tfrac{b}{c}\right)^2 = 1 \quad \Longrightarrow \quad a^2 + b^2 = c^2$$

One can get such a rational point by forming a line with positive rational slope through the point $P = (-1, 0)$ and intersecting the line with the circle. Using methods from high school algebra, you can see that if the line intersects the circle in one rational point, the other point will also have rational coordinates.[1] Hence, it was known early on that there are infinitely many Pythagorean triples (details are in Cuoco & Rotman 2013).

There are several *algebraic* methods for generating Pythagorean triples as well. One method uses complex numbers and the observation that

$$x^2 + y^2 = (x + yi)(x - yi) \tag{18.1}$$

The sum of two squares can thus be written as the product of complex number and its complex conjugate. So, if you want *integers* a and b so that $a^2 + b^2$ is a perfect square, you might write the sum of these two squares as

$$a^2 + b^2 = (a + bi)(a - bi)$$

and try to make each factor on the right-hand side the square of a complex number with integer real and imaginary parts. And it's within the scope of high school mathematics to show that if $a + bi = (r + si)^2$, then $a - bi = (r - si)^2$. So, for example,

$$\begin{aligned} (3 + 2i)^2 &= 5 + 12i \quad \text{and} \\ (3 - 2i)^2 &= 5 - 12i. \end{aligned}$$

So,

$$\begin{aligned} 5^2 + 12^2 &= (5 + 12i)(5 - 12i) \\ &= (3 + 2i)^2(3 - 2i)^2 \\ &= ((3 + 2i)(3 - 2i))^2 \\ &= \left(3^2 + 2^2\right)^2 \\ &= 13^2, \end{aligned}$$

[1]In most high school curricula, students find the intersections of a line and a circle, given their equations. This leads to a quadratic equation, and if the equations of the line and circle have rational coefficients, so will the resulting quadratic equation. If one solution to this equation is rational, the other one will be too, because the sum of the solutions is the (negative of) the coefficient of the linear term.

and (5, 12, 13) is a Pythagorean triple. This gives an easily programmed algorithm for generating Pythagorean triples:

(1) pick a complex number $a + bi$ with a, b integers (and $a > b$, for reasons that will be apparent if you try some numerical examples)
(2) write $(a + bi)^2$ as $c + di$
(3) then $(c, d, \sqrt{c^2 + d^2})$ is a Pythagorean triple.

This gives high school teachers a useful tool for their lesson planning, and if you carry this algorithm out for a generic complex number $a + bi$, you'll see a usual formula for pythagorean triples that shows up in the literature [see Cuoco and Rotman (2013), for example].

A little notation and formalism will help put this method into a more general setting and will show its connection to abstract algebra. Complex numbers of the form $a + bi$ where a and b are integers are called *Gaussian integers*. The set of all Gaussian integers is denoted by $\mathbb{Z}[i]$, because $\mathbb{Z}$ (*Zahlen*) usually denotes the system of ordinary integers that is the focus of much of arithmetic in school; so $\mathbb{Z}[i]$ is obtained from $\mathbb{Z}$ by "adjoining i." Both $\mathbb{Z}$ and $\mathbb{Z}[i]$ are endowed with two operations. The properties of addition and multiplication that allow one to calculate with integers also hold in $\mathbb{Z}[i]$—order doesn't matter in addition or multiplication, multiplication distributes over addition, and so on. Formally, both systems are examples of *commutative rings*, and, in fact, $\mathbb{Z}$ is a subring of $\mathbb{Z}[i]$. What's relevant here is that we used implicitly the algebra of $\mathbb{Z}[i]$ by factoring $a^2 + b^2$ into a Gaussian integer times its complex conjugate, and then we made each of the factors perfect squares in $\mathbb{Z}[i]$, ensuring that their product was a perfect square. A little more work shows that every Pythagorean triple is obtained in this way.

Up One Level of Abstraction

This can all be put in a more general setting, one that applies to other contexts. If $z = a + bi$ is any complex number, we can denote its complex conjugate $a - bi$ by $\overline{z}$. A calculation shows that

(1) $\overline{z + w} = \overline{z} + \overline{w}$,
(2) $\overline{zw} = \overline{z}\,\overline{w}$,
(3) $z = \overline{z}$ if and only if z is a real number, and
(4) $z\overline{z} = a^2 + b^2$

When z is a Gaussian integer, this last equation conjures up images of Pythagorean triples, a connection that can be pushed. The map $z \mapsto z\overline{z}$ shows up often enough to be named: It's called the *complex norm* map, denoted (here) by N: $N(z) = z\overline{z}$. Naming the norm allows us to encapsulate it and investigate its properties. In particular, item 4 above tells us that the search for Pythagorean triples amounts to the search for Gaussian integers with perfect square norms.

And another direct calculation shows that the norm is *multiplicative*:

$$N(zw) = N(z)\,N(w),$$

but my favorite proof of this fact uses the analogous property of conjugation (the conjugate of the product is the product of the conjugates):

$$\begin{aligned} N(zw) &= zw\,\overline{zw} \\ &= zw\,\overline{z}\,\overline{w} \\ &= z\,\overline{z}\,w\,\overline{w} \\ &= N(z)\,N(w) \end{aligned}$$

It then follows that

$$N(z^2) = (N(z))^2$$

for all complex numbers z, and, in particular, for Gaussian integers. This equation says that *if we want $N(z)$ to be a square, just take z to be a square*. Table 18.1 was generated with a computer algebra system and shows $(r+si)^2$ and the corresponding norm.

The table contains many interesting patterns, so that it could be a basis for a high school investigation of its own (look, for example, at the imaginary parts of the numbers as you go down a column).

Heron Triangles

The method in the previous section can be modified to produce triangles with *rational* side-length and area. Such triangles can be scaled to produce Heron triangles. Suppose, in Fig. 18.1, that the lengths of the sides and the area are all rational numbers and $m\angle BCA = \theta$.

Since the area of the triangle is rational and is equal to $\frac{1}{2}ab\sin\theta$, $\sin\theta$ is a rational number.

Table 18.1 $(r+si)^2$ and the resulting norm

	$s=1$	$s=2$	$s=3$	$s=4$
$r=2$	$3+4i$, 5			
$r=3$	$8+6i$, 10	$5+12i$, 13		
$r=4$	$15+8i$, 17	$12+16i$, 20	$7+24i$, 25	
$r=5$	$24+10i$, 26	$21+20i$, 29	$16+30i$, 34	$9+40i$, 41
$r=6$	$35+12i$, 37	$32+24i$, 40	$27+36i$, 45	$20+48i$, 52
$r=7$	$48+14i$, 50	$45+28i$, 53	$40+42i$, 58	$33+56i$, 65
$r=8$	$63+16i$, 65	$60+32i$, 68	$55+48i$, 73	$48+64i$, 80

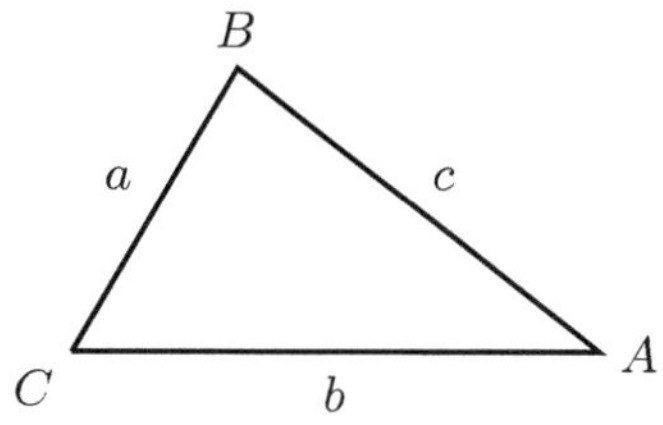

Fig. 18.1 a, b, c, and the area are all rational

By the law of cosines,

$$c^2 = a^2 + b^2 - 2ab\cos\theta$$

Since a, b, and c are rational, $\cos\theta$ is also a rational number.

And since

$$\sin^2\theta + \cos^2\theta = 1,$$

$(\cos\theta, \sin\theta)$ is thus a rational solution to

$$x^2 + y^2 = 1$$

In other words, $(\cos\theta, \sin\theta)$ is a rational point on the unit circle. The methods of the previous section can thus be used to find such rational points—they are in 1–4 correspondence with Pythagorean triples. For example, the triple $(3, 4, 5)$ corresponds to the four rational points $(\pm\frac{3}{5}, \pm\frac{4}{5})$.

In order to find a (rational) Heron triangle, then, it's enough to find a rational point $(\cos\theta, \sin\theta)$ on the unit circle, and then to find three positive rational numbers (a, b, c) so that

$$c^2 = a^2 + b^2 - 2ab\cos\theta$$

The triangle with sidelengths a, b, and c will then have area $\frac{1}{2}\sin\theta$, a rational number.

So, the question comes down to this: Given a rational point $(\cos\theta, \sin\theta)$, how can one find rational numbers a and b so that $a^2 + b^2 - 2ab\cos\theta$ is the square of a rational number? Complex numbers enter again.

Let $\alpha = -\cos\theta + i\sin\theta$. Then $\overline{\alpha} = -\cos\theta - i\sin\theta$. Two facts follow from this definition:

$$\alpha + \overline{\alpha} = -2\cos\theta \quad \text{and}$$

$$\alpha \cdot \overline{\alpha} = 1$$

Consider the ring $\mathbb{Q}[\alpha]$ constructed just as we constructed the Gaussian integers: the set of all complex numbers of the form $a+b\alpha$ where a and b are rational numbers and $\mathbb{Q}$ is the field of rational numbers. If $z = a + b\alpha$, then $\overline{z} = a + b\overline{\alpha}$, and

$$\begin{aligned} N(z) = z\,\overline{z} &= (a + b\alpha)(a + b\overline{\alpha}) \\ &= a^2 + ab(\alpha + \overline{\alpha}) + b^2\alpha\,\overline{\alpha} \\ &= a^2 + ab(-2\cos\theta) + b^2 \cdot 1 \\ &= a^2 - 2ab\cos\theta + b^2 \end{aligned}$$

This is precisely the expression that we want to make equal to the square of a rational number. So, we want to find $z = a + b\alpha$ such that $N(z)$ is the square of a rational number.

But recall that norm is multiplicative:

$$N(zw) = N(z)N(w)$$

and that

$$N(z^2) = (N(z))^2$$

for all complex numbers z, and in particular, for our special complex numbers of the form $a + b\alpha$. This equation says (again) that *if we want $N(z)$ to be a square, just take z to be a square.*

So, the algorithm works like this:

(1) Pick a rational point on the unit circle, $(\cos\theta, \sin\theta)$ $(0 < \theta < \pi)$.
(2) Let $\alpha = -\cos\theta + i\sin\theta$.
(3) Pick any number z of the form $r + s\alpha$ where r and s are rational numbers, and square it (this will make the norm a square).
(4) Write z^2 in the form $a + b\alpha$ using the relation

$$\alpha^2 + 2\cos\theta\,\alpha + 1 = 0$$

Then the triangle with sidelengths a and b and included angle θ will have a rational number as its third sidelength and a rational number as an area (it will be $\frac{1}{2}ab\sin\theta$). This triangle can be scaled to produce a Heron triangle.

For example:

(1) Pick the point $(\frac{3}{5}, \frac{4}{5})$, so that $\theta = \cos^{-1}\left(\frac{3}{5}\right)$.
(2) $\alpha = -\frac{3}{5} + \frac{4}{5}i$. Note that

$$\alpha^2 + 2(\tfrac{3}{5})\alpha + 1 = \alpha^2 + (\tfrac{6}{5})\alpha + 1 = 0$$

(3) Let $z = 2 + \alpha$ $(r = 2, s = 1)$.

(4) Then

$$\begin{aligned} z^2 &= (2+\alpha)^2 \\ &= 4 + 4\alpha + \alpha^2 \\ &= 4 + 4\alpha + \left(-\tfrac{6}{5}\alpha - 1\right) \\ &= 3 + \tfrac{14}{5}\alpha \end{aligned}$$

So, the triangle whose sides have length 3 and $\frac{14}{5}$ and included angle $\cos^{-1}\left(\frac{3}{5}\right)$ can be scaled to produce a Heron triangle. The third side can be found by taking the norm of $2+\alpha$ (or by the law of cosines) to be $\frac{13}{5}$. Scaling by 5, the triangle whose sidelengths are 15, 14, and 13 is a Heron triangle.[2]

And there's another method that would appeal to geometers. We're looking for rational points that satisfy

$$c^2 = a^2 - 2ab\cos\theta + b^2$$

These correspond to rational points on the conic defined by

$$1 = x^2 - 2xy\cos\theta + y^2 \tag{18.2}$$

The graph is an ellipse. So, just as in the Method of Diophantus, take a line with rational slope (greater than 1, say) through $(0, -1)$ and intersect it with the ellipse. The second point will also be rational. This parametrizes all the rational points on the conic and all Heron triangles.

Eisenstein Triples

Speaking of ellipses, this "secant and conic" method has other applications to task design. In the introduction, I mentioned the (5, 7, 8) triangle as an example of the meta-problem of constructing scalene integer sided triangles with a 60° angle.

Applying the law of cosines to the triangle in Fig. 18.2, we have

Fig. 18.2 A nice triangle

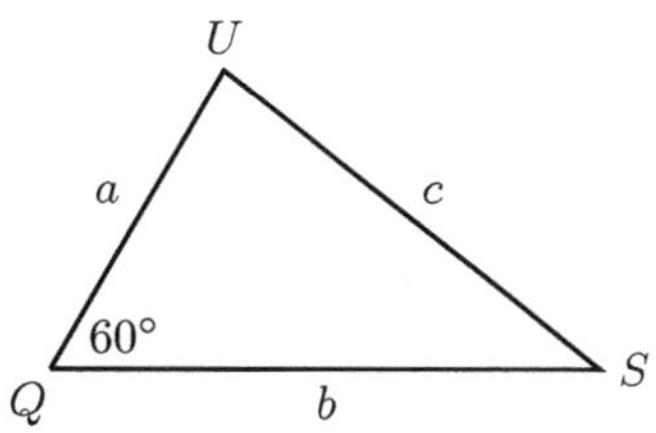

[2]Over a dozen years ago, Bowen Kerins and friends published another method in Kerins (2003).

$$\begin{aligned} c^2 &= a^2 + b^2 - 2ab \cos 60^\circ \\ &= a^2 + b^2 - 2ab\,\tfrac{1}{2} \\ &= a^2 - ab + b^2 \end{aligned}$$

Such triples (a, b, c) are called Eisenstein triples, and the same equation rears its head again. We want integers (a, b, c) so that

$$a^2 - ab + b^2 = c^2. \tag{18.3}$$

Dividing by c^2,we have

$$\left(\tfrac{a}{c}\right)^2 - \left(\tfrac{a}{c}\right)\left(\tfrac{b}{c}\right) + \left(\tfrac{b}{c}\right)^2 = 1. \tag{18.4}$$

A little hindsight suggests a version of the equation that hints of Diophantus—we're looking for rational points on the graph of

$$x^2 - xy + y^2 = 1. \tag{18.5}$$

This graph is an ellipse. A secant through $(-1, 0)$ with rational slope (and that intersects the conic in the first quadrant) generates a rational point and hence a solution to our meta-problem—more details are in Cuoco and Rotman (2013).

And just as before, there's an algebraic analogue to this method. Just as we can form the ring of Gaussian integers $\mathbb{Z}[i]$ by adjoining a fourth root of unity, i, to $\mathbb{Z}$ (that is, a solution to $x^4 - 1 = 0$), so we can form the ring of *Eisenstein integers* by adjoining the *cube* roots of unity—the solutions to $x^3 - 1 = 0$. Because

$$x^3 - 1 = (x - 1)(x^2 + x + 1),$$

the three cube roots of unity are

$$\{1, \frac{-1 + i\sqrt{3}}{2}, \frac{-1 - i\sqrt{3}}{2}\}.$$

Let

$$\omega = \frac{-1 + i\sqrt{3}}{2},$$

and consider $\mathbb{Z}[\omega] = \{a + b\omega \mid a, b \in \mathbb{Z}\}$. This is a ring (the *Eisenstein integers*) with structural similarities to $\mathbb{Z}$ and $\mathbb{Z}[i]$ (details are in Cuoco & Rotman 2013). In particular, because

$$\begin{aligned} \omega + \overline{\omega} &= -1 \quad \text{and} \\ \omega\overline{\omega} &= 1, \end{aligned}$$

Table 18.2 $(r + s\omega)^2$ and the resulting norm

	$s = 1$	$s = 2$	$s = 3$	$s = 4$
$r = 2$	$3 + 3\omega,\ 3$			
$r = 3$	$8 + 5\omega,\ 7$	$5 + 8\omega,\ 7$		
$r = 4$	$15 + 7\omega,\ 13$	$12 + 12\omega,\ 12$	$7 + 15\omega,\ 13$	
$r = 5$	$24 + 9\omega,\ 21$	$21 + 16\omega,\ 19$	$16 + 21\omega,\ 19$	$9 + 24\omega,\ 21$
$r = 6$	$35 + 11\omega,\ 31$	$32 + 20\omega,\ 28$	$27 + 27\omega,\ 27$	$20 + 32\omega,\ 28$
$r = 7$	$48 + 13\omega,\ 43$	$45 + 24\omega,\ 39$	$40 + 33\omega,\ 37$	$33 + 40\omega, 37$
$r = 8$	$63 + 15\omega,\ 57$	$60 + 28\omega,\ 52$	$55 + 39\omega,\ 49$	$48 + 48\omega,\ 48$
$r = 9$	$80 + 17\omega,\ 73$	$77 + 32\omega,\ 67$	$72 + 45\omega,\ 63$	$65 + 56\omega,\ 61$
$r = 10$	$99 + 19\omega,\ 91$	$96 + 36\omega,\ 84$	$91 + 51\omega,\ 79$	$84 + 64\omega,\ 76$

a direct calculation shows that

$$\begin{aligned} N(z) &= (a + b\omega)(\overline{a + b\omega}) \\ &= (a + b\omega)(a + b\overline{\omega}) \\ &= a^2 - ab + b^2. \end{aligned}$$

Hence, the same mantra applies:

To make $N(z)$ a square in $\mathbb{Z}$, make z a square in $\mathbb{Z}[\omega]$.

Once again, we have a method for generating Eisenstein triples (see Table 18.2). See Cuoco and McCallum (2018a) for a full treatment.

Where This Sits in the Bigger Picture

Similarly, the solutions to $x^n - 1 = 0$ are the "nth roots of unity." One such nth root of unity is

$$\zeta_n = \cos\frac{2\pi}{n} + i\sin\frac{2\pi}{n}$$

So, $\zeta_4 = i$ and $\zeta_3 = \omega$. The ring $\mathbb{Z}[\zeta_n]$ is formed in the same way as the Gaussian and Eisenstein integers are formed. In Cuoco and Rotman (2013), we discuss in detail how these structures played important roles in the development of abstract algebra and in early attempts to prove Fermat's Last Theorem.

Interlude

Before going on, I should say that none of these applications was part of my undergraduate preparation. All of them were learned on the job, while I was teaching and studying. But they fit perfectly in undergraduate courses in algebra and in mathematics education; they are part of the profession-specific applications of algebra that teachers find so useful in the design of their lessons. And they are also part of an algebraic heritage that leads to problems at the forefront of what's known in mathematics.

For example, every Pythagorean triple defines a triangle whose area is an integer. But there are also right triangles with *rational* side-lengths and with integer area—for example, the triangle whose sides have length

$$\left(\frac{3}{2}, \frac{20}{3}, \frac{41}{6}\right)$$

has area 5. The classification of such triangles is an open problem, one that has been around for centuries and is still the object of current research in number theory (more details are in Cuoco & Rotman 2013). Here we have a problem whose statement is accessible to high school students and that shows how algebra is an alive and thriving discipline. Understanding the problem and some methods used to investigate it surely belongs in mathematics teacher preparation.

Polynomial Interpolation

The previous section exploited the structure and properties of a particular ring ($\mathbb{Z}[i]$ or $\mathbb{Z}[\omega]$) to build solutions to various versions of the "meta-problem problem." In this section, I'll look at two more applications of abstract algebra.

Many high school curricula contain a unit on "fitting functions to tables." For polynomial functions, there are two methods from algebra that situate the topic in more general settings. I was about to write "modern algebra" instead of "algebra," but both of the methods have a long pedigree, going back centuries. And yet both are useful tools that arise in many modern settings and that exploit the structural similarities between ordinary integers and polynomials with coefficients in a field.

Newton's Difference Formula

Suppose we are looking for a polynomial that fits this table:

n	$f(n)$
0	1
1	−1
2	11
3	49
4	125
5	251
6	439
7	701

A common first step is to look at the table of successive differences. In the table below, the Δ stands for "difference"—it indicates the difference between consecutive terms. For example, the "76" in the table below is equal to $125 - 49$:

n	$f(n)$	Δ	Δ^2	Δ^3	Δ^4	Δ^5	Δ^6	Δ^7
0	1	−2	14	12	0	0	0	0
1	−1	12	26	12	0	0	0	
2	11	38	38	12	0	0		
3	49	76	50	12	0			
4	125	126	62	12				
5	251	188	74					
6	439	262						
7	701							

By the way difference tables are built, *every* entry in the interior of the table is the sum of its "up and over." $11 = -1 + 12$, $38 = 38 + 12$, $262 = 188 + 74, \ldots$.

So, one can take any entry in the $f(n)$ column, replace it by its up and over, replace these two numbers by *their* up and overs, and keep moving up the table towards the first row. For example, with $f(3)$, take the up-and-overs all the way to the top:

$$\begin{aligned} f(3) &= 49 \\ &= 11 + 38 \\ &= (-1 + 12) + (12 + 26) = -1 + 2 \cdot 12 + 26 \\ &= (1 + -2) + 2 \cdot (-2 + 14) + 14 + 12 = 1 + 3 \cdot (-2) + 3 \cdot 14 + 12 \end{aligned}$$

Look again, with some emphasis added and details suppressed:

$$\begin{aligned} f(3) &= \mathbf{1} \cdot 49 \\ &= \mathbf{1} \cdot 11 + \mathbf{1} \cdot 38 \\ &= \mathbf{1} \cdot -1 + \mathbf{2} \cdot 12 + \mathbf{1} \cdot 26 \\ &= \mathbf{1} \cdot 1 + \mathbf{3} \cdot (-2) + \mathbf{3} \cdot 14 + \mathbf{1} \cdot 12 \end{aligned}$$

Fig. 18.3 Up and over

n	$f(n)$	Δ	Δ^2	Δ^3
0	1	–2	14	12
1	–1	12	26	12
2	11	38	38	12
3	49	76	50	12
4	125	126	62	12
5	251	188	74	
6	439	262		
7	701			

I was quite blown away when I first saw this (again, on the job). Just to make sure, take a look:

$$\begin{aligned}
f(4) &= \mathbf{1} \cdot 125 \\
&= \mathbf{1} \cdot 49 + \mathbf{1} \cdot 76 \\
&= \mathbf{1} \cdot (11 + 38) + \mathbf{1} \cdot (38 + 38) \\
&= \mathbf{1} \cdot 11 + \mathbf{2} \cdot 38 + \mathbf{1} \cdot 38 \\
&= \mathbf{1} \cdot (-1 + 12) + \mathbf{2} \cdot (12 + 26) + \mathbf{1} \cdot (26 + 12) \\
&= \mathbf{1} \cdot -1 + \mathbf{3} \cdot 12 + \mathbf{3} \cdot 26 + \mathbf{1} \cdot 12 \\
&= \mathbf{1} \cdot (1 + -2) + \mathbf{3} \cdot (-2 + 14) + \mathbf{3} \cdot (14 + 12) + \mathbf{1} \cdot (12 + 0) \\
&= \mathbf{1} \cdot 1 + \mathbf{4} \cdot (-2) + \mathbf{6} \cdot 14 + \mathbf{4} \cdot 12 + \mathbf{1} \cdot 0
\end{aligned}$$

More generally, if seems as if we have, for any non-negative integer n,

$$f(n) = \binom{n}{0} \cdot 1 + \binom{n}{1} \cdot (-2) + \binom{n}{2} \cdot 14 + \binom{n}{3} \cdot 12$$

Replacing the binomial coefficients with their algebraic equivalents, we have:

$$f(n) = 1 \cdot 1 + n \cdot (-2) + \frac{n(n-1)}{2} \cdot 14 + \frac{n(n-1)(n-2)}{6} \cdot 12$$

And a little algebra turns this into a nice cubic:

$$f(n) = 2n^3 + n^2 - 5n + 1$$

A check shows that this function agrees with the table.

The general proof follows via a generic calculation, and the result goes back at least to Isaac Newton:

Newton's Difference Formula: Suppose we have a table with inputs $0 \ldots m$:

Input	Output	Δ	Δ^2	Δ^3	$\ldots$	Δ^m
0	a_0	a_1	a_2	a_3	$\ldots$	a_m
1						
2						
3						
4						
5						
6						
$\vdots$						
m						

If f is a function that agrees with the table, then for $0 \leq n \leq m$:

$$f(n) = \sum_{k=0}^{m} a_k \binom{n}{k}$$

The theorem talks about a function defined on non-negative integers, but one can extend the domain of the binomial coefficients $\binom{n}{k}$ via the formulation

$$\binom{n}{k} = \frac{n(n-1)(n-2)(n-3)\ldots(n-k+1)}{k!}$$

The right-hand side makes sense for any element n of a field, and leads to a set of polynomials $\binom{x}{k}$ (for non-negative integer k) that have found applications in many fields. Folks around Boston call $\binom{x}{k}$ the "Mahler basis" for polynomials, because Kurt Mahler used them in the 1960s to interpolate another family of functions defined on rings quite different from $\mathbb{Z}$. Writing a polynomial, not as a linear combination of the various powers x^k, but as a linear combination of the Mahler polynomials $\binom{x}{k}$, often reveals hidden algebraic and combinatorial treasures. Some of the many examples are in Graham et al. (1992) and Cuoco (2005).

Lagrange Interpolation

When the input data don't follow a simple linear pattern, there's another method that highlights the structural similarity between $\mathbb{Z}$ and the ring of polynomials in one variable with coefficients in a field.

Students are often asked to find a polynomial that agrees with a table like this:

Input	Output
−3	12
2	22
3	72
−4	−26

On the surface, the problem of fitting data seems to have little to do with abstract algebra and number theory. But this problem fits right into the theory of commutative rings. Here's how.

We can reformulate the problem like this: Suppose we can find polynomials g, h, k, and ℓ satisfying

$$\begin{array}{cccc} g(-3) = 1 & h(-3) = 0 & k(-3) = 0 & \ell(-3) = 0 \\ g(2) = 0 & h(2) = 1 & k(2) = 0 & \ell(2) = 0 \\ g(3) = 0 & h(3) = 0 & k(3) = 1 & \ell(3) = 0 \\ g(-4) = 0 & h(-4) = 0 & k(-4) = 0 & \ell(-4) = 1. \end{array}$$

Setting $f = 12g + 22h + 72k - 26\ell$, we have a polynomial that fits the original table. Now, the Factor Theorem from second year algebra shows that g is divisible by the linear polynomials $x-2$, $x-3$, and $x+4$. Since they are irreducible in $\mathbb{Q}[x]$, they are pairwise relatively prime; hence, g is divisible by their product: there is a polynomial A such that

$$g(x) = A(x-2)(x-3)(x+4).$$

In fact, we can choose A to be a constant that makes $g(-3) = 1$: set

$$1 = g(-3) = A(-3+2)(-3-3)(-3+4);$$

that is, $A = 1/30$ and $g(x) = \frac{1}{30}(x-2)(x-3)(x+4)$. Similarly,

- $h(x) = B(x+3)(x-3)(x+4)$ and $h(2) = 1$ implies $B = -1/30$, so
$$h(x) = -\tfrac{1}{30}(x+3)(x-3)(x+4)$$
- $k(x) = C(x+3)(x-2)(x+4)$ and $k(3) = 1$ implies $C = 1/42$, so
$$k(x) = \tfrac{1}{42}(x+3)(x-2)(x+4)$$
- $\ell(x) = D(x+3)(x-2)(x-3)$ and $\ell(-4) = 1$ implies $D = -1/42$, so
$$\ell(x) = -\tfrac{1}{42}(x+3)(x-2)(x-3).$$

Now putting $f = 12g + 22h + 72k - 26\ell$, we have, after simplification:

$$f(x) = 2x^3 + 4x^2 - 8x + 6.$$

You can check that f matches the table.

This method is called *Lagrange Interpolation*, and it applies to any finite set of input–output pairs. Careful readers will get the feeling that this looks a great deal like the ideas behind the Chinese Remainder Theorem in arithmetic; the ideas are applied here to polynomials rather than integers.

More precisely, the Chinese Remainder Theorem provides a method for classifying all integers that leave prescribed remainders when divided by a set of relatively prime divisors [see Cuoco and Rotman (2013), for example]. And the remainder theorem from second year algebra implies that we are looking here for a polynomial f so that

$$f(-3) = 12 \Leftrightarrow \text{the remainder when } f(x) \text{ is divided by } (x+3) \text{ is } 12$$

$$f(2) = 22 \Leftrightarrow \text{the remainder when } f(x) \text{ is divided by } (x-2) \text{ is } 22$$

$$f(3) = 72 \Leftrightarrow \text{the remainder when } f(x) \text{ is divided by } (x-3) \text{ is } 72$$

$$f(-4) = -26 \Leftrightarrow \text{the remainder when } f(x) \text{ is divided by } (x+4) \text{ is } -26.$$

So, we have the same problem, but in a different ring—the ring $k[x]$ where k is a field. And the *proof* of the Chinese Remainder Theorem carries over almost word for word to this setting. Digging deeper, we see that this similarity is due to the structural similarities between the two rings—both are principal ideal domains (Cuoco & Rotman 2013).

Interlude

Once again, we have genuine applications of abstract algebra to the work of teaching high school mathematics.

- Newton's difference formula provides a real application to the many, many references to Pascal's triangle that run throughout precollege education. It can also be used to tie together many of the results that live in the folklore of school mathematics—that constant third differences imply a cubic fit, for example. And the use of the Mahler basis leads to coherent and elegant developments for the various summatory polynomials (the sum of the first n values of the kth powers of integers) that are often used in calculus class.
- Lagrange interpolation and associated ideas from high school algebra can be used to classify all polynomial functions that agree on a finite set of data. This often gives high school teachers a nice "trick" (but it's not a trick) to find many functions that match a table, allowing for a topic we call "fooling the test makers"—many standardized tests (used to) ask students to "find *the* rule" that agrees with a table.
- And the fact that the method is the Chinese Remainder Theorem for polynomials gives an example of the structural similarities between the two main systems in school mathematics. Here's a pair of problems from EDC (2014).

5: At Sasha's party. Tony presents the following puzzle: "I'm thinking of a number. If I divide it by 3, the remainder is 2. If I divide it by 5, the remainder is 3. If I divide it by 7, the remainder is 1. What's my number?"

a: What number might Tony be thinking of?
b: Is there more than one integer that fits Tony's puzzle? If so, name two of them. If not, explain why.

6: Later that night, Derman takes the floor and presents the following puzzle: "I'm thinking of a polynomial. If I divide it by $x - 3$, the remainder is 16. If I divide it by $x - 5$, the remainder is 42. If I divide it by $x - 7$, the remainder is 84. What's my polynomial?"

a: What polynomial might Derman be thinking of?
b: Is there more than one polynomial that fits Derman puzzle? If so, name two of them. If not, explain why.

Parting Thoughts

There are many more examples like this, algebra applied to the work of teaching. The same holds for other branches of mathematics; analysis connects to geometry, linear algebra connects to probability—the list is long. Most undergraduate mathematics courses include applications to professions in which mathematics is used, and this chapter adds to the call to include the teaching profession as one of the sites for this kind of applied mathematics.

But there is only so much time in 4 years of college. The CBMS report on teacher preparation (CBMS 2012) details some of the ways that teachers can continue their study of "mathematics applied to teaching" after they enter the profession. Indeed, inservice professional development provides opportunities for years of professional learning, a chance to fine-tune the uses of mathematics in teaching.

For the past several decades, I've been working together with high school mathematics teachers in a variety of collaborations. I'm convinced that there is an untapped treasure of applications that lives in what teachers know and do, just waiting to be mined for ideas for teacher preparation programs.

References

CBMS. (2012). *The mathematical preparation of teachers–II*. Providence RI and Washington DC: American Mathematical Society and Mathematical Association of America.
Cuoco, A. (2000). Meta-problems in mathematics. *College Mathematics Journal, 31*, 5.
Cuoco, A. (2005). *Mathematical connections: A companion for teachers and others*. Washington, DC: MAA.

Cuoco, A., & McCallum, W. (2018a). The double continuity of algebra. In G. Kaiser et al. (Eds.), *Invited Lectures from the 13th International Congress on Mathematical Education* (pp. 49–63). ICME-13 Monographs. Hamburg, Germany: ICMI.

Cuoco, A., & McCallum, W. (2018b). Curricular coherence in mathematics. In Y. Li et al. (Eds.), *Mathematics matters in education*. Advances in STEM Education. Springer International Publishing.

Cuoco, A., & Rotman, J. (2013). *Learning modern algebra from early attempts to prove fermat's last theorem*. Washington, DC: MAA.

EDC. (2014). *The CME Project*, a four year high school program. Pearson.

Graham, R., Knuth, D., & Patashnik, O. (1992). *Concrete mathematics*. Reading, MA: Addison Wesley.

Kerins, B. (2003). Gauss, Pythagoras, and Heron. *Mathematics Teacher, 96*, 350–357.

Lang, S. (1967). *Algebra*. Reading, MA: Addison Wesley.

Wasserman, N. (2015). Unpacking teachers' moves in the classroom: Navigating micro- and macro-levels of mathematics complexity. *Educational Studies in Mathematics, 90*(1), 75–93.

Chapter 19
Impact of Abstract Algebra on Teachers' Understanding of and Approaches to Instruction in Solving Equations

Eileen Murray and Erin E. Baldinger

In 2012, the College Board of the Mathematical Sciences (CBMS) published their second report focused on teacher education. The *Mathematical Education of Teachers II* (*MET-II*; CBMS, 2012) intends to be a resource for undergraduate teachers of mathematics, specifically those who are responsible for teaching prospective PreK–12 mathematics teachers. Among the main goals of the report were to provide updated recommendations for teacher preparation and professional development in mathematics. In the recommendations for teacher preparation, the report advocated that future high school mathematics teachers take advanced mathematics courses, such as abstract algebra. Currently, among 4-year institutions with secondary pre-service teaching certification programs, 89% of all mathematics departments require their students to take abstract algebra (Blair, Kirkman, & Maxwell, 2013). Despite this requirement and the recommendations from the *MET-II*, there is a need for more research on how knowledge of university-level mathematical content is employed in teaching practice, as well as how secondary teachers view their university courses as relevant to their teaching (Zazkis & Leikin, 2010).

This chapter presents a study investigating how a workshop focused on abstract algebra impacted teachers' understanding of secondary content and pedagogy. By discussing how the workshop impacted the ways teachers understand the mathematics they will teach and how they will teach it, our goal is to better understand how knowledge of the connections between advanced and secondary

E. Murray (✉)
Department of Mathematical Sciences, College of Science and Mathematics, Montclair State University, Montclair, NJ, USA
e-mail: murrayei@montclair.edu

E. E. Baldinger
Department of Curriculum and Instruction, College of Education and Human Development, University of Minnesota, Saint Paul, MN, USA

N. H. Wasserman (ed.), *Connecting Abstract Algebra to Secondary Mathematics, for Secondary Mathematics Teachers*, Research in Mathematics Education,
https://doi.org/10.1007/978-3-319-99214-3_19

content may be related to the teaching and learning of mathematics. That is, we hope to advance our understanding of the specific abstract algebra content that could provide secondary teachers with the knowledge they need to make best use of mathematical connections in ways that not only improve their own understanding of algebra, but also allow them to make pedagogical decisions that can help students develop deeper understandings of algebra.

We begin the chapter by describing what we mean by connections in terms of mathematical content in abstract algebra and ways of thinking about and engaging with that content. We tie this to instructional practice to explain how we envision the influence that understanding mathematical connections can have on a teacher's practice. We then present the nature of the workshop and discuss the design and redesign of the workshop materials. We close by describing how teachers' involvement impacted the participants' mathematical understandings, as well as their views on how they might teach students about particular content. Specifically, we describe how instruction in and exposure to algebraic structures and their properties influences teachers' understanding of and approaches to teaching students about the mathematical properties used to solve equations in secondary algebra classrooms.

Background

Most teacher preparation programs in the USA require abstract algebra for prospective secondary teachers. Unfortunately, the standard topics of groups, rings, and fields are not often well connected to secondary mathematics within these courses (Cuoco & Rotman, 2013). This lack of connecting content being learned to secondary mathematics could be due to facilitators' insufficient knowledge of the connections themselves (Hodge, Gerberry, Moss, & Staples, 2010). Even so, if the goal is to build prospective teachers' content knowledge in a way that supports their teaching, abstract algebra facilitators should infuse links between advanced and secondary mathematics (Blömeke & Delaney, 2012). So, while most university facilitators may consider abstract algebra relevant for secondary teachers because the content is inherently related, we argue that the connections between abstract algebra and secondary mathematics that directly influence teaching should be explicitly discussed. Such discussions can be precisely about the content (e.g., polynomials and polynomial rings), but should also include discussions of specific ways of knowing and thinking about mathematical connections (e.g., precision of language). In the subsequent sections, we discuss four ways in which we view connections that may help bridge the gap between abstract algebra and secondary mathematics.

Algebraic Structures and Their Properties

Mathematicians and mathematics educators have discussed content connections between abstract algebra and secondary mathematics for decades (e.g., Arcavi, 1995; Barbut, 1987; Cuoco, 2001; Ferrini-Mundy, Floden, & McCrory, 2005; McCrory, Floden, Ferrini-Mundy, Reckase, & Senk, 2012). Some examples include developing the complex numbers by using the central technique of reducing modulo a prime (Cuoco, 1997), proving the fundamental theorem of algebra and relating the theorem to finding roots of quadratics in school mathematics (Baldinger, 2013), or understanding inverse function as an object rather than the "undoing" of an action (Wright, Murray, & Basu, 2016). While many more examples exist, we focus on the role of algebraic structures and their properties because of their pervasiveness in the school curriculum and the historic difficulties students face understanding these concepts.

Inverse and identity are two critical focal points in both abstract algebra and school mathematics. Students in lower secondary school (Bryant, Christie, & Rendu, 1999), and even young children (Gilmore & Spelke, 2008; Rasmussen, Ho, & Bisanz, 2003), can reason with inverse additive relationships. This supports the idea that the concept of inverse can be considered a fundamental human activity and does not depend on specific computational understanding, which provides an interesting backdrop when one considers cases of inverses and identities in university-level mathematics, such as abstract algebra (Plaxco, 2015). Even so, we cannot assume that understanding inverses and identities can be easily developed in new contexts for students. In fact, research points to issues students may have with understanding, for example, the identity element around quantifiers or unfamiliar operations. Specifically, students could inappropriately borrow from known groups, or operations, and rely on the assumption that zero or one would be the only possibilities for an identity (Melhuish, 2015).

For instance, lower secondary school students' perception of number systems is challenged with the introduction of negative rational numbers and mathematical operations on these numbers (Wasserman, 2016). During this time, subtraction of one positive number from itself (e.g., $8 - 8$) tends to be reconsidered as the equivalent operation of adding one positive number with its opposite (e.g., $8 + -8$). Teachers can help students make sense of the relationship between addition and subtraction through the properties of operations (e.g., commutativity, associativity of addition, additive inverses), and the connection that subtraction can be regarded as finding an unknown addend (CBMS, 2012).

Students can also struggle transitioning from the concept of inverse operation to inverse element, especially with respect to inverse function (Wright et al., 2016). The introduction of the function concept and the binary operation composition compounds the difficulties of coordinating inverse operation and inverse element. As composition is developed as a binary operation associated with a set of functions, inverse elements become inverse functions, which may seem different than more familiar inverse operations. Knowledge of abstract algebra may help teachers understand these various applications and conceptions of identity and inverse. This knowledge may additionally help teachers and students develop their conceptual understanding of inverse as "'objects' in a set based on connecting inverse function to abstract algebra" (Wasserman, 2016, p. 37). Finally, attention to identity and inverse may assist teachers in understanding that the concepts of multiplicative inverse, additive inverse, inverse matrix, and inverse function are examples of the same idea (CBMS, 2012).

Connections are about more than just mathematical content. They also include specific ways of knowing and thinking about mathematics. In 2001, the National Research Council (NRC) published *Adding it Up: Helping Children Learn Mathematics* (NRC, 2001). The purpose of the book was "to address the conflicts in current proposals for changing school mathematics by giving a more rounded portrayal of the mathematics children need to learn, how they learn it, and how it might be taught to them effectively" (p. xiv). Since this time, the book has informed much of the discussion around the teaching and learning of mathematics, including the *MET-II* publication. In the next three sections, we discuss ways of engaging in mathematical ideas. These categories are overlapping and interconnected, but provide a way for us to look more broadly at the ways of knowing and doing mathematics, as both teachers and learners.

Language in the Mathematics Classroom

In order for students to develop mathematical proficiency, they need experience exploring mathematical properties, justifying solutions, and analyzing problems (NRC, 2001). We know this because mathematical knowledge includes understanding the discipline itself, including the nature of discourse in mathematics. In order for students to learn how to communicate mathematically, they must be provided opportunities "to offer solutions, make claims, answer questions, and provide explanations" (NRC, 2001, p. 345). One of the main objectives for discourse in the classroom should be to help students develop understanding of key mathematical ideas. Beyond content benefits, mathematical discourse in the classroom "provides opportunities to emphasize and model mathematical reasoning and problem solving and to enhance students' dispositions toward mathematics" (NRC, 2001, p. 346).

An important component of mathematical discourse is precision of language. According to Barwell (2005), when it comes to mathematical language, "Any ambiguity, that is, any possibility of more than one interpretation for a mathematical expression arises from sloppy use of language rather than any uncertainty of mathematical ideas" (p. 118). Barwell argues that as students explore mathematical ideas, they are better able to participate in mathematical practices and further develop their mathematical thinking. In this way, ambiguity "acts as an important resource for students and teachers, serving as a means of articulating between thinking and discourse" (p. 125). However, it is also true that as "students encounter algebraic ideas, they discover the value of precise language and of working with clear definitions" (NRC, 2001, p. 280).

As secondary mathematics teachers engage in learning advanced content, such as abstract algebra, they are afforded the opportunity to communicate about new mathematical ideas as a learner. By engaging in the activity of communicating mathematically, prospective and practicing teachers may reconsider how to talk about mathematical ideas and the relevance of mathematics by connecting approaches, content, and principles.

In addition to communicating mathematically as learners and doers of mathematics, teachers can also think about their use of language in the classroom as it pertains to discourse. In managing classroom talk in meaningful ways, teachers make judgments about questioning, scaffolding, and prompting students' participation. Researchers have explored the nature (and significance) of classroom discourse for mathematics teaching and learning (e.g., Herbel-Eisenmann & Cirillo, 2009). As teachers begin to help students engage as a community of learners, they may find that the purpose of discourse within their classrooms allows both the teacher and the students to learn more mathematics, since "[s]tudent conversation is focused on exploring and explaining mathematical ideas and their connections" (Manouchehri & St. John, 2006, p. 546).

One way in which teachers can promote student engagement in, and conversations about, mathematical ideas is through questioning. However, in order to use questioning effectively, it is important to begin with what the students know, while also having clear goals for what is to be learned (Stein & Smith, 2011). According to Stein and Smith, teachers can prepare for instruction by adequately anticipating student responses to tasks. In this way, teachers can apply questions at appropriate moments that help students explore mathematical meanings, prompt students to explain their thinking, and generate discussion among students. These types of questioning techniques scaffold students' thinking in a way that helps them "think harder and more deeply about the ideas at hand" (p. 62). As secondary mathematics teachers engage in learning advanced content, they are afforded the opportunity to deepen their own understanding of mathematical concepts, such as algebraic structures and properties. This understanding may aid teachers in being better able to anticipate student responses (and difficulties), which in turn can guide instructional actions, such as questioning.

The notions about mathematical discourse and precision of language are relevant for both teaching and learning mathematics. As people learn mathematics, and in the process of doing mathematics, teachers and students may better understand the importance of "refining ill-formed ideas through the use of precise language" (CBMS, 2012, p. 68). The focus on language is also supported by the types of examples and tasks used during instruction, as the breadth and depth of what is done in the classroom can either highlight or hide these important differences.

Use of Examples and Tasks

Using multiple examples and non-examples is an important practice for mathematics teaching and learning. In fact, examples and non-examples are "associated with conceptualization and definitions, and serve to highlight critical features of a concept; as well as counter-examples that are associated with claims and their refutations" (Zodik & Zaslavsky, 2008, p. 165). Therefore, examples and non-examples can aid in learning about mathematical ideas through conceptualizations, generalizations and abstractions.

The importance of tasks in the mathematics classroom and for instruction is also highlighted in the *MET-II* report, which points to the need for teachers to understand how the choice of numbers, manipulatives, or problem contexts may impact student learning (CBMS, 2012). Teachers should understand how to pose tasks that are able to focus on particular mathematical ideas. Recent research has shown that students can be provided with opportunities to use mathematical practices (e.g., generalization) and can see the intended objects of learning better when they are presented with a series of tasks that vary in an intentional way (Barlow, Prince, Lischka, & Duncan, 2017). Hence, facilitating teacher learning about mathematics and the curriculum may help teachers provide high-quality lessons using carefully selected examples and tasks. Such lessons can advance students' conceptual understanding and support students' higher-order thinking (Murray, 2011).

In order for students to develop all five strands of mathematical proficiency (conceptual understanding, procedural fluency, strategic competence, adaptive reasoning, and productive disposition (NRC, 2001)), a teacher's ability to choose appropriate content, representations, tasks, and materials is based on an understanding that effective mathematics learning is in part a function of students' engagement in tasks. Mathematics tasks used in secondary instruction directly influence students' opportunities to learn. In fact, "[r]esearchers have recently taken a closer look at instruction by investigating the choice and use of academic tasks. Tasks are central to students' learning, shaping not only their opportunity to learn but also their view of the subject matter" (NRC, p. 335).

For example, consider the use of the properties of number used in secondary algebra. A teacher may assume that students already understand the properties of commutativity or associativity, and therefore, might not think to include examples or tasks specifically highlighting the properties. However, if a teacher has had the opportunity to explore properties of non-commutative groups during an abstract algebra course, they may be more inclined to emphasize the non-triviality of commutativity and help students see how it is not a property of the operation alone.

> That is to say, if one only considers subtraction and/or division for examples of non-commutativity, one might miss the point that commutativity is a characteristic of particular group structures. Because both subtraction and division are also not associative, they do not exemplify a group structure, and thus do not help illuminate the connection between this property, the operation, and the set. (Murray, Baldinger, Wasserman, Broderick, & White, 2017, pp. 5–6)

The point in being aware of how examples and tasks impact student learning is critical in helping students develop appropriate mathematical reasoning skills.

Habits of Mind

In considering ways of thinking and engaging in mathematics, we take into account mathematical habits of mind (Cuoco, Goldenberg, & Mark, 1996) and standards for mathematical practice (e.g., National Governors Association Center for Best Practices & Council of Chief State School Officers, 2010). These habits of mind include the ability to describe and understand examples and non-examples of mathematical concepts, as well as communicating mathematically. However, when we consider habits of mind, we also consider other ways of participating in mathematics, such as looking for patterns, giving precise descriptions, making conjectures, and attending to precision. We use the term mathematical habits of mind to encompass all of these different mathematical practices. Such habits of mind cut across levels and areas of mathematical study (Murray et al., 2017). By including other important ways of thinking about mathematics beyond those discussed in the previous two sections, we include habits of mind (generally) as another way of connecting advanced and secondary mathematics.

Research points to the importance of developing mathematical habits of mind in teacher education. For example, Baldinger (Chap. 11; 2015) illustrated how pre-service teachers engaged in mathematical practices during an abstract algebra course could improve their capacity to enact these practices with respect to secondary content. For the purposes of this work, we consider how engaging in mathematical habits of mind in abstract algebra are similar to and different from engaging in those same habits in secondary mathematics, particularly as they relate to effective instruction.

When secondary teachers have the opportunity to tackle challenging problems in abstract algebra, they sometimes stretch beyond their comfort level and recall the importance of mathematical habits of mind in developing conceptual understanding. These types of experiences may remind teachers to support their own students in using habits of mind in their own learning. Additionally, teachers may use their experiences in abstract algebra to help develop their knowledge about the nature of mathematics, including the forms of argument, justification, and proof for both abstract and school algebra. For example, "simplifying algebraic expressions rely on the properties of rings of polynomials, including distributivity of multiplication over addition, additive inverses, and associativity" (Ferrini-Mundy et al., 2005, p. 32). We argue that to support teachers in developing students' mathematical habits of mind, it is important to help teachers develop deep understandings of the underlying algebraic structures and to see how such structures and their properties connect to various topics within secondary mathematics.

Workshops

To explore how a workshop focused on abstract algebra impacts teachers' understanding of secondary mathematics and approaches to instruction, we conducted two workshops with preservice and in-service teachers, both facilitated by the first author. The first workshop, our pilot, took place over one 2.5-h class session for prospective and practicing teachers enrolled in a mathematics master's degree program with a concentration in mathematics education. The learning goals were to consider the importance of mathematical properties in solving equations and the connection between these properties and algebraic structures. The second workshop was a professional development workshop with four in-service teachers (Conor, Dylan, Orlaith, and Aidan; all names are pseudonyms) from an urban high school in the northeastern USA. The workshop consisted of four 3-h sessions held on consecutive days prior to the beginning of the school year. Our findings are based on analysis of the professional development workshop.

In the pilot and the professional development workshops, participants started with secondary content, worked with abstract algebra concepts, and ended with an activity situated in a secondary classroom scenario. The secondary content in the workshops focused on solving linear and quadratic equations. The abstract algebra content focused on groups, rings, and fields, with particular attention to properties in these structures, including zero divisors. Participants worked in small groups to solve problems, and shared their solutions and reasoning in whole class discussions.

Each session concluded by returning to a classroom scenario. The purpose of using classroom contexts to frame some of the conversation was motivated by research on the use of vignettes and cases with prospective and practicing teachers (Hauk, Toney, Jackson, Nair, & Tsay, 2014; NRC, 2001). The goal of presenting vignettes to help teachers reflect on practice is to illustrate how effective teaching should extend beyond exact transmission of facts or understanding of information to include the mathematical and social background and experiences of the students in the room (Hauk et al., 2014). In fact, research has shown that teachers' learning in professional development workshops that employ cases from real classrooms can help them develop a deeper understanding of mathematical structures and connections among mathematical ideas, as well as gain a greater range of ways to represent mathematical ideas (NRC, 2001).

Because of the difference in duration between the pilot and professional development workshops, there were notable changes made between the two workshops. One major difference was the amount of time spent on abstract algebra content. The professional development workshop had more time to discuss the abstract algebra content, which was divided into two units: the first unit focused on groups, and the second unit focused on rings and fields. Together, these two units took 3 out of the 4 days of the professional development workshop. The second major difference was that, in the professional development workshop, we incorporated scripting, or "Lesson Play," activities (Zazkis, 2017), rather than just conversing about a classroom situation. This decision was informed by our experiences in the pilot workshop with the nature of the participants' conversation when they discussed their own work versus exploring student work. We also considered research, which suggested that lesson play activities were effective for preparing for instruction and providing insights into participants' mathematical understandings (Zazkis & Zazkis, 2014). We hypothesized that using scripting activities would shift teachers' thinking during discussions of teaching practice from focusing primarily on planning to more intentionally imagining responses to student contributions and attending to student learning.

Workshop Units

The unit on groups began by considering a student's work solving a linear equation (see Fig. 19.1). Participants reflected on the nature of the work and then engaged in a conversation about the mathematical properties the student used in the student's solution strategy. The idea was to elicit thinking about what mathematical ideas students use when solving simple equations. This supported the focus on exploring connections between abstract algebra and solving equations in secondary school because classic algorithms for solving linear equations require repeated use of the field axioms (e.g., associativity, existence of inverse and existence of identity), as well as other theorems (e.g., addition or multiplication property of equality) (McCrory et al., 2012). For the unit on rings and fields, the initial activity was also student work, but this time accompanied by a short explanation from the student about his method (see Fig. 19.2). After independently considering the student's work and explanation, participants discussed the validity of the method and the properties employed in the solution strategy. The objectives were to have the teachers: (1) discuss the logic, (2) try to connect this example to the work done to define groups, and (3) to motivate two new algebraic structures (rings and fields).

After the initial classroom scenarios, participants engaged in exploring abstract algebra content. The unit on groups included developing the group axioms using symmetries (cf., Larsen, 2009, 2013), formally defining a group, and exploring examples and non-examples of groups. The abstract algebra content concluded with solving equations using the group S_3. In the unit on rings and fields, participants were provided definitions of these algebraic structures. After a brief discussion about the properties of rings and fields and the relationship of rings and fields to groups, participants first explored the set of integers modulo n, $\mathbb{Z}_n = \{0, 1, 2, \ldots, n - 1\}$, where each element represents the equivalence class consisting of all those integers having a remainder of r when divided by n. Using this set, teachers generated Cayley tables for $\mathbb{Z}_5$ and $\mathbb{Z}_6$ under both addition and multiplication, and discussed the patterns they observed in the tables. Participants then worked on solving linear and quadratic equations in $\mathbb{Z}_5$ and $\mathbb{Z}_6$, and discussed the differences in solutions based on the algebraic structure (see Fig. 19.3).

Both units ended with a scripting activity (see Appendix 1). The purpose of the activity was to see how the teachers incorporated the abstract algebra concepts

Groups Warm-Up

During a lesson on solving multistep equations, Mr. Hatton asked his students to solve the equation $4(5x - 11) = 16$. While walking around the class looking at what the students were writing, he noticed the following strategy:

$$\frac{1}{4} \cdot 4(5x - 11) = 16 \cdot \frac{1}{4}$$

$$5x - 11 = 4$$

$$5x = 15$$

$$x = 3$$

What mathematical properties are used for each individual step?

Fig. 19.1 Group unit warm-up

Rings and Fields Warm-Up

Ms. Murray asked her students to solve the following quadratic equation:

$$3x^2 - 6x - 24 = 0.$$

As Jibran explains his solution, he says, "I added 24 to both sides and divided by 3."

$$3x^2 - 6x = 24$$

$$x^2 - 2x = 8$$

$$x(x - 2) = 8$$

He then concludes, "The only numbers that are 2 apart and multiply to be 8 are 2 and 4, and -2 and -4, so x has to be 4 or -2." Students agreed that 4 and -2 work when you substitute them into the original equation, but they were unsure about his method.

- What mathematical properties did Jibran use to solve this equation?
- What would you say to Jibran about his method? Explain your thinking.

Fig. 19.2 Rings and fields unit warm-up

and ideas into their simulated interactions with students. After completing both units, the participants responded in writing to reflection questions. Their responses provided further information about how their experiences with abstract algebra may have impacted the participants' understanding of mathematical content, connections between abstract algebra and secondary mathematics, and how these understandings may impact their future instruction.

Solving Equations in $\mathbb{Z}_5$ and $\mathbb{Z}_6$

Solve the following equations in $\mathbb{Z}_5$ and $\mathbb{Z}_6$. Describe the mathematical properties you use for each step in your solution. Think about how solving equations in this context is similar to/different from what your students do when they solve equations.

1. $2(x+3) = 1$
2. $4x = 2$
3. $x^2 + 3x + 2 = 0$
4. $(x+1)^2 = (x+1)(x+3)$

Fig. 19.3 Solving equations in $\mathbb{Z}_5$ and $\mathbb{Z}_6$

Table 19.1 List of codes

Theme	Codes
Mathematical Content	Algebraic structures • Building structures with particular properties and understanding the elements of the sets (including when the elements are the same) • Understanding binary operations and how they "act" with different sets • Defining structures with particular properties
	Properties • Understanding what properties are, what they do, and the characteristics of properties • Understanding properties in different structures • Understanding the ways in which properties are used to answer questions (e.g., solving equations)
Pedagogical Practice	Language
	Examples and tasks
	Habits of mind

Data Collection and Analysis

Each day of the professional development workshop was video recorded and transcribed for analysis. Other data collected included posters and other materials used during the workshops, and all of the teachers' written work. We used thematic analysis (Braun & Clarke, 2006) to identify the main features of the workshop. Two major topics emerged in the analysis: mathematical content and pedagogical practices, each with several contributory codes (see Table 19.1).

We used these themes to divide the transcripts into episodes. Each episode consisted of a section of transcript about a particular topic. An episode ended when the focus of the conversation changed. Researchers coded independently and then resolved any disagreements through discussion.

After coding all of the transcripts, we applied the same set of codes to teachers' written reflections. These reflections occurred at two time points during the workshop. Teachers responded to the first set of questions after completing the rings and fields packet on Day 3, and responded to the second set of questions at the end of the workshop on Day 4. The first set of questions focused explicitly on connecting algebraic structures to solving equations: (1) What does it mean to think about solving equations of different algebraic structures (group, ring, or field)?; (2) How is this similar/different to how you think about solving equations?; and (3) In what ways, if any, have these discussions impacted your thinking about instruction of solving equations for students? The second set of questions focused more on the properties of identity and inverse: (1) What does it mean to think about an identity or inverse in groups, rings, and fields?; (2) How is this similar/different to how you think about identity and inverse in secondary mathematics?; and (3) What ways, if any, have these discussions impacted your thinking about instruction of the identity and inverse for students? We used the teachers' responses to these questions to provide evidence of the potential impact that the different ideas raised in the professional development workshop might eventually have on their teaching.

In total, there were 168 (139 content; 29 pedagogical practice) coded episodes. Episodes were associated with exactly one theme, and within each theme, episodes could receive multiple codes. One hundred and forty-four episodes came from transcripts of the workshops (129 content; 15 pedagogical practice), and 24 episodes came from participants' reflections (10 content, 14 pedagogical practice).

Influence of Workshops

Both the focus on mathematical content and the discussions about pedagogical practices contributed to shifting teacher thinking regarding how they might teach solving equations. In the following sections, we first consider episodes where mathematical content related to algebraic structures, binary operations, and defining structures were evident in participants' engagement in the workshop activities and their written reflections. Then we explore episodes related to pedagogical practice, focusing on language, examples and tasks, and habits of mind. Though there were far fewer episodes related to pedagogical practice, they remained a persistent topic throughout the data. Taken together, the attention to mathematical content and pedagogical practices helped inform the way teachers were thinking about how to improve their own teaching practice.

Mathematical Content

Algebraic Structures

During the workshop activities, participants had multiple opportunities to work with foundational content. Through the activities, participants worked on determining elements of various sets with particular properties and understanding the elements in those sets. They also worked on developing their understanding of binary operations and how they "act" with different sets. The participants grappled with these different sets and operations in a way similar to how they would normally work with real numbers. They explored sets of symmetries as they unpacked the definition of a group. They also explored the similarities and differences between $\mathbb{Z}_5$ and $\mathbb{Z}_6$, while learning about the distinctions between rings and fields.

We coded 43 episodes from the workshop transcripts and written reflections, as relating to work on algebraic structures. In 15 of those episodes, participants worked on carefully constructing various algebraic structures and understanding the elements in those structures. In particular, they worked on identifying when elements are the same. For example, in one of the first activities focused on groups, the participants were working with symmetries of an equilateral triangle. For each symmetry, participants wrote a verbal description, drew a diagram, and created a symbol to represent the symmetry. The participants were trying to determine how to combine symmetries and which combinations would be equivalent. In the episode below, we see Dylan and Orlaith considering which combinations of symmetries might be equivalent to the 240° rotation.

Dylan: I mean eventually you'll do the same as if you rotated it [the triangle] two times [120 degrees each time] . . . if you—you end up getting the same thing.
Orlaith: So, that would be our 240?
Dylan: Yeah.
Orlaith: Same as 240?
Dylan: Yeah, but it's a different way of getting there instead of rotating.

In a similar way, as the participants worked with modular arithmetic in the unit on rings and fields, they discussed the meaning of elements of $\mathbb{Z}_5$.

Facilitator: So, you said the numbers are going to be zero, one, two, three, four.
Dylan: Yeah, those are remainders.
Facilitator: So, can you explain what the element two is in that set?
Dylan: Well, so for example, if you're seven divided by five with the remainder of two—
Conor: One remainder two.
Dylan: —so it would be two, seven. Oh, maybe—
Conor: So, is seven a number in the set or is two the number?
Dylan: No, two would also be a number cause technically two . . . Wait, would two be a number?—
Orlaith: [inaudible]
Dylan: —Right. Yeah. It's still a remainder of two. Two divided by five the remainder is two.
Conor: Yeah.
Dylan: So, that will be—so, it will be 2, 7, 12, 17, 22, etc.

Through this discussion, Dylan explicitly stated which elements would be equivalent to two in $\mathbb{Z}_5$. The teachers continually returned to this work of understanding the elements in sets throughout the units on groups, rings, and fields.

In 25 episodes, the participants were working on building their understanding of binary operations. They were pushed to move beyond their understanding of operations on the real numbers and explored how operations "act" with different sets. One way this happened was through exploring non-commutative contexts, which challenged the ways the teachers thought about the properties of the operations. For example, the teachers explored symmetries of equilateral triangles, and found that different combinations of rotations and reflections could produce equivalent results.

Facilitator:	So, we see that we have a lot of equivalent pairs, right? So, if I look at this, right? I see that R_{120} [120-degree rotation] with F_C [reflection over median through vertex C] is equal to F_A. But the other things equal to F_A are R_{240} with F_B and then R_{360} with F_A. These are all the same. So, we have like these are all equivalent to F_A.
Orlaith:	Uh huh.
Facilitator:	That's why we see this like equivalent diagonal there [in the Cayley table].
Conor:	Hm.
Orlaith:	And it's whichever way you do it, right?
Facilitator:	Well, that's a good—what do you mean whichever way you do it?
Orlaith:	So, if you look at R_{120} first and then F_C, it's gonna be F_A. Now, let's look at F_C first.
Facilitator:	Okay.
Orlaith:	And R_{120}. So no it's not the same.
Facilitator:	F_B.
Orlaith:	So, it's going to be F_B.
Facilitator:	So, what does that tell you about the way we're combining the symmetries?
Conor:	Is it appropriate to say that they're not commutative?
Orlaith:	Yes.

Relatedly, they examined whether or not operations such as subtraction and division could be used for defining a group using real numbers.

Facilitator:	Yes, real numbers with subtraction, is it a group?
Conor:	Yes.
Dylan:	Well, I don't necessarily know—
Facilitator:	How do you know?
Orlaith:	Yeah. I would say that cause you're looking at—
Dylan:	—I don't necessarily know if it's associative though cause you do a minus b is not the same as saying or if you do the quantity of a minus b minus c that's not necessarily gonna be the same as your...
Orlaith:	That's true. Uh huh.
Conor:	Ohhhh!
Facilitator:	Because those two things are the same.
Dylan:	So, might not be associative at least.

Written reflections also captured how teachers were thinking about binary operations. For example, when reflecting on identity and inverse, Aidan wrote, "It means to determine what operations we are dealing with. Identity or inverse mean different things based on what mathematical concept." Conor wrote, "Groups, rings, and fields have different properties that govern them. Therefore, they have different operators that are valid or invalid." Orlaith wrote, "How you think about it is determined by what operation you are looking at. You may or may not have inverse or identities, therefore may or may not have groups." These reflections all show that the focused work on binary operations that occurred during the workshop helped shape how the participants interpreted the meaning of identities and inverses.

Finally, in nine of the episodes, the teachers were defining structures with different sets. They explored the formal definitions of groups, rings, and fields and made determinations about whether given sets, such as $\mathbb{Z}_5$ and $\mathbb{Z}_6$, were groups, rings, or fields under addition or multiplication. Conor noted this important feature when he wrote that "groups, rings, and fields have different properties that govern them."

Properties

Mathematical properties emerged as one of the greatest areas of focus during the workshop, due to how the teachers were interacting with the workshop materials. This focus was also prevalent in the teachers' reflections. There were three main ways in which properties were taken up. First, there was work on understanding what the properties are for groups (closure, associativity, identity, and inverse); on understanding how the properties work; and on understanding how the properties relate to one another. Second, there was work exploring how the properties behaved in various structures, including if commutativity or associativity held in certain contexts, or noting how the properties are the same or different in various structures. This kind of work caused the teachers to revisit their understandings of properties, with which they were already familiar, and deepen the way they thought about them. Finally, there was work understanding how the properties might be used to answer questions, such as solving equations. We coded 110 episodes from the workshop and written reflections as relating to mathematical properties.

Understanding properties appeared in 34 episodes. When working on problems together, participants often paused to re-define the properties for themselves, especially the definitions for associativity and the identity and inverse elements. For example, on the third day of the workshop, after having discussed the identity and inverse in several examples, the participants had to revisit these elements for the integers modulo n as they were building and exploring the Cayley tables for addition and multiplication in $\mathbb{Z}_5$ and $\mathbb{Z}_6$.

Dylan:	So, when you're adding something to zero.
Facilitator:	Right.
Dylan:	So, it would be like one plus zero is itself, one.
Orlaith:	Oh! O.K. Yes.
Dylan:	Two plus zero is itself, two.
Orlaith:	O.K.
Dylan:	So, you're not looking at the actual solution but you're looking at the—the set of zero, like the—the whole column of zero and the whole row of zero.
Conor:	Oh! So, that's what you're looking at the column and row of zero to make sure that when you add zero to something you get that same...
Dylan:	You get itself.
Orlaith:	You get the same thing.
Dylan:	Yes.
Conor:	That's how we're checking to make sure the additive identity holds. Got it.

The participants also explored the ways in which the properties interacted with one another. This focus on defining properties was a significant emphasis for the classroom scenarios, where the teachers had to describe what properties were being used in each step of a student's solution for solving an equation. For example, in the first scripting activity (see Appendix), the participants looked at a conversation about solving $3(x-2)^2 = 6(x-2)(x+5)$. One hypothetical student suggested dividing by the term $x - 2$, after the equation had been simplified to $(x-2)^2 = 2(x-2)(x+5)$. At first, the participants did not see a problem with this. With some prompting by the facilitator (to discuss which property the student was using in order to simplify the equation by first dividing by three and then dividing by $x - 2$), they began to understand how this could be problematic when the term did not have a multiplicative inverse, namely when $x = 2$. At the end of the workshop, Dylan reflected on the importance of the definitions of properties in his writing about inverse and identity. He wrote, "This is something I have always struggled with. In order to think of identity or inverse, one must really have a clear idea and definition of each. We worked on today, that depending on your definition, changes everything."

By far the most common focus in the workshops and reflections was understanding how the properties work in different structures. This came up in 67 episodes. Teachers explored the properties in symmetric groups, in $\mathbb{Z}_5$ and $\mathbb{Z}_6$, and in the real numbers. They had multiple conversations trying to unpack the similarities and differences regarding the identity element in different contexts. For example, when determining the elements for the set of symmetries of the equilateral triangle with composition as the binary operator, the participants thought about what properties the 360° rotation would have.

Facilitator:	Do we have a zero here?
Orlaith:	Yes, our zero is three—R_{360}.
Dylan:	Hm.
Orlaith:	Well, actually, it's a one.
Dylan:	Yeah.
Facilitator:	So, what's the difference between zero and one?
Dylan:	One is . . .
Orlaith:	There's no change.
Facilitator:	Cause yesterday we said it was a one.
Orlaith:	Yeah.
Facilitator:	And today we're saying it's a zero. So, what's the difference between . . .
Orlaith:	No, it's not a zero. The—If it was a z—If it was a zero then anything times—well, anything union 360 would be zero.
Dylan:	Not 360. Yeah.
Orlaith:	Right. Not 360, so it really is a one.
Dylan:	Yeah.
Facilitator:	So, when you say it really is a one, what is it acting like?
Orlaith:	Identity.
Facilitator:	Identity.
Conor:	Uhm. Identity.

Facilitator:	What kind of identity? It's really interesting to me that you're saying zero is coming into the mix here. So like one is what identity? What kind of identity is one?
Dylan:	A whole. In probability terms it's the one whole.
Orlaith:	Unit. Unit identity. I'm not sure exactly.
Facilitator:	Yeah. So, what I'm pushing you to think about is like when we say something is the identity.
Orlaith:	Uh huh.
Facilitator:	Right?
Dylan:	It's itself.
Facilitator:	But what does that mean? Like we say that we have an identity for the set, so what—what does that mean? Why do you call this the identity?
Orlaith:	Cause it doesn't make any change.
Facilitator:	Doesn't make any change.
Orlaith:	It doesn't change anything.
Facilitator:	So, when I combine it with anything else, I just get that thing I combine it with.
Orlaith:	Right.

The properties were also explored by examining the Cayley tables of different groups, which provided a concrete way to investigate the properties of a particular algebraic structure. Reflecting on different algebraic structures, Conor wrote,

> My mentality, like a student taking Algebra II (the subject I teach), we solely focused on the set of all real numbers when solving equations. Therefore, because $\mathbb{R}$ is a field, my thinking was oriented to the ability to use more operators. And, more specifically, those operators that apply to $\mathbb{R}$ and fields. So solving equations for me has traditionally been in that context.

The properties were influential in how Conor was starting to think about how abstract algebra might connect to his own teaching.

The main goal of the workshop was to consider mathematical properties used when solving equations and their connections to algebraic structures. Therefore, it was not surprising that we found 38 episodes focused on the use of properties to solve equations. The participants had the opportunity to solve equations in a variety of contexts, such $\mathbb{Z}_5$ and $\mathbb{Z}_6$. In addition, the participants connected properties and solving equations to their school contexts when they discussed solving equations over the real numbers through the scripting activities. Dylan, reflecting on solving equations over different algebraic structures, wrote:

> As I previously mentioned, I had to think deeply and consider alternative methods. Similar—in terms of you are trying to find solutions. Different—in terms of how you find solutions. Solving equations like this took me out of my "comfort zone" but in a good way. I have a better understanding of solving equations.

Orlaith wrote, "It is different because it shows that there are different ways that equations can be solved with the same answers being obtained." These reflections indicate how the teachers were beginning to make sense of multiple approaches to solving equations, and how those approaches were related to the underlying properties and algebraic structures. The focus on using properties to solve equations, along with the time spent exploring properties in different contexts, supports the idea that the workshop materials had their intended effect; namely, encouraging teachers to delve deeper into how the properties are used as tools, and how the structures involved have an impact.

Pedagogical Practices

In addition to the work on content, participants also explicitly attended to pedagogical practices, which might serve as a connection between abstract algebra and secondary content. We examined the instances of attention to language, examples and tasks, and general habits of mind in the workshops and in the teachers' reflections. In total, we coded 29 episodes as relating to these three categories, with some episodes having multiple codes.

Language

Attention to language occurred in 13 episodes. During the workshops, the teachers would stop and reflect on how they might talk with their students about a particular idea. For example, in a discussion about the elements in a symmetry group, the teachers noted that a rotation of 480° would be equivalent to a 120° rotation, and need not be included in their list of elements.

Facilitator:	Right? But if I did a 480 rotation, would that be the same as any of these?
Conor:	Yeah.
Facilitator:	It would be. So, it's an equivalent symmetry.
Conor:	Right.
Facilitator:	So, do we need it?
Orlaith:	We don't need it.
Facilitator:	We don't need it.
Orlaith:	But, I mean, if you're teaching students you can make that point.
Facilitator:	Right, understanding that there are these equivalencies. Same thing like you have equivalent fractions, right?
Orlaith:	Right.
Facilitator:	One half is two fourths, is three sixths, right?
Conor:	Sure.
Orlaith:	Yeah.
Facilitator:	So, it's this idea that we have these equivalent ideas in mathematics.

In this discussion, we see how the teachers begin to make the connection between equivalent elements in this abstract context (symmetries of an equilateral triangle) and equivalencies in school content, through being attentive to how they talk to students about equivalencies. The facilitator bolsters that connection by drawing the teachers' attention to the topic of equivalent fractions. The connection here focuses on how a teacher might talk about equivalence with her students in the future.

Dylan reflected on this at the end of the workshop in his writing about inverse and identity. He wrote, "I have always thought of the word inverse as something undoing something. After this week, I need to be careful with my math vocabulary.

This PD has pushed me into being more precise with my words." In her reflections on inverse and identity, Orlaith wrote, "It makes me think about how important knowing [and] emphasizing definitions [and] understanding of definitions are to teaching [secondary] students." These two reflections illustrate the ways in which the teachers were reexamining their approach to language in their teaching.

This focus on language was also driven by the facilitator bringing explicit attention to language issues throughout the workshop. For example, the participants were working on solving equations in the context of a symmetric group. As they were discussing the properties used that enabled them to solve the equation, the facilitator paused the discussion to have them reflect on how naming these properties (e.g., associative, commutative) is often left implicit when solving equations with students. The teachers agreed that they were guilty of this as the facilitator helped the teachers reflect on their use of imprecise language such as "canceling" when they were actually talking about the use of inverses.

Facilitator:	Or what else do you say all the time? It "cancels."
Dylan:	Uh huh.
Orlaith:	Yeah, we do say it cancels. [...]
Conor:	They cancel, I don't know.
Dylan:	I rarely use cancel.
Facilitator:	I hear it all the time. Right? These cancel. We—I heard it here yesterday.
Dylan:	It adds up to zero, divides out to one, and that's what I use. (Laughs)
Facilitator:	Right.
Orlaith:	Yeah, you say it divides out.
Facilitator:	But, when you say zero and one, you're also slightly hiding this idea of identity in numbers, too. Cause like you just say "adds to zero," right? Assuming that they understand that zero is the additive identity and so then-
Conor:	Right.
Facilitator:	Right? So, it's like we make all these assumptions that our students understand these things.
Conor:	Right. Yea, but if you were to stop every time you're solving an equation, you wouldn't say "oh, well these two... are... because zero is the additive identity." You're like—You're not—You're not gonna stop in the middle of it.
Facilitator:	No, but you might do it at the beginning.

In this conversation, we see how the facilitator also highlighted the use of language during teaching, and how particular words, such as cancel, can be ambiguous for students. The teachers recognized this ambiguity as being related to the use of the word for different underlying mathematical ideas (e.g., multiplicative versus additive inverse). Even so, we see the teachers continue to struggle with how they might alter their current teaching to attend to the ambiguity, while keeping the content accessible to their students.

Examples and Tasks

Attention to examples and tasks came up in six of the episodes. One episode occurred after a discussion about whether or not subtraction and division are associative. Orlaith made a connection to the warm-up problem, in which the teachers had analyzed a student's solution of an equation. In the first step, the student had multiplied both sides of the equation by $\frac{1}{4}$, and the teachers had initially discussed this as an instance of the division property of equality.

Orlaith:	So, that's why looking at that first problem that we had, we multiplied by one fourth instead of dividing by [four]. And then would that have something also to do with that divided by $(x + 2)$?
Facilitator:	Yes, exactly because if I am constantly thinking about it as division then, yeah, I can divide. Why can't I divide?
Orlaith:	Right.
Facilitator:	But if I'm thinking in terms of multiplication and trying to find the multiplicative inverse, I realize, oh wait a minute, that could be zero, zero doesn't have a multiplicative inverse, then suddenly that is not a possibility.
Conor:	Hm.
Orlaith:	Wow!
Facilitator:	I know right?
Orlaith:	This is like very ...
Conor:	This really does affect our instruction!

This dialogue shows how the teachers were thinking about the relationship between the properties and the tasks they might use with students. Conor's comment at the end also suggests their recognition of how studying abstract algebra might impact their instruction.

Dylan reflected on how he might change the examples or tasks he used in his class. In his reflection on identity and inverse, he wrote:

> I will be lesson planning completely different. Our first unit is all about solving and inverse operations. I must change how I approach my introduction throughout this unit because I want my students to not have misconceptions as they approach higher levels.

He saw a connection to his teaching practice, recognizing that his first unit of the year would focus on inverse operations. He began to imagine alternative tasks he might use with his students to enhance their understanding about the role that identity and inverse elements play in inverse operations and in solving equations.

Habits of Mind

In coding for habits of mind, we looked for episodes about mathematical practices beyond language and examples and tasks. These general habits of mind included connecting representations or general problem solving strategies. These ideas came up in 13 episodes in which both the participants and the facilitator generated ideas connected to habits of mind. For example, as they worked on completing the script for the classroom scenario in which hypothetical students debated whether both sides of an equation can be divided by $x - 2$, the participants considered how to support students in developing habits of mind through discussion. The facilitator reinforced observations on supporting mathematical practices by sharing her own experiences with connecting representations.

Facilitator:	And so I can do that numerically but I can also do that graphically thinking about these two things [$x(x - 2)$ and 8] as two separate functions and when

	these two functions are equivalent. Which is something I think important for our students especially in Algebra I and Algebra II is helping them make connections between multiple representations. That's like a huge thing in the standards...
Orlaith:	Yes.
Conor:	For sure.
Facilitator:	...is for us to be able to associate these multiple representations.

Teachers and the facilitator were able to contribute important discussion points around habits of mind through the activities, especially in terms of the scripting activities and classroom scenarios.

Habits of mind were actually more prevalent in their reflections than in their conversations, showing that attention to habits of mind may be something the teachers picked up on in the workshops that would impact their teaching. All four of the teachers made related comments. For example, Orlaith wrote:

> It gives a basis for why we can go about solving equations in specific ways and will assist in allowing students to explore how to solve equations. It will also allow for more thought provoking questions to be raised which will hopefully allow for students gaining a deeper understanding of math and how the principles are integrated.

Dylan wrote, "The biggest take away is to never hold down a student. There are multiple pathways to solve the same problem. Never hinder a student's thinking process." Aidan wrote, "Solving an equation is not only about applying procedural steps to isolate the variable and come up with the solution. But to think deeply about the concept by exploring its methods and misconceptions, to also establish connections." Conor wrote, "I realize the importance of generalizing the definitions as they apply to different sets of numbers."

Though there were fewer episodes related to pedagogical practices than there were related to content, spending an extended amount of time on the content may be part of the reason participants had "a-ha" moments about things they do in the classroom and the way they might approach the content. These experiences seemed to help make the pedagogical connections to language, examples and tasks, and habits of mind stronger and more salient for the teachers. At the conclusion of the workshop, participants were more aware of connections between advanced algebra and secondary mathematics than they had been at the beginning and they each began to reconsider their own practice in different ways.

Discussion

The goal of the workshop was to engage teachers in an exploration of connections between abstract algebra and secondary mathematics. In particular, we were interested in better understanding how instruction in and exposure to algebraic structures and their properties might influence secondary teachers' understanding of and approaches to instruction focused on solving equations. We found that working through abstract algebra tasks problematized the participants' understanding of the properties of real numbers, including associativity, commutativity, inverse, and identity. This was evident in the participants' need to revisit basic definitions of the properties and how the properties could be recognized in various contexts. We found that teachers greatly benefitted from working to make sense of the mathematical properties of groups, rings, and fields repeatedly through different contexts. These many opportunities seemed to deepen participants' conceptual understanding. The need for repeated experiences with these concepts could be a result of the participants not having seen enough examples to be able to generalize the mathematical objects as an algebraic structure, or because of the difficulty of transferring mathematical knowledge (Bransford et al., 2006; Bransford & Schwartz, 1999). Regardless, it is important for teachers to have a productive struggle with mathematical content they may feel comfortable with in order for them to gain deeper conceptual understanding (CBMS, 2012). This deeper understanding can guide teachers in their efforts to confront student thinking and handle misunderstandings.

As one example, participants in this study often had to work through several examples to begin to generalize the related properties and algebraic structures. Exploring different examples was important to the generalization process. One potential implication is that if one of the goals of an abstract algebra course is related to helping prospective and practicing teachers generalize ideas about solving equations (a primary aim of our workshop), then they likely need to solve equations on several different structures. In a typical abstract algebra course, while examples of different algebraic structures are frequently discussed, time is not necessarily taken to solve equations on those different structures. If we want prospective and practicing teachers to develop generalizations about solving equations, it seems that experience solving equations on different structures could be productive, especially by having them identify how solving equations on those structures is similar or different. Such insights about solving equations more broadly might help teachers connect to the secondary content of solving equations with real numbers and prompt further pedagogical reflections.

The influence of the workshop on the teachers' thinking about instruction is evident, mostly through the written reflections at the conclusion of day 3. The participants highlighted the importance of emphasizing mathematical properties, using specific language during instruction, and making connections across content. Additionally, as the participants worked through the abstract algebra activities, they were able to demonstrate higher levels of reasoning and began to be aware of how their own use of language was critical in justification and argumentation. This development is comparable to students' abilities to demonstrate improved levels of reasoning when they are encouraged to reflect and communicate their mathematical thinking "both in their precision of language and in their use of arguments" (NRC, 2001, p. 285). The workshop placed the teachers in the position of learners by presenting abstract algebra tasks focused on foundational concepts relevant to secondary mathematics. This gave them the mental space to consider how the explicit inclusion of associativity, commutativity, inverse, and identity in their teaching could impact students' understanding of solving equations.

One limitation of this study is that the data used to explore the influence on approaches to instruction focused on solving equations is based on teachers' self-reflections. Without looking at teachers' actual classroom practice, we cannot claim that these influences changed instruction. Nevertheless, by engaging in these activities, the teachers were at least made aware of possible strategies they might employ to strengthen student understanding of solving equations.

Conclusion

As mathematicians and mathematics teacher educators, we understand the vertical connections between abstract algebra and secondary mathematics, both in terms of content and mathematical practices. We also know that abstract algebra is a challenging subject for most undergraduates, making it difficult for prospective and practicing teachers to connect the advanced and secondary content in meaningful ways. Therefore, for these four participants to make strides in their understanding of the content and to perceive impacts on their teaching is a positive result from the workshop. Even so, we need to better understand how we might be able to get secondary teachers to think through advanced content either as prospective teachers or in their practice to really understand the foundational concepts in algebra that span the grade levels.

Based on our experiences, we recommend that teachers engage in sustained ways with key mathematical ideas. Evidence from this study indicates that revisiting key concepts multiple times is important to improved understanding. It is through an improved understanding that teachers may make changes in practice—a superficial understanding is unlikely to lead to truly sustained changes. Not only do teachers need to understand the mathematics, they need to understand it in a way that makes it worthwhile to change routine teaching practices. For example, deeply understanding what an inverse is ought to influence the language a teacher uses when talking about inverses in different school contexts. The deep understanding developed through sustained attention with abstract examples should motivate the teacher to use more precise language because they understand why that language matters and how it might influence student understanding.

We also recommend including scenarios or scripting tasks as touchstones for teachers to highlight the connections between abstract algebra and their teaching practice. The teachers regularly referred back to the scenarios and scripting activities throughout the workshop. They used the scripts as ways to talk about their own mathematical thinking and to better understand each other's thinking. The workshop also provided space for the teachers to discuss their own practice, and, in doing so, they were able to identify for themselves how the ideas from abstract algebra might be relevant to their teaching. In many ways, the recommendation to include teaching scenarios in an abstract algebra workshop parallels the recommendation that secondary mathematics teaching include the use of problems set in "real life" contexts. In this case, the "real life" context is teaching mathematical ideas, and embedding abstract algebra work in those "real life" teaching contexts can help teachers recognize their application in practice.

Appendix 1: Scripting Tasks

Groups

Given below is the beginning of a conversation between a teacher, Mrs. Fryling, and two students, Samara and Francisco. Your task is to extend this imaginary interaction in a form of dialogue between the teacher and students. Include explanations and/or examples that you would use.

Mrs. Fryling:	Class, today we're going to solve $3(x-2)^2 = 6(x-2)(x+5)$. Who has an idea about how to start?
Samara:	I was thinking about dividing both sides by 3.
Mrs. Fryling:	Ok, if we do that, what will we get?
Samara:	We'd get $(x-2)^2 = 2(x-2)(x+5)$.
Mrs. Fryling:	Ok. What next?
Samara:	Now we could divide both sides by $(x-2)$.
Francisco:	You can't divide by $(x-2)$.
Samara:	Why not? If you can divide both sides by 3, why can't you divide by $(x-2)$?

Rings and Fields

Given below is the beginning of a conversation between a teacher, Mr. Freeman, and his students. Your task is to extend this imaginary interaction in a form of dialogue between the teacher and students. Include explanations and/or examples that you would use.

Mr. Freeman:	Let's look at the equation $x^2 - 4 = 0$. What do you think we can do to determine the values for x that satisfy this equation?
Avi:	I think you should add four to both sides to get $x^2 = 4$.
Mr. Freeman:	How does that help?
Zareen:	Because we know that two squared and negative two squared is 4. So we have our two answers, 2 and -2.
Mr. Freeman:	Then what if we started with $x^2 + 4 = 0$?
Avi:	We can just do the same thing, and then take the square root of both sides to find the answer.
Zareen:	That doesn't make any sense! You can't square a number to get -4!

References

Arcavi, A. (1995). Teaching and learning algebra: Past, present, and future. *Journal of Mathematical Behavior, 14*(1), 145–162.

Baldinger, E. E. (2013). Connecting abstract algebra to high school algebra. In M. Martinez & A. Castro Superfine (Eds.), *Proceedings of the 35th annual meeting of the North American Chapter of the International Group for the Psychology of Mathematics Education* (pp. 733–736). Chicago, IL: University of Illinois at Chicago.

Baldinger, E. E. (2015). Pre-service secondary teachers learning to engage in mathematical practices. In T. G. Bartell, K. N. Bieda, R. T. Putnam, K. Bradfield, & H. Dominguez (Eds.), *Proceedings of the 37th annual meeting of the North American Chapter of the International Group for the Psychology of Mathematics Education* (pp. 608–615). East Lansing, MI: Michigan State University.

Barbut, E. (1987). Abstract algebra for high school teachers: An experiment with learning groups. *International Journal of Mathematical Education in Science and Technology, 18*(2), 199–203.

Barlow, A. T., Prince, K. M., Lischka, A. E., & Duncan, M. D. (2017). Developing algebraic reasoning through variation in the U.S. In R. Huang & Y. Li (Eds.), *Teaching and learning mathematics through variation* (pp. 321–339). Rotterdam: Sense Publishers. https://doi.org/10.1007/978-94-6300-782-5_18

Barwell, R. (2005). Integrating language and content: Issues from the mathematics classroom. *Linguistics and Education, 16*(2), 205–218. https://doi.org/10.1016/j.linged.2006.01.002

Blair, R., Kirkman, E. E., & Maxwell, J. W. (2013). *Statistical abstract of undergraduate programs in the mathematical sciences in the United States: Fall 2010 CBMS survey*. Providence, RI: American Mathematical Society.

Blömeke, S., & Delaney, S. (2012). Assessment of teacher knowledge across countries: A review of the state of research. *ZDM Mathematics Education, 44*(3), 223–247. https://doi.org/10.1007/s11858-012-0429-7

Bransford, J. D., & Schwartz, D. L. (1999). Rethinking transfer: A simple proposal with multiple implications. *Review of Research in Education, 24*(1), 61–100. https://doi.org/10.3102/0091732X024001061

Bransford, J. D., Stevens, R., Schwartz, D. L., Meltzoff, A., Pea, R. D., Roschelle, J., et al. (2006). Learning theories and education: Toward a decade of synergy. In P. A. Alexander & P. H. Winne (Eds.), *Handbook of educational psychology* (2nd ed., pp. 209–244). Mahwah, NJ: Lawrence Erlbaum.

Braun, V., & Clarke, V. (2006). Using thematic analysis in psychology. *Qualitative Research in Psychology, 3*(2), 77–101. https://doi.org/10.1191/1478088706qp063oa

Bryant, P., Christie, C., & Rendu, A. (1999). Children's understanding of the relation between addition and subtraction: Inversion, identity, and decomposition. *Journal of Experimental Child Psychology, 74*(3), 194–212. https://doi.org/10.1006/jecp.1999.2517

Conference Board of the Mathematical Sciences. (2012). *The mathematical education of teachers II. Issues in mathematics education*. Providence, RI and Washington, DC: American Mathematical Society and Mathematical Association of America Retrieved from https://www.cbmsweb.org/archive/MET2/met2.pdf

Cuoco, A. (1997). Constructing the complex numbers. *International Journal of Computers for Mathematical Learning, 2*, 155–186.

Cuoco, A. (2001). Mathematics for teaching. *Notices of the AMS, 48*(2), 168–174.

Cuoco, A., Goldenberg, E. P., & Mark, J. (1996). Habits of mind: An organizing principle for mathematics curricula. *Journal of Mathematical Behavior, 15*(4), 375–402. https://doi.org/10.1016/S0732-3123(96)90023-1

Cuoco, A., & Rotman, J. R. (2013). *Learning modern algebra: From early attempts to prove Fermat's last theorem*. Washington, DC: Mathematical Association of America.

Ferrini-Mundy, J., Floden, R. E., & McCrory, R. (2005). Knowledge for teaching school algebra: Challenges in developing an analytic framework. In *American Education Research Association Annual Meeting*, Montreal, Quebec, Canada.
Gilmore, C. K., & Spelke, E. S. (2008). Children's understanding of the relationship between addition and subtraction. *Cognition, 107*(3), 932–945. https://doi.org/10.1016/j.cognition.2007.12.007
Hauk, S., Toney, A., Jackson, B., Nair, R., & Tsay, J.-J. (2014). Developing a model of pedagogical content knowledge for secondary and post-secondary mathematics instruction. *Dialogic Pedagogy: An International Online Journal, 2*, 16–40. https://doi.org/10.5195/dpj.2014.40
Herbel-Eisenmann, B. A., & Cirillo, M. (Eds.). (2009). *Promoting purposeful discourse: Teacher research in mathematics classrooms*. Reston, VA: National Council of Teachers of Mathematics.
Hodge, A. M., Gerberry, C. V., Moss, E. R., & Staples, M. E. (2010). Purposes and perceptions: What do university mathematics professors see as their role in the education of secondary mathematics teachers? *Primus, 20*(8), 646–663. https://doi.org/10.1080/10511970902810733
Larsen, S. P. (2009). Reinventing the concepts of group and isomorphism: The case of Jessica and Sandra. *Journal of Mathematical Behavior, 28*(2–3), 119–137. https://doi.org/10.1016/j.jmathb.2009.06.001
Larsen, S. P. (2013). A local instructional theory for the guided reinvention of the group and isomorphism concepts. *Journal of Mathematical Behavior, 32*(4), 712–725. https://doi.org/10.1016/j.jmathb.2013.04.006
Manouchehri, A., & St. John, D. (2006). From classroom discussions to group discourse. *Mathematics Teacher, 99*(8), 544–551.
McCrory, R., Floden, R. E., Ferrini-Mundy, J., Reckase, M. D., & Senk, S. L. (2012). Knowledge of algebra for teaching: A framework of knowledge and practices. *Journal for Research in Mathematics Education, 43*(5), 584–615. https://doi.org/10.5951/jresematheduc.43.5.0584
Melhuish, K. M. (2015). *The design and validation of a group theory concept inventory*. Unpublished doctoral dissertation, Portland State University, Portland, OR.
Murray, E. (2011). *Implementing higher-order thinking in middle school mathematics classrooms*. Unpublished doctoral dissertation, University of Georgia, Athens, GA.
Murray, E., Baldinger, E. E., Wasserman, N. H., Broderick, S., & White, D. (2017). Connecting advanced and secondary mathematics. *Issues in the Undergraduate Mathematics Preparation of School Teachers, 1*(August), 1–10.
National Governors Association Center for Best Practices, & Council of Chief State School Officers. (2010). *Common core state standards for mathematics*. Washington, DC: Author.
National Research Council. (2001). Adding it up: Helping children learn mathematics. In J. Kilpatrick, J. Swafford, & B. Findell (Eds.). Washington, DC: National Academies Press. https://doi.org/10.17226/9822
Plaxco, D. (2015). *Relating understanding of inverse and identity to engagement in proof in abstract algebra*. Unpublished doctoral dissertation, Virginia Polytechnic Institute and State University, Blacksburg, VA.
Rasmussen, C., Ho, E., & Bisanz, J. (2003). Use of the mathematical principle of inversion in young children. *Journal of Experimental Child Psychology, 85*(2), 89–102. https://doi.org/10.1016/S0022-0965(03)00031-6
Stein, M. K., & Smith, M. (2011). *Five practices for orchestrating productive mathematics discussions*. Reston, VA: National Council of Teachers of Mathematics.
Wasserman, N. H. (2016). Abstract algebra for algebra teaching: Influencing school mathematics instruction. *Canadian Journal of Science, Mathematics and Technology Education, 16*(1), 28–47. https://doi.org/10.1080/14926156.2015.1093200
Wright, M., Murray, E., & Basu, D. (2016). Impact of abstract algebra knowledge on the teaching and learning of secondary mathematics. In T. P. Fukawa-Connelly, N. E. Infante, M. Wawro, & S. Brown (Eds.), *Proceedings of the 19th Annual Conference on Research in Undergraduate Mathematics Education* (pp. 481–495). Pittsburgh, PA: Mathematical Association of America.

Zazkis, R. (2017). Lesson Play tasks as a creative venture for teachers and teacher educators. *ZDM Mathematics Education, 49*(1), 95–105. https://doi.org/10.1007/s11858-016-0808-6

Zazkis, R., & Leikin, R. (2010). Advanced mathematical knowledge in teaching practice: Perceptions of secondary mathematics teachers. *Mathematical Thinking and Learning, 12*(4), 263–281. https://doi.org/10.1080/10986061003786349

Zazkis, R., & Zazkis, D. (2014). Script writing in the mathematics classroom: Imaginary conversations on the structure of numbers. *Research in Mathematics Education, 16*(1), 54–70. https://doi.org/10.1080/14794802.2013.876157

Zodik, I., & Zaslavsky, O. (2008). Characteristics of teachers' choice of examples in and for the mathematics classroom. *Educational Studies in Mathematics, 69*(2), 165–182.

Chapter 20
Blue Skies Above the Horizon

Ami M. Mamolo and Peter D. Taylor

Introduction

> How can secondary teacher education approaches, ones that use abstract algebra as it might be applied in secondary teaching situations, inform the way we think about connecting abstract algebra to secondary mathematics, for secondary mathematics teachers?

This is the question we were asked to consider when framing our commentary to the chapters included in this section of the volume, *Connecting Abstract Algebra to Secondary Mathematics, for Secondary Mathematics Teachers*. As we discussed this question, and how we might address it, we found ourselves posing further questions, and we frame our commentary around those. Specifically, we will focus on the following two broad questions:

1. *Connections*. In what ways does abstract algebra connect to secondary school mathematics and how can we understand these connections in terms of teachers' disciplinary knowledge?
2. *Approaches*. How can we then build on these connections to support the development of teachers' disciplinary knowledge?

A. M. Mamolo (✉)
Faculty of Education, University of Ontario Institute of Technology, Oshawa, ON, Canada
e-mail: ami.mamolo@uoit.ca

P. D. Taylor
Department of Mathematics and Statistics, Queen's University, Kingston, ON, Canada

N. H. Wasserman (ed.), *Connecting Abstract Algebra to Secondary Mathematics, for Secondary Mathematics Teachers*, Research in Mathematics Education,
https://doi.org/10.1007/978-3-319-99214-3_20

Connections

Each of the chapters presented in this section exemplifies ways in which abstract algebra may be connected and applicable to secondary school mathematics teaching, with a particular emphasis on connections amongst mathematical content and practices. The authors present different perspectives for why they promote their particular connections: (1) to make the work of teachers easier, (2) to enrich mathematical understanding for pupils, (3) to support teacher responses to contingency, and (4) to highlight relationships amongst mathematical ideas. Somewhat implicit in all of this are the authors' underlying intentions for secondary school mathematical experiences. We have our own views on this, and we articulate some of them here, to help contextualize our discussion of the ideas raised within each chapter. As teacher educators, the overarching goals we hold for secondary school mathematics inform our priorities when structuring experiences for prospective secondary teachers. How we support prospective teachers in developing practices that are in line with our goals relates to our understanding of teachers' disciplinary knowledge, and vice versa. Our understanding of teachers' disciplinary knowledge informs the support we can give prospective teachers in developing practices that are in line with our goals. Thus, our response to the question "in what ways does abstract algebra connect to secondary mathematics and how can we understand these connections in terms of teachers' disciplinary knowledge?" is broken down into the following subsections:

- Goals for secondary mathematics: Blue skies and feet on the ground
- Teachers' disciplinary knowledge: A view of mathematical horizons
- Analyzing connections

Goals for Secondary Mathematics: Blue Skies and Feet on the Ground

A detailed discussion of our goals for secondary mathematics would take us well outside the limits of this chapter, so instead, we offer highlights. First and foremost, our position is that secondary students should be engaging with meaningful, relevant, interesting, and challenging activities and problems. We acknowledge that this is a bit of a "blue sky" goal—many teachers only rarely are able to tackle problems of that kind. Administrators, parents, and students can be reticent toward approaches that diverge from established routines and practices, and too often those practices rely on bite-sized exercises that reinforce rules, processes, and calculations. Nevertheless, our blue sky goals are motivated by pragmatic considerations—we want students to appreciate mathematics, its structure, its relevance, its beauty, and how it connects to their scholarly, professional, and personal life trajectories. For this to work, their teachers must also have such an appreciation.

There are challenges. Disparities continue to exist amongst mathematical practices engaged in at school and mathematical practices needed by mathematicians, those that are applicable to various professions, and those that are valuable for informed citizenry. Cuoco, in his introduction (Chap. 18), speaks of a career-long personal agenda to find ways to close the (huge) gap between school mathematics and mathematics as it is practiced by mathematics professionals. Boaler (2016) and Taylor (2018) both declare that the examples we work with in class should be ones that are of interest to a mathematician—and we interpret this broadly to include various "types" of applied mathematicians. This view, and more general versions of it, goes back a long way, for example, to Whitehead (1929) and Dewey (1934), both of whom emphasized the quality of the student experience. This draws our focus much more to doing rather than knowing. Or if you like, knowledge is wonderful, provided you are doing something interesting with it, which is not always the case in school mathematics.

Let us state an operational version of this. Whenever we, as teachers of mathematics, meet with students at any level, we should bring, along with the knowledge we are offering, an activity or problem that is interesting and that has meaning in the students' lives. Papert (1972) makes this point:

> The important difference between the work of a child in an elementary mathematics class and that of a mathematician is not in the subject matter (old fashioned numbers versus groups or categories or whatever) but in the fact that the mathematician is creatively engaged in the pursuit of a personally meaningful project. In this respect a child's work in an art class is often close to that of a grown-up artist (p. 249).

The analogy between mathematics and art is compelling to us, and we note that several researchers have emphasized the creative and aesthetic character of mathematics (e.g., Barabe & Proulx, 2017; Boaler, 2016; Gadanidis, Borba, Hughes, & Lacerda, 2016; Raymond, 2018; Sinclair, 2006; Taylor, 2018). For us, this reaches back to the theme of Dewey's (1934) *Art and Experience*, that the aesthetic experience is jointly constructed between painter and viewer, performer and audience, that both are called to be artists in a shared experience. We imagine a parallel ideal, where both teacher and student are called to be mathematicians in a shared experience of doing interesting mathematics. To realize such an ideal, teachers need to have the courage and the knowledge to be more independent in how they interpret and enact curricula, so as to work with material that really interests and excites them and their pupils. They need an appreciation for the nature of mathematics, its structure, its relevance, its beauty. We see abstract algebra as a wonderful context in which to foster such an appreciation, and one that has important implications for teachers' disciplinary knowledge.

Teachers' Disciplinary Knowledge: A View of Mathematical Horizons

Mathematical knowledge required for teaching has been widely discussed, with attention focusing on knowledge in teaching, for teaching, and of teachers (e.g., Adler & Ball, 2009; Ball, Thames, & Phelps, 2008; Davis & Simmt, 2006; and many, many others). As part of their extensive work on teacher knowledge, Ball and colleagues introduced the construct of Knowledge at the Mathematical Horizon (KMH), which was described as knowledge which "engages those aspects of the mathematics that... illuminate and confer a comprehensible sense of the larger significance of what may be only partially revealed in the mathematics of the moment" (Ball & Bass, 2009, p. 5). While there is no consensus on how to define KMH, it has been characterized by key elements of a teachers' mathematical knowledge that extend beyond curricular content, such as, knowledge of mathematical structures, practices, and values (Ball & Bass, 2009). Evidence suggests that KMH is an integral part of teachers' disciplinary knowledge, and it has been positively linked to teachers' abilities to respond in the moment to classroom interactions (Fernandez & Figueiras, 2014; Jakobsen, Thames, Ribeiro, & Delaney, 2012; Zazkis & Mamolo, 2011), to plan and extend lessons (Wasserman & Stockton, 2013), and in anticipating and responding to student learning (Mamolo & Pali, 2014).

Our view of the horizon aligns with the perspective of Zazkis and Mamolo (2011), who suggest that KMH is intimately related to a teacher's focus of attention and his or her ability to flexibly shift attention, such that relevant properties, generalities, or connections, which embed particular mathematical content in a greater structure, are accessed in teaching situations. Specifically, KMH is conceptualized as a teacher's knowledge of the horizon of a mathematical object, and draws on Husserl's philosophical notions of inner and outer horizon (Follesdal, 2003). That is, when an individual attends to an object, he or she will focus on particular features of that object—e.g., if you think of a circle, your attention might be focused on its size and shape—while other features of that same object will lie in the periphery—e.g., the equation of that circle, or its position in space. These peripheral features, all of which are specific to the object of thought, lie within the object's inner horizon—they are "aspects of an object that are not at the focus of attention, but that are also intended" (Zazkis & Mamolo, 2011, p. 9). Further in the periphery are "features which are not in themselves aspects of the object, but which are connected to the world in which the object exists" (*ibid*, p. 9), and as such comprise the object's outer horizon. With respect to our circle, the outer horizon includes structures, such as trigonometric identities, conics, and geodesics, as well as ways of working mathematically with circles.

Metaphorically, knowledge of the horizon depends on our location in the terrain. A high vista can offer a broad view of the terrain—where we've been, where we can go, ways we can get there, and potential obstacles along the way. In contrast, anyone who has ever been lost in the woods knows the challenges of navigating your way without such a view. Zazkis and others have argued that

Table 20.1 Interpreting components of KMH

KMH components (Ball & Bass, 2009)	Our interpretation of components	Connections to inner and outer horizons
Mathematical environment surrounding current "location"	Knowledge of how the current subject matter relates to previously learned and future concepts, within and across specific grades	Is influenced by focus of attention (inner horizon) and understanding of the lay of the land (outer horizon)
Major disciplinary ideas and structures	Knowledge of the underlying structural components of mathematics, such as connections between seemingly disparate content	Structure embeds specific content within the greater mathematical world (outer horizon)
Key mathematical practices	Including conjecturing, generalizing, and proving	Ways of engaging in specific practices (inner horizon) within a greater mathematical world (outer horizon)
Core mathematical values and sensibilities	Including precision, axiomatic thinking, and questioning conventions	Ways of being within the greater mathematical world (outer horizon)

studying advanced mathematics can help teachers acquire a broad view by providing access to these "higher vistas." Our conceptualization of KMH illustrates how advanced mathematical knowledge can influence teachers' view of the terrain through a connection we make to the components of horizon knowledge introduced by Ball and Bass (2009). Specifically, Ball and Bass (2009) describe four major components of KMH: (1) a sense of the mathematical environment surrounding the current "location" in instruction; (2) major disciplinary ideas and structures; (3) key mathematical practices; and (4) core mathematical values and sensibilities. These components connect to the notions of inner and outer horizons, as described in Table 20.1. These connections speak most clearly to instances when the object of thought corresponds to mathematical content (e.g., circles, functions), however similar connections exist when the object of thought is more abstract (e.g., proving, recursive thinking, axiomatic thinking). In particular, we observe that knowledge of an object's outer horizon can influence what about that object is in view and what lies in the periphery (inner horizon). This, in turn, influences the view of the mathematical landscape, which embeds that object within a greater mathematical world (outer horizon), as well as provides a map for "instructional locations."

Applying these ideas to examples from the chapters allowed us to analyze some of the ways in which abstract algebra can be connected to, and is applicable for, secondary mathematics teaching. We highlight three examples here—depicted in Figs. 20.1, 20.2, and 20.3—and then discuss further connections in the following subsection.

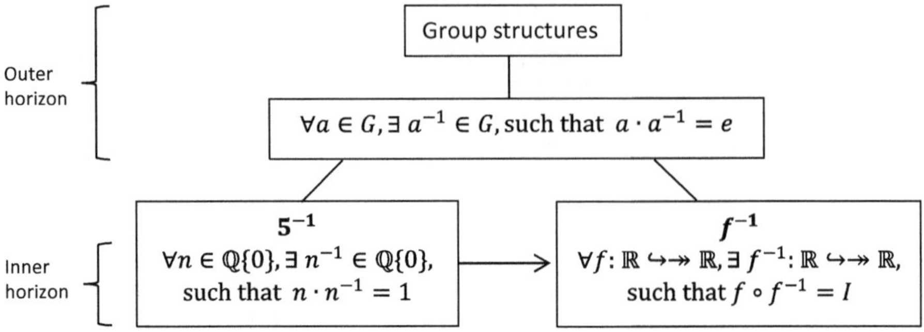

Fig. 20.1 Connecting examples of inverses via structures in the horizon

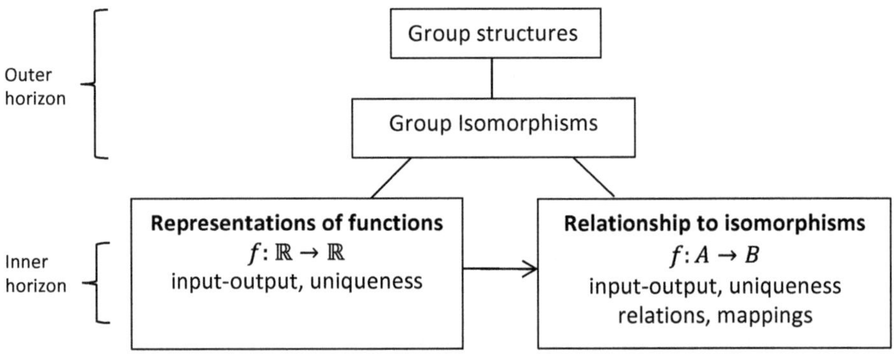

Fig. 20.2 Shifting focus from representations to relationships via broadened horizons

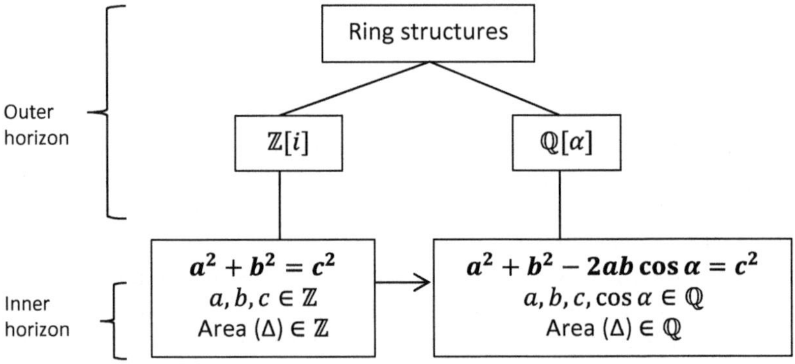

Fig. 20.3 Connecting Pythagorean triples with Heron triangles via structures in the horizon

Figure 20.1 offers an example of how KMH embeds specific examples in a broader context via knowledge of group theory. The example comes from Zazkis and Marmur (Chap. 17), who noted that knowledge of group theory is helpful in appreciating "reciprocal," as a specific instance of "inverse," as it draws attention to the structural similarities of the two. This is in contrast with their observation that many prospective secondary teachers view the concepts as distinct and context-specific. The analysis in Fig. 20.1 focuses on the number 5^{-1}, with some of its intended features lying in the peripheral inner horizon, including, for example, that 5 is a rational non-zero number, its reciprocal exists, and if you multiply 5 by its reciprocal, the product is 1. In the outer horizon are the group-theoretic structures which embed this specific instance in a more general and abstract setting, including the closure property of groups, and the existence of inverse and identity elements. These structures overlap with the outer horizon of f^{-1} and form a pathway for connecting the two examples. In the periphery or inner horizon of f^{-1}, we have depicted what we believe would be likely intended by secondary teachers, namely that the function is a bijection which acts on real numbers.

Figure 20.2 also considers functions and illustrates how KMH can occasion a shift in attention such that different properties of an object fall into view. The example comes from Wasserman and Galarza (Chap. 16), who characterized changes in secondary teachers' portrayal and emphases of functions after their engagement with the Function Module. Prior to the module, teachers emphasized multiple representations of functions (e.g., equations, tables, graphs) and restricted their examples to numerical ones. Figure 20.2 positions representations of functions at the focus of attention, and includes, in the inner horizon, the restriction of f to the set of real numbers and its "every input has a unique output" characterization. The Function Module introduced group isomorphisms and seemed to broaden teachers' awareness of how to think about functions within a greater context. That is, they drew on more abstract examples of functions, they positioned functions as a special kind of more general relations, and they stopped attending to representations of functions, and started attending to relationships between functions and isomorphisms. We highlight this shift in attention, but are careful not to place a value-judgment on it. It is not clear from the chapter whether a shift in attention from function representations to relationships with isomorphisms is a useful shift, even if it is an applicable shift. The usefulness would depend on how the material was enacted with students, and we return to this idea later on.

Our third example, depicted in Fig. 20.3, stems from Cuoco's rich and interesting exploration of Pythagorean triples and Heron triangles. In his Chap. 18, Cuoco develops a method for solving "one of the oldest meta-problems [which] involves the search for Pythagorean triples" (p. 385), and then modifies it to produce Heron triangles. The method abstracts from the particulars of the meta-problem to address it via ring theoretic structures, which are themselves at different levels of abstraction. Cuoco introduces ring structures from $\mathbb{Z}[i]$ to address the search for Pythagorean triples. Contextualizing the ring $\mathbb{Z}[i]$ in a more general and abstract setting allows connections to be made to other contexts. This allows Cuoco to construct the ring $\mathbb{Q}[\alpha]$, whose structures are analogously applied in the search for

Heron triangles. Figure 20.3 illustrates how KMH can forge links between different areas (and grades) of curriculum, via knowledge of ring theory.

Analyzing Connections: Wherefore, Abstract Algebra?

In the preceding subsection we analyzed examples from the chapters through a lens of KMH in order to illustrate how abstract algebra knowledge can inform a teacher's understanding of secondary school mathematics. It is worth questioning whether other areas of advanced mathematics would be similarly applicable, and we suggest that in many cases, more mathematics would not lead to better mathematics. Framing the discussion in terms of KMH helped us articulate for ourselves what in particular about abstract algebra makes it stand out in a crowded chest of mathematical treasures. Zazkis and Marmur suggest that it is precisely the abstract nature of algebraic structures which is so applicable: "While abstraction is regarded as a source of difficulty, it nonetheless possesses the potential to make the algebraic construct relevant and applicable for a large variety of concrete examples and mathematical topics" (p. 365). More than that though, it is structure itself that really stands out for us. It is the explicit attention to structure (e.g. of groups, fields, rings) that is of central importance in the study of abstract algebra—we compare and contrast structures, exemplify and extend them, investigate implications, push boundaries, and play with relationships, all with a sort of directness and cohesion that are not so clearly visible in other areas, such as real analysis.

In the language used to develop KMH, one could say that in abstract algebra, major disciplinary structures become the focus of attention, and key mathematical practices, values, and sensibilities are developed through engaging with these structures. If clear connections to secondary school mathematics can be made (either by or for the individual), then enriched understandings of the mathematical horizon can emerge. Knowledge of abstract algebra can:

- Enrich an object's outer horizon by linking that object to major disciplinary structures and ways of working with them
- Broaden an object's inner horizon by shedding light on previously unknown or unacknowledged properties of the object
- Occasion a shift in attention such that different or more general properties of the object come into view

The applicability of horizon knowledge in general and abstract algebra knowledge in particular, depends on how an individual connects this knowledge to teaching situations. We highlight some connections amongst abstract algebra content and secondary school content, as well as connections amongst abstract algebra understanding and decision-making in teaching situations.

Connections Amongst Abstract Algebra Content and Secondary School Content

A quick review of select curricula shows some direct connections between abstract algebra content and secondary content—vectors, their properties, and performing operations on them are included in, for example, the US Common Core Standards (2010), The National Curriculum in England (2014), the Australian Curriculum (2012), and the Ontario Curriculum (2007), which is used in much of eastern Canada (curricula for central, western, and northern Canada do not address vectors). In looking for other content-specific connections, and given the prominence of functions in various secondary mathematics curricula, it was not surprising to find that three of the four chapters in this section identified ways that abstract algebra could be applicable to working with functions. Wasserman and Galarza identified binary operations and group isomorphisms as examples of functions, and noted that working with binary operations and group isomorphisms could foster "a broader understanding of function," which could in turn "help exemplify nuances within and boundaries around the idea of functions as more than merely symbolic rules" (p. 341). Zazkis and Marmur looked at inverse functions and seemingly invertible functions and noted that "an internalized recognition of the existence of an algebraic structure of mathematical objects (including objects outside the algebraic domain) may aid teachers in their thought process" (p. 373) when clarifying student confusion or designing tasks. They also note that a group theoretic understanding of secondary content "brings the discussion to a higher level of abstraction, where different ideas exemplify the same structure" (p. 371). Structural similarities between rings were exploited by Cuoco when considering fitting functions to tables; his discussion of Newton's Difference Formula and Lagrange Interpolation highlights how the reformulation of problems with abstract algebra structures can shed new light on familiar content. Group, ring, and field structures were also applied to solving equations, number properties and operations, and "meta-problems" involving the measures of triangles.

Connections Amongst Abstract Algebra Understanding and Teaching Decisions

In considering different ways in which the understanding of abstract algebra may connect to teaching situations, we restrict our attention to four aspects related to the mathematical work of teaching:

- *Planning* includes such things as unit and lesson preparation, structuring of courses, and establishing assessment approaches
- *Task design* includes the creation, development, or acquisition, of specific tasks for learning, consolidation, assessment, practice, and so on

- *Norm enactment* includes how the teacher embodies and fosters social and socio-mathematical norms in the classroom, such as expectations for justifying statements, posing questions, or precision in defining terms
- *In-the-moment responses* include teachers' reactions to unanticipated mathematical ideas or utterances offered by students

The applicability of abstract algebra to each of these aspects of teaching was well exemplified by the chapters in this section. In planning activities, there were considerations of which definitions to use, what emphases to place, and which properties to address (Chap. 16); in designing tasks, careful choices in numerical examples were made so as not to "cloud the underlying method" (Chap. 18, p. 384); norms related to problem solving became more salient (Chap. 19); and in-the-moment responses were shaped and influenced by a sense of underlying structure (Chap. 17). Further, these four aspects of the mathematical work of teaching are all clearly interconnected, and each one can influence all of the others. For instance, the norms recognized and valued by a teacher can influence the planning and design of course materials in terms of what and how content is addressed, and affect how moments of contingency are addressed. Murray and Baldinger speak to such interconnectedness with their discussion of how an abstract algebra workshop helped teachers value the importance of precise language in defining concepts, justification and argumentation. The participants in this study were given "the mental space" to consider how valuing precision in their teaching "could impact students' understanding of solving equations" (p. 424). Engaging in workshop activities exposed teachers to the "possible strategies they might employ to strengthen student understanding of solving equations" (p. 425) in their lesson planning, task design, and in response to student questions. In-the-moment responses can serve as catalysts for rethinking planning and design, while also promoting and fostering normative standards of the classroom. Zazkis and Marmur provide an example of how an unexpected student question triggered a reconsideration of the original task and a redesign of the lesson for future teachers. They state: "In both the original and adapted version of the task, basic properties of group theory served as a guide for the instructor's responses and as a basis for further mathematical inquiry" (p. 371).

Approaches

Following the claim of Ball and Bass (2009, p. 11) that "we do not know how horizon knowledge can be helpfully acquired and developed," we sought ideas from the chapters to inform our thinking about this issue. We have thus far presented our view that abstract algebra has the potential to be a useful (and we might add beautiful, fun, enticing) context through which horizon knowledge can be enriched. However, we have not yet addressed the question of "how," and for us this boils down to a look at teaching approaches. For this, we consider the different roles that faculties of education and departments of mathematics each play in preparing secondary teachers. This helps paint a picture of where teachers might reasonably acquire and develop their horizon knowledge. We then look at specific approaches exemplified in the chapters, and highlight similarities and differences in approaches, as they might apply to the different demographics of prospective and practicing teachers. We conclude this section and chapter by applying what we have learned and offer a set of examples that build off of ideas from the chapters of this section. Thus, our exploration of approaches draws on our discussion of ways that abstract algebra connects to secondary mathematics. Specifically, we address the question of "*how can we build on these connections to support the development of teachers' disciplinary knowledge?*" via the following subsections:

- A vision of teacher preparation: Where from, abstract algebra?
- Structuring educational approaches: Broadening horizons
- Extending connections: Blue skies above the horizon

A Vision of Teacher Preparation: Where from, Abstract Algebra?

Murray and Baldinger cite recommendations from the CBMS (2012) report that advocate for the inclusion of courses in advanced mathematics, such as abstract algebra, in the preparation of future secondary mathematics teachers. They note further that "among four-year institutions with secondary pre-service teaching certification programs, 89% of all mathematics departments require their students to take abstract algebra" (p. 403). The implication seems to be that there is recognition that abstract algebra "can sure help" teachers, as Zazkis and Marmur quipped. However, many teacher education programs do not include content-specific requirements for admission, nor does the pool of secondary teacher candidates include students only (or perhaps even mostly) from departments of mathematics. In Canada, for example, a secondary mathematics teacher candidate can gain entry to a bachelor of education program with only three 6.0 credit mathematics courses from their undergraduate studies, and these courses tend to be Calculus I and II, and Statistics. It is important to note that the content and teaching approaches that are common, and even possible, in (typically) large-scale calculus and statistic courses, are significantly different from what is common or possible when studying abstract algebra. Further, there seems to be little impetus for prospective teachers to study abstract algebra in university: calculus and statistics have more apparent connections to school curricula, they have "friendlier" reputations amongst students, and there are often more supports for student learning in these courses. Thus, we agree with Zazkis and Marmur that "it is reasonable to assume that future teachers studying at university are not likely to focus on group theory as an important topic in support of their future career" (p. 365).

We mention this because one of the first reactions we had to the examples discussed in these chapters was that, for better or worse, very few of the problems, activities, or approaches illustrated would be appropriate in a bachelor of education program. The typical focus on "methods" in faculties of education almost precludes the introduction of new (advanced, non-curricular) content, even in content-focused applications, such as analyzing student error patterns. In our experiences with colleagues from various faculties of education, it has been fairly standard to find attitudes ranging from indifference to open hostility toward advanced mathematical knowledge. As such, we suggest that making space for abstract algebra in a bachelor of education program would be a tough sell. Indeed, the authors of this section's chapters might agree with us: the modules discussed in Wasserman and Galarza were developed for a masters-level abstract algebra course, Cuoco's examples stemmed from years of teaching experiences and a personal affinity for mathematics encountered during graduate studies, Murray and Baldinger developed content-specific workshops for practicing teachers, and Zazkis and Marmur exemplified how a teacher educator's knowledge of abstract algebra can help direct prospective teachers' attention to important mathematical structures within school curriculum.

It seems to us that the departments of mathematics might be a more hopeful place where teachers could be exposed to abstract algebra concepts and their connections to secondary school teaching. We recall a conversation with a colleague who taught a masters-level mathematics course called *Abstract Algebra for Teachers*. When asked how "abstract algebra for teachers" was different from "abstract algebra," the response was "Actually, I hadn't thought about that." So, we thought about it. And in reflecting on our own experiences with the subject matter, and with the ideas raised in the chapters, we suggest that first of all, there needs to be a difference, and secondly, there ought not to be much difference at all!

The need for a difference comes from a difference in course objectives and a difference in learners' focus of attention. First, we look at course objectives: For the most part, courses in abstract algebra offered in undergraduate mathematics programs have mathematical content as a primary objective. Thus, they tend to treat topics in a comprehensive manner and move with reasonable speed through the material. Their choice of topics is also significantly influenced by areas of current research interest; for example, the beginning course that all mathematics majors take, typically offered in the second or third year of undergraduate studies, focuses on rings and fields, leaving group theory to later more specialized courses that many majors do not take. On the other hand, a course aimed at teachers could have the development of mathematical thinking and structured play as primary objectives, and comprehensive coverage would not be so important. Indeed, one could imagine a course with a collection of wonderful activities around groups and rings that would *not* start students on the road to a PhD in mathematics, but *would* give them a sense of the power of working with structure, of using aesthetic principles as a guide to the way forward (Sinclair, 2006), and most importantly, how guided play with concrete objects can lead to an understanding of abstract structures.

Having said all of that, the more we interact with undergraduate students in an honours mathematics program, both while they are students and after they graduate, the more we have come to believe that the course that we have just described for teachers would be right for the capacities and needs of these students as well. This is certainly the case for those students who wind up in the general business/industry environment, including the specialized STEM areas, but we feel that it would also be the case for those few who *do* go on to a PhD in mathematics, as these students will presumably already have the capacity to extract theoretical results from mathematically rich particular examples. The key here is to use "low-floor, high-ceiling" activities (Gadanidis et al., 2016; Boaler, 2016) that give all students welcome access and invite more ambitious students to probe more deeply.

To be clear, the courses we are suggesting here are about mathematics, not pedagogy. It is their pedagogical style, focusing on mathematical thinking and investigation, that will set them apart from most of what is currently offered in programs for math majors. We feel that all students would welcome that type of course. Having said *that*, experience with our colleagues in mathematics departments has led us to feel that they may resist such a change in pedagogy. Although wonderful exceptions exist, a comprehensive linear development remains the normative choice for both textbooks and classrooms.

Regarding learners' focus of attention, we have in mind two possible ways teachers may study abstract algebra in support of their career—as prospective teachers in an undergraduate mathematics course, or as practicing teachers in a graduate mathematics education course. In the former case, the primary learning objective would likely align with the primary course objective of the undergraduate program and focus on mathematical content. Thus, their attention would be on their own personal scholarly growth. In the latter case, the primary learning objective of the teacher-student is not solely on personal scholarly growth, but also includes the scholarly growth of their students. In other words, the two sets of learners are differently motivated in their choice of courses. These differences have implications for educational approaches in abstract algebra, as we discuss in the next section.

Structuring Educational Approaches: Broadening Horizons

In our work with prospective teachers, we have seen differences in how they tackle a problem. Some will stare at it with little sense of how to begin, of how to even think of what needs to be done. Others seem to be able to see that "this is a special case of that," or "this belongs over there," or "if I am to show this, I'm going to have to show that first." These students have a grasp of the structure of the problem that allows them to start moving. Recalling metaphors for horizon, it's analogous to finding yourself in the middle of a large unfamiliar city with instructions that you are to be somewhere at a certain time. Having an aerialmap makes a huge difference, particularly if you have experience in reading the map, for example, in how the different modes of potential transport are color-coded, and how you can transfer from one to the other. We suggest that the study of abstract algebra can help develop such a map for the mathematical landscapes of secondary school and further, can foster the ways of working with that map to understand how to get around. That is, the study of abstract algebra can develop the capacity to identify and work with the structure of a problem, and this is a valuable component of a teacher's KMH.

When examining the chapters in this section for ideas of how to structure educational approaches for learning abstract algebra, we found both explicit and implicit suggestions. Cuoco's chapter, although it did not specifically address educational approaches, nevertheless exemplified important considerations. Through the author's own rich view of the horizon, we note an importance in understanding definitions and properties, as well as in connecting different methods to solve the same problem. Cuoco's understanding of mathematical structure is exemplified in his comparisons of algebraic and geometric methods, as well as in connecting methods at different levels of abstraction. For each idea addressed, it is clear that he has a sense of "where this sits in the bigger picture" (p. 393). We found many good problems in this chapter that could be very useful, depending on how they are framed. More on this idea in a bit.

Of the educational approaches explicitly addressed in the chapters, we note the importance of Mason's (2002) "experience of disturbance," as described by Zazkis and Marmur. A disturbance can compel an individual to rethink previous knowledge, to seek out new ideas, and it can elicit a shift in attention, and can enhance awareness—in short, it can broaden horizons. We also appreciated the emphasis placed on question-posing by Murray and Baldinger. They developed questions to explicitly draw teachers' attention toward the similarities and differences of secondary mathematics and abstract algebra. Explicit connections amongst secondary mathematics and abstract algebra were also advocated for by Wasserman and Galarza. We suggest that explicit attention to these connections is valuable for both undergraduate mathematics students, as well as practicing teachers, but for different reasons. The former group would benefit from the opportunity to build on what they have already learned, and it could go a long way for making the material more accessible, without necessarily hampering the agenda of professors hoping to educate the next generation of abstract algebra researchers. The benefits for practicing teachers were discussed by Wasserman and Galarza, who drew on the instructional model developed by Wasserman, Fukawa-Connelly, Villanueva, Mejia-Ramos, and Weber (2017) and Wasserman, Weber, and McGuffey (2017) in the design of their modules. The instructional model "is composed of two parts: building up from and stepping down to practice" (p. 338). The authors elaborate that, "[i]n between building up from and stepping down to practice, the advanced mathematics topics are taught by the instructor in ways true to its advanced nature with formal rigorous treatment" (p. 338). Such a model intends to help teachers appreciate the relevance and role of advanced mathematical knowledge in their teaching practice.

The need to consider how we frame abstract algebra problems, particularly for practicing teachers, relates to the teacher-students' learning objectives mentioned above: their interest is shared between personal growth and the potential to foster their students' growth. Thus, we suggest that what will compel teachers to learn abstract algebra will be different from what will compel undergraduate students, and as such, problems need to be framed differently. Perhaps most significantly, there will be a difference in what might constitute a disturbance of experience – for practicing teachers, that disturbance will likely need to come from classroom practice.

Extending Connections: Blue Skies Above the Horizon

In concluding this commentary chapter, we repeat our main premise: that abstract algebra can provide a wonderful context in which to develop and nurture a genuine appreciation of mathematics. Our engagement with material from the chapters in this section elicited a broadening of our horizons and inspired some blue-sky thinking about valuable ways to engage with mathematics that connect to and extend some of the examples presented. We use this opportunity to offer instructional ideas, which are motivated by our desire to foster interesting, meaningful, and valuable mathematical engagement for secondary teachers.

Order of Operations

Zazkis and Marmur broach this topic. It is a huge idea in mathematics, and indeed the whole notion of linearity, $f(a + b) = f(a) + f(b)$, is about interchanging order of operations, as is the fundamental theorem of calculus. In the task Zazkis and Marmur present, they discuss expressions, all alternating multiplication and division, such as

$$a \div b \times c \div d$$

which were given to prospective teachers. The teachers were told that a student did the calculation by computing the two divisions first, followed by the multiplication, and were asked whether this was correct. We felt that this task could make mathematics seem more confusing and harder than it really is, though our concerns might have been assuaged if we had been provided with more details about exactly how the task was scaffolded and what discussions might have preceded it. Nevertheless, the task offers an interesting context in which to foster important mathematical sensibilities, such as recognizing, challenging, and exploring within the constraints of mathematical conventions. The expression can invite one to first ask what the conventions are and second to explore how many different answers might be obtained with all possible conventions. We connect this to the kind of mathematical thinking that Mason (2001) refers to as searching for freedom within constraint. This could lead to rich mathematical discussions, as well as motivate why mathematicians generally avoid the use of the symbol $\div$, preferring that students get into the habit of writing expressions such as

$$\frac{a}{b} \cdot \frac{c}{d}$$

where the conventions are clear.

Seemingly Invertible Functions

Another task presented by Zazkis and Marmur defined a pair of functions f and g from $\mathbb{R}$ to $\mathbb{R}$ to be "seemingly invertible" if

$$f\,(g(x)) = x$$

for every x in $\mathbb{R}$. The question given to the teachers was whether this implied that g is invertible. This task certainly has mathematical meaning and could potentially connect with a number of different kinds of functions and transformations. Similar to their other example, we found ourselves wondering more about the details of how the authors engaged teachers in the investigation. For example, in our first reading of the paper, we were unsure how we might manage to get the teacher candidates that we have experience with, to come up with the exponential-logarithm example, as it seems to be unintuitive at first. More generally, our feeling is that teacher candidates, along with most graduates of first-year calculus courses, are uncomfortable "playing" with functions of a real variable.

Here we give an example of how we might envision the scaffolding of the task. We would first make sure that the students knew what an invertible function was, and being wary of their facility with formal notation, we would ask them to present the definition with a simple diagram and a finite domain. We would expect a diagram such as Fig. 20.4.

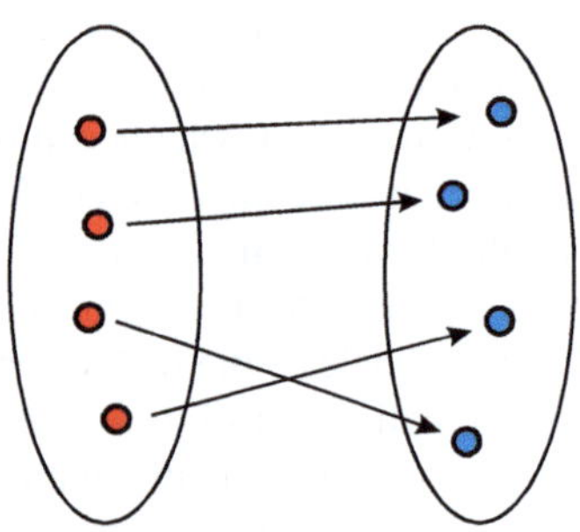

Fig. 20.4 A simple example of an invertible function

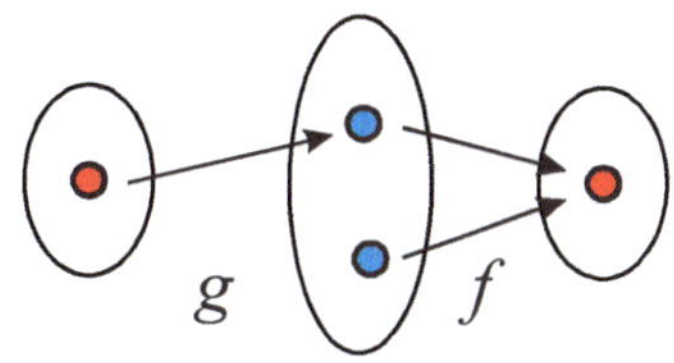

Fig. 20.5 The simplest example in which $f(g(x)) = x$ but g is not invertible

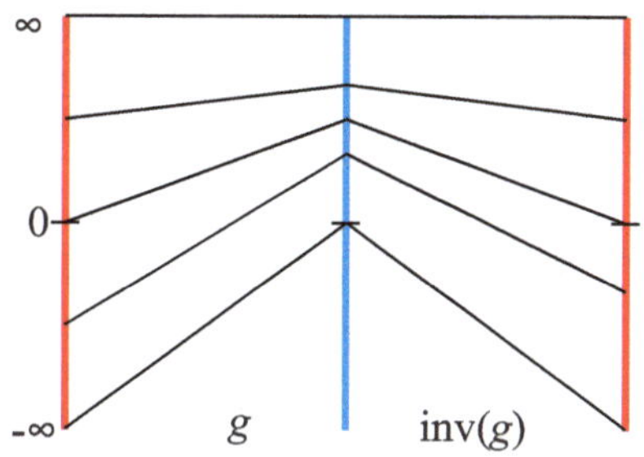

Fig. 20.6 A generalization of Fig. 20.5 to real number line domains

Then, we would ask them to explore the question of the invertibility of g using the arrow representation with the smallest possible domain sets that give an example in which $f(g(x)) = x$ with g not invertible. We would expect them to come up with the counterexample in Fig. 20.5.

Of course our final objective is to explore the example on the real number line and an interesting task at this point would be to ask the students to adapt the setup of Fig. 20.5 to the case in which both domains would be $\mathbb{R}$. We feel that this could lead to an interesting small group activity. It seems clear that the two blue points in the middle set would need to be "half" the real number line each, while the red point would be the entire line. We expect that this would lead to the diagram of Fig. 20.6.

This diagram might well lead the students to the exponential and logarithmic forms, possibly via the square and square root functions, as discussed in the chapter. Of course, we still have the question of what f should do with the negative part of the middle domain and it might easily be seen from this setup that it doesn't really matter.

Math and Music

Wasserman and Galarza present a task that exploits the observation that the keys on a piano work in a *mod* 12 manner, and they introduce the "distance from C" operation as a context in which to discuss isomorphic groups. This example seemed to us to be a bit limited, and perhaps contrived, in that it missed some important and natural learning opportunities.

What is significant about the remarkable number of interactions between mathematics and the real world is that they have the capacity to give us new and powerful insights into the structure of the phenomenon being modeled. This is certainly the case for the structure of the musical scale and there are a number of activities we might give students that then could provide insights into this structure. In Fig. 20.7,

a

0	1	2	3	4	5	6	7	8	9	10	11
12	13	14	15	16	17	18	19	20	21	22	23
24	25	26	27	28	29	30	31	32	33	34	35
36	37	38	39	40	41	42	43	44	45	46	47
48	49	50	51	52	53	54	55	56	57	58	59
60	61	62	63	64	65	66	67	68	69	70	71
72	73	74	75	76	77	78	79	80	81	82	83
84	85	86	87	88	89	90	91	92	93	94	95
96	97	98	99	100	101	102	103	104	105	106	107

The integers arranged in **mod(12)** columns.

b

C	C#	D	E♭	E	F	F#	G	G#	A	B♭	B
16.35	17.32	18.35	19.45	20.60	21.83	23.12	24.50	25.96	27.50	29.14	30.87
32.70	34.65	36.71	38.89	41.20	43.65	46.25	49.00	51.91	55.00	58.27	61.74
65.41	69.30	73.42	77.78	82.41	87.31	92.50	98.00	103.8	110.0	116.5	123.5
130.8	138.6	146.8	155.6	164.8	174.6	185.0	196.0	207.7	220.0	233.1	246.9
261.6	277.2	293.7	311.1	329.6	349.2	370.0	392.0	415.3	440.0	466.2	493.9
523.3	554.4	587.3	622.3	659.3	698.5	740.0	784.0	830.6	880.0	932.3	987.8
1047	1109	1175	1245	1319	1397	1480	1568	1661	1760	1865	1976
2093	2217	2349	2489	2637	2794	2960	3136	3322	3520	3729	3951
4186	4435	4699	4978	5274	5588	5920	6272	6645	7040	7459	7902

The frequencies of the piano keys in cycles/second (using what is called the "even-tempered scale"). The standard piano range is A0 = 27.50Hz to C8=4186Hz. "Middle C" is usually taken to be at 261.6 Hz.

Fig. 20.7 (**a**) The integers arranged in *mod*(12) columns. (**b**) The frequencies of the piano keys in cycles/second (using what is called the "even-tempered scale"). The standard piano range is A0 = 27.50 Hz to C8 = 4186 Hz. "Middle C" is usually taken to be at 261.6 Hz

we give an example that works with the multiplicative structure of the piano keys. The two tables are not only the same size, but they also have a parallel arithmetic structure. First of all, there is a clear one-to-one correspondence based on position, that is 0 corresponds to 16.35 and 67 corresponds to 784.0. Let's calls this mapping the function F. Thus $F(67) = 784.0$ Some questions we could ask include:

1. What happens algebraically as you go along the rows of the frequency table?
2. What happens algebraically as you go down the columns of the frequency table?
3. Find a formula for $F(n)$ in terms of n.

In fact there is much treasure to be harvested from the questions posed in this example, particularly if there's a keyboard available. The point is that if we are going to bring music into the school classroom, we should bring it fully in, play with it, and seek to understand how it works. In many ways, that is what mathematicians do best[1].

[1] For those who might be interested in a more detailed discussion, the following link presents such an activity: http://www.mast.queensu.ca/~math9-12/musical%20magic%20of%2012.html.

Table of Differences

In his polynomial interpolation example, Cuoco starts with a sequence that might have emerged from a particular exploration, and shows how the successive-difference method can tell the student whether the sequence can be generated by a polynomial, and in this case, the number of steps until the differences are constant will tell us the degree of the polynomial. Many students have seen this constant difference argument, often in working with quadratic or cubic polynomials, but given these constant differences, some work remains in finding the coefficients of the polynomial. By carefully tracking these calculations, Cuoco produces an elegant formulation of the polynomial in terms of the combinatorial coefficients. He suggests that these coefficients often reveal "combinatorial treasures" hidden in the original example. We agree. In fact, we have used the well-known problem about the number of regions, *R*, in a circle, formed from all chords between *n* points on the circle (in which chords intersect at distinct points), with secondary school students. The investigation leads to $R = \binom{n}{0} + \binom{n}{2} + \binom{n}{4}$, which can be obtained nicely by Cuoco's difference method, and indeed provides combinatorial treasures for discussion as well as opportunities to discuss inductive proof approaches.

References

Adler, J., & Ball, D. (Eds.). (2009). Knowing and using mathematics in teaching. *For the Learning of Mathematics, 29*(3), 1–56.

Australian Curriculum. (2012). Retrieved from: https://www.australiancurriculum.edu.au/senior-secondary-curriculum/mathematics/

Ball, D. L., & Bass, H. (2009). With an eye on the mathematical horizon: Knowing mathematics for teaching to learners' mathematical futures. Paper presented at the 43rd Jahrestagung der Gelleschaft fur Didaktic der Mathematik. Oldenburg, .Germany. Retrieved May 15, 2011, from: www.mathematik.uni-dortmund.de/ieem/BzMU/BzMU2009/BzMU2009-Inhalt-fuer-Homepage.htm

Ball, D., Thames, H. M., & Phelps, G. (2008). Content knowledge for teaching. *Journal of Teacher Education, 59*(5), 389–407.

Barabe, G., & Proulx, J. (2017). Revolutionner l'enseignment des mathematiques: Le projet visionnaire de Seymour Papert. *For the Learning of Mathematics, 37*(2), 25–29.

Boaler, J. (2016). *Mathematical mindsets: Unleashing students' potential through creative math, inspiring messages and innovative teaching*. San Francisco, CA: Jossey-Bass.

Common Core State Standards in Mathematics (CCSS-M). (2010). Retrieved from: http://www.corestandards.org/the-standards/mathematics

Conference Board of Mathematical Sciences (CBMS). (2012). *The mathematical education of teachers II*. Washington, DC: American Mathematical Society and Mathematical Association of America.

Davis, B., & Simmt, E. (2006). Mathematics-for-teaching: An ongoing investigation of the mathematics that teachers (need to) know. *Educational Studies in Mathematics, 61*(3), 293–319.

Department for Education. (2014). *The national curriculum in England: Key stages 3 and 4 framework document*. Retrieved from: https://assets.publishing.service.gov.uk/government/uploads/system/uploads/attachment_data/file/381754/SECONDARY_national_curriculum.pdf

Dewey, J. (1934). *Art as experience*. New York, NY: Putnam.

Fernandez, S., & Figueiras, L. (2014). Horizon content knowledge: Shaping MKT for a continuous mathematical education. *REDIMAT, 3*(1), 7–29.

Follesdal, D. (2003). Husserl, Edmund (1859-1938). In E. Craig (Ed.), *Routledge encyclopaedia of philosophy* (2nd ed.). London: Routledge. Retrieved July 19, 2010, from: www.rep.routledge.com/article/DD029

Gadanidis, G., Borba, M., Hughes, J., & Lacerda, H. (2016). Designing aesthetic experiences for young mathematicians: A model for mathematics education reform. *International Journal for Research in Mathematics Education, 6*(2), 225–244.

Jakobsen, A., Thames, M. H., Ribeiro, C. M., & Delaney, S. (2012). Delineating issues related to horizon content knowledge for mathematics teaching. Paper presented at the Eighth Congress of European Research in Mathematics Education (CERME-8). Retrieved from: cerme8.metu.edu.tr/wgpapers/WG17/WG17_Jakobsen_Thames_Ribeiro.pdf

Mamolo, A., & Pali, R. (2014). Factors influencing prospective teachers' recommendations to students: Horizons, hexagons, and heed. *Mathematical Thinking and Learning, 16*(1), 32–50.

Mason, J. (2002). *Researching your own practice: The discipline of noticing*. London: Routledge Falmer.

Mason, J. (2001). Mathematics as a constructive enterprise. In J. Winter (Ed.), *Proceedings of the British Society for Research into Learning Mathematics (Vol. 21, Num. 3)* (pp. 52–57). Southampton: British Society for Research into Learning Mathematics.

Ministry of Education. (2007). *The Ontario curriculum grades 11 and 12: Mathematics*. Retrieved from: http://www.edu.gov.on.ca/eng/curriculum/secondary/math1112currb.pdf

Papert, S. (1972). Teaching children to be mathematicians versus teaching about mathematics. *International Journal of Mathematical Education in Science and Technology, 3*(3), 249–262.

Raymond, K. (2018). M is not just for STEM: How myths about the purposes of mathematics education have narrowed mathematics curricula in the United States. *Education Sciences, 8*(2), 47.

Sinclair, N. (2006). *Mathematics and beauty*. New York, NY: Teachers College Press.

Taylor, P. (2018). Teach the mathematics of mathematicians. *Education Sciences, 8*(2), 56.

Wasserman, N., Fukawa-Connelly, T., Villanueva, M., Mejia-Ramos, J. P., & Weber, K. (2017). Making real analysis relevant to secondary teachers: Building up from and stepping down to practice. *PRIMUS, 27*(6), 559–578.

Wasserman, N., Weber, K., & McGuffey, W. (2017). Leveraging real analysis to foster pedagogical practices. In A. Weinberg, C. Rasmussen, J. Rabin, M. Wawro, & S. Brown (Eds.), *Proceedings of the 20th annual conference on Research in Undergraduate Mathematics Education (RUME)* (pp. 1–15). San Diego, CA: RUME.

Wasserman, N. H., & Stockton, J. C. (2013). Horizon content knowledge in the work of teaching: A focus on planning. *For the Learning of Mathematics, 33*(3), 20–22.

Whitehead, A. N. (1929). *Aims of education*. New York, NY: The Free Press.

Zazkis, R., & Mamolo, A. (2011). Reconceptualizing knowledge at the mathematical horizon. *For the Learning of Mathematics, 31*(2), 8–13.

Index

N. H. Wasserman (ed.), *Connecting Abstract Algebra to Secondary Mathematics, for Secondary Mathematics Teachers*, Research in Mathematics Education,
https://doi.org/10.1007/978-3-319-99214-3

MIX
Papier aus verantwortungsvollen Quellen
Paper from responsible sources
FSC® C105338

If you have any concerns about our products,
you can contact us on
ProductSafety@springernature.com

In case Publisher is established outside the EU,
the EU authorized representative is:
Springer Nature Customer Service Center GmbH
Europaplatz 3, 69115 Heidelberg, Germany

Printed by Libri Plureos GmbH
in Hamburg, Germany